Introduction to Risk Analysis

A Systematic Approach to Science-Based Decision Making

Daniel M. Byrd III
C. Richard Cothern

Government Institutes
An imprint of
The Scarecrow Press, Inc.
Lanham, Maryland • Toronto • Oxford
2005

Published in the United States of America
by Government Institutes, an imprint of The Scarecrow Press, Inc.
A wholly owned subsidiary of
The Rowman & Littlefield Publishing Group, Inc.
4501 Forbes Boulevard, Suite 200
Lanham, Maryland 20706
http://govinst.scarecrowpress.com

PO Box 317
Oxford
OX2 9RU, UK

British Library Cataloguing in Publication Information Available

Library of Congress Cataloging-in-Publication Data

Byrd, Daniel M.
Introduction to risk analysis : a systematic approach to science-based decision making / Daniel M. Byrd III, C. Richard Cothern.
p. cm.
Includes bibliographical references and index.
ISBN: 978-0-86587-696-5
Technology—Risk assessment. 2. Environmental risk assessment. 3. Decision making.
I. Cothern, C. Richard. II. Title.

T174.5.B97 2000
363.7'072—dc21 99-086190

The paper used in this publication meets the minimum requirements of American National Standard for Information Sciences—Permanence of Paper for Printed Library Materials, ANSI/NISO Z39.48-1992. Manufactured in the United States of America.

One of the authors (CRC) wishes to dedicate this volume to his seven grandchildren: Ellen Grace, Hannah Elizabeth, Julia Catherine, Kara Lynn, Kristin Elizabeth, Kelsey Christine, and Noah Richard.

Summary Contents

Contents

Chapter Two

Functions, Models, and Uncertainties 41

Chapter Three

Chapter Four

Chapter Six

Chapter Seven

Chapter Ten

Chapter Eleven

Chapter Twelve

Chapter Thirteen

List of Figures and Tables

Preface

We, the authors, first met while working at the U.S. Environmental Protection Agency (U.S. EPA) 20 years ago. Because the U.S. EPA operates under several regulatory statutes that make decisions based on risk, this agency has developed and incorporated many risk-based analyses and guidelines that we have discussed over the years. Other federal agencies (e.g., the Consumer Product Safety Commission, the Occupational Safety and Health Administration, the Food and Drug Administration, and the Nuclear Regulatory Commission) also assess health, safety, and environmental risks. We subscribe to this approach and have been involved in it for almost two decades.

We offer this volume for those entering the field of environmental risk who need to understand how all the varying and diverse pieces are used and how they fit together. The book is structured to be used as a text by taking the chapters in sequence. However, many of the chapters stand on their ownand can be read and understood independently.

This volume is the first comprehensive and integrated volume to provide professionals with an introduction to all aspects of the field of risk analysis. The three major areas of risk analysis: risk assessment (exposure assessment and dose-response curves), risk communication (it needs to be two-way) and risk management (including values and ethics) are presented in an integrated way along with many other related topics.

Risk is a quantifiable concept. The quantitative nature of risk is introducedin chapter two and carried through the rest of this book. In chapter two we start with graphing data and carefully take the reader through logarithmic thinking, statistics for pedestrians, uncertainty, risk estimating, models, and finally curve fitting, which leads to risk characterization (in individuals and populations). This allows the reader to fully understand how risk numbers are developed and how to do the same for new data.

In addition to those topics mentioned above we have included others to complete the coverage of all relevant topics, including epidemiology, the sister science to toxicology; comparative risk analysis, to relate different and diverse risks; and ecological risk, equally important to human health risk. Finally, we present some selected case studies to allow the new professional a glimpse of how this all works in the real world.

Daniel M. Byrd III
C. Richard Cothern
Washington, D.C.
September 1999

About the Authors

Daniel M. Byrd III received both B.A. (1964) and Ph.D. (1971) degrees from Yale University. After postdoctoral work at the University of Maryland, Johns Hopkins University, and the National Cancer Institute, he taught pharmacology and conducted independent research into mechanisms and dosimetry of chemotherapeutic drugs at Roswell Park Memorial Institute and at the University of Oklahoma. He subsequently held positions in the Office of Chemical Control, the Office of Pesticide Programs, the Carcinogen Assessment Group, and the Science Advisory Board at the U.S. Environmental Protection Agency (EPA), which recognized his performance with several Outstanding Performance Evaluations, a Special Act Citation, and a Silver Medal for outstanding managerial leadership and initiative. The American Board of Toxicology conferred Diplomate status on him in 1982. He also has managed scientific and medical committees for three trade associations, the Distilled Spirits Council, the Halogenated Solvents Industry Alliance, and the American Petroleum Institute. He is the author of more than forty scientific articles and one hundred regulatory publications. He holds memberships in eleven professional societies.

Since 1988, Dr. Byrd has directed Consultants in Toxicology, Risk Assessment and Product Safety (CTRAPS), a scientific support firm that helps clients acquire, interpret, and use biomedical information. In addition to regulatory toxicology and risk assessment experience, CTRAPS has expertise in applying information theory, operations research techniques, and simulation models in several regulatory areas, including agricultural import-export assessments, Clean Air Act requirements (residual risk determinations), food contaminant analyses, food additive petitions, testing and risk assessment guidelines, pesticide registrations, premanufacturing notifications, medical device applications, regulatory risk analysis, Safe Drinking Water Act rulemakings, and testing rule compliance.

Dr. Byrd advocates the use of population distributions for risk assessment purposes. He served as a panel member in the 1982 review of EPA's exposure assessment methods, which evolved into the 1986 guidelines. He was a charter member of the Delivered Dose Workgroup of the American Industrial Health Council, a focal point for the development of physiological-pharmacokinetic models in industry and their application to integrate exposure pathways using biomarker measurements, a technical approach that CTRAPS currently provides for its clients.

In 1998, Dr. Byrd chaired a peer review of proposals in response to the National Exposure Research Laboratory's solicitation about multipathway human exposure and source-to-dose relationships. He recently commented on the EPA's Exposure Assessment

Handbook, noted problems with EPA's pesticide exposure methods before the Tolerance Reassessment Advisory Committee, and observed the SAB's reviews of the National Human Exposure Assessment Survey and of integrated exposure models.

C. Richard (Rick) Cothern has spent the past 19 years with the U.S. Environmental Protection Agency serving the Center for Environmental Statistics, Science Advisory Board (as Executive Secretary), and the Offices of Drinking Water (international expert on radioactivity and risk assessment) and Pollution Prevention and Toxic Substances. He is professor of engineering management and chemistry at the George Washington University and a professor of technology and management at the University of Maryland's University College. At the George Washington University he is an Associate Professor of Chemistry and Professor of Engineering Management. At Hood College he is an Adjunct Professor of Chemistry. Previously, he taught high school physics and undergraduate and graduate level physics for seventeen years at the University of Manitoba and at the University of Dayton (as Assistant and Associate Professor of Physics).

Dr. Cothern's research concerns nuclear spectroscopy, environmental physics, and surface analysis. He has authored over 100 scientific articles including many related to public health, the environment, risk assessment, photoelectron spectroscopy, surface analysis, and low-energy nuclear spectroscopy. He has written and edited 14 books on such diverse topics as science and society, energy and the environment, trace substances in environmental health, lead bioavailability, environmental arsenic, environmental statistics and forecasting, risk assessment and radon, and radionuclides in drinking water. He received his B.A. from Miami University (Ohio), his M.S. from Yale University, and his Ph.D. from the University of Manitoba.

Introduction to Risk Analysis

Risk Analysis

Probability of Future Loss

Risk analysis is a controversial subject. Specific analyses often examine emotion-laden, socially important topics. Contending groups with conflicting ethical values intentionally distort information about risk in their attempts to prevail. Often, the stakes are high. Recently, the United States has seen many regulations based on risk analyses with large economic stakes, including the standards for radon in homes, particulate matter and ozone in the air, microbes and disinfectants in drinking water, and additives and pesticides in foods.

Problems of language also bedevil risk analysis. Because of terminology difficulties, arguments break out between experts talking past each other. Often, each speaker unknowingly debates a different topic. Stan Kaplan, one of the founders of modern risk analysis, informally defines two kinds of semantic confusion. In the first kind, a single word means several different things; in the second kind, several different words mean the same thing. In risk analysis, you can often detect sources of confusion by applying Kaplan's definitions.

Some linguists and educators believe that you do not understand a word unless you can define it. Yet, some explanations of risk analysis fail to explain what risk is. In this book, we try to untangle risk analysis by defining key terms and enforcing these definitions throughout the text. Some sidebars describe controversies about terms that bedevil the current practice of risk analysis.

Risk Defined

We define *risk* as "the probability of a future loss." Even expert practitioners of risk analysis have not universally accepted this definition. It encompasses a wide range of current activities, ranging from deaths to earthquakes to bankruptcies. The definition is sufficiently general and consistent with our overall approach. We definitely mean to imply that risk analysis resembles other forecasting activities, such as actuarial practices in the life insurance industry or picking stocks.

By defining risks as probabilities, we have transferred dispute from one term to another. Probability also is a controversial notion. In this book we define probability as an expression of the confidence that an event will occur. Do you think the Dodgers will lose the World Series next year? By convention, a high probability, such a 0.9, means *nine chances out of ten (9/10)*; it indicates a highly likely event. A low probability, such as 0.000009, means *nine chances out of one million (9/1,000,000)*; it indicates an unlikely event. To understand risk analysis, you have to learn a little about the probability calculus. We briefly summarize it in the section called "Uncertainty" in this chapter.

Predictive Nature of Risk

Our definition of risk emphasizes its predictive nature. To us, regulatory risk analysis has much in common with forecasting the weather or predicting next year's gross national product. Thus, we require that each statement of risk have an accompanying, explicit statement of the undesirable outcome or expected loss—the thing you do not want—as a necessary footnote that travels everywhere with the probability by implication. We might believe that the chance of bad weather tomorrow is 0.3 (three out of ten), but eventually we need to know whether the bad weather will be snow or rain, and whether we will get two feet of snow or two inches of rain. Similarly, we might believe that the chance of the economy declining next year is 0.007 (seven out of a thousand), but we also want to know whether the decline will be a recession or a depression.

Nature of Risk

The top experts, members of the Society for Risk Analysis, have discussed the concept of risk extensively and not been able to agree on a definition. The debate mostly divides over only one difference: whether risk has of the nature of a probability or that of a utility. A utility is the probability of an event times the value of the event. For example, if you hold one ticket out of ten thousand in a fair lottery, and if the winning ticket is worth one millon dollars, then your ticket is worth one hundred dollars ($1,000,000/10,000 = $100). If you believe that every undesirable event has a price, thinking about risks as utilities makes good sense, but for many analysts, loss outcomes are difficult to value, sometimes even incommensurate. For example, stating a dollar value for a case of kidney cancer is problematic and seldom necessary. The analyst just states the probability of a case of kidney cancer. Trying to add cases of kidney cancer to other outcomes, like lottery tickets, seems silly, even if you express the loss as the cost of a case of kidney cancer. In this book, we require that analysts state the undesirable outcome and the probability that it will occur, but not necessarily multiply the two values together. Leaving them stated separately is acceptable.

Do we mean that the gross national product will decline by some amount, or are we discussing a 10 percent decrease in the Dow Jones average or 2 percent seasonally adjusted increase in unemployment rates? Risks clearly have precise meanings with accompanying quantitative units. You may hear risks expressed in terms like "a probability of one hundred automobile accidents per year from riding on expressways." As we show below, this statement is incomplete.

Unfortunately, risk analysts often talk in shorthand about risk. (Sometimes, they even get sloppy or make mistakes.) We omit units or defining frameworks, because we believe everyone listening understands the units. The statement "a risk of one hundred automobile accidents per year from riding on expressways" does not fully express the conditions relevant to the risk. Instead, we might add, "given a specific definition of population riding on expressways in 1992 in the United States."

Attaching an expression of the outcome creates a conditional probability; that is, the risk occurs only under certain conditions. Once we know the denominator, the automobile-riding population on expressways during 1992 in the U.S., we also can express the risk as a numerical probability: 100/6,700,000, or 0.000015 accidents per person per year. (For more information about the units of risk, see the section called "Units" later in this chapter.) We also understand something of the relevance of the estimate. We can include a definition of expressways or the number of miles driven. Do you believe that circumstances in 1910 resemble those in 1999? Do you define expressways as four-lane roads or as limited-access roads? If you are trying to understand your personal risk from driving to work on an expressway, this information becomes particularly useful.

Virtually every event has an associated risk. Try to think of something that does not have a risk. Did you imagine sitting in your armchair, reading a book? Think again. Suppose your house is catching on fire, and that you will not notice the flames until it is too late to avoid losing the house and getting burned. The risk is very small, but it is finite. If we wanted to, we could make predictions about the risk of increased home damage per hour spent reading. We could even attempt to validate these predictions against national fire statistics or other data.

Universality of Risk

All activities, objects, and processes have risks. Thus, risk analysis is ubiquitous, an activity that is pervasive at an informal level. Everyone does it while making everyday decisions. Most of us are familiar with examples from gambling. We do not advocate gambling, even as a way to learn about risks, because over time, most games of chance lead to an expected loss. (The gambling house or bookmaker takes out something. So, over the long term, you get back less than you bet.) However, when your money is at stake, based on your estimate of a future probability, the elements of risk become particularly obvious.

Expert practitioners of risk analysis have a wide spectrum of activities to analyze. Risk is associated with any object, such as the risk that a boiler will explode or that an earthquake will occur. The content may change between applications, but the techniques and ideas are surprisingly similar. In this book, we emphasize practical outcomes, or "risks in the real world," mostly from health and safety examples, because we have expertise in this area. Much regulatory attention focuses on health and safety, particularly human health risks. To gain insight into this subject, we will refer later in this chapter to the U.S. life table, a compendium that describes the experiences of the U.S. population with death (mortality).

Predicting an event that commonly occurs, like a cold, that you expect soon, but that usually leads to small losses, intuitively differs from forecasting a rare event, far off in the future, that might lead to a huge loss, such as a meteor striking the earth and eliminating most biological species. The differences become more understandable, when you break the comparison into smaller pieces: the probability, the outcome, and the amount of time until the outcome of each event.

Structure of Risk Analysis

Problems of language and definition so cloud the field of risk analysis that a uniform terminology does not exist to describe its structure. Engineers use the term *risk assessment* very differently from scientists. Many confuse the practice of safety assessment with risk assessment. (*See* Chapter Seven) Suppose that you have a specific problem involving risk. You might want to obtain an in-depth understanding of the risk, figure out some ways to control it, and learn how to explain what you know to others. In this book, we similarly divide the overall process of risk analysis into three components: *risk assessment*, *risk management*, and *risk communication*.

Risk Assessment

To us, *risk assessment* is the process of characterizing a risk. It involves estimating the probability, usually a mathematical process, and specifying the conditions that accompany the outcome. It usually incorporates descriptive data and scientific theories. So, it resembles policy analysis. Some risk assessors even think that it is a branch of policy analysis. Others describe risk assessment as science-based inference or fact-based opinion. We put it into the same category as forecasting earthquakes or bankruptcies. Risk assessors ask what can go wrong, how likely is the bad outcome, how long will it take before it occurs, and what might be the importance of the loss, if it does.

Risk Management

In contrast to the process of risk assessment, *risk management* is the process of deciding what to do about risk. A risk manager looks at the available options to control the probability of loss, the time until the loss, and the magnitude of the loss. The risk manager always has options of doing nothing, involving stakeholders, publicizing the risks, or obtaining more information. Risk managers usually attempt to make decisions that will reduce risks.

A risk manager will ask what we can do about a risk, and what tradeoffs do the different control options involve? The process usually begins by determining the options and estimating the risks under each option, but the risk manager must evaluate a wider array of information, including economic costs, technical feasibility, social acceptance, legal conformance, political perceptions, regulatory objectives, and enforceability. Balancing these different concerns, especially when the information is uncertain or incomplete, is a difficult task. Some analysts describe risk management as science-based decision making under conditions of uncertainty.

Usually, the risk manager has an array of tools to support this difficult task. One of the tools is *decision analysis*, a mature branch of operations research, and risk analysts can apply this highly structured, coherent approach. Given opportunity and information, a risk manager can balance the risk of one option against the risk of another. Still another tool the risk manager may use is an analysis of the economic costs and benefits of each option, giving preference to the option with the lowest cost-benefit ratio. In addition, the risk manager may select the option with the lowest ratio of risk to benefits, by directly comparing risks with economic data. Theoretically, a risk manager could use any or all of *risk-risk*, *cost-benefit*, and *risk-benefit* approaches to explore the best way to manage a risk.

Separating Risk Assessment from Risk Management

Some stakeholders deeply involved in government regulation have wanted government agencies to separate risk assessment and risk management activities, not just conceptually, but even physically. Their idea was to send all of the risk assessors to some remote location, where isolation would protect them from political influence. Otherwise, the reasoning goes, risk assessors are too vulnerable to demands from their bosses to produce estimates that support or justify politically motivated preferences.

The high tide of the separation idea came in the early-to-middle 1980s. Congress funded a Committee on Institutional Means at the National Academy of Sciences. This committee produced a landmark book, *Risk Assessment in the Federal Government: Managing the Process*, otherwise known as the "Red Book" because of the color of its cover." Science courts" are based upon the same idea of separating risk assessment from risk management. In them, assessors could appeal to wise scientists, who would approve or reject risk assessments. The supporting interest groups do not want risk assessors involved in making political decisions. Yet, few have more to say about risk management options

Practically, many constraints may prevent full exploration, including the unimportance of the decision, lack of time, legal restrictions, and sparseness of data and funds to support the process.

than the people who prepared the risk assessment. In contrast, government observers are all too familiar with government scientists distorting risk assessments to prevent managers from "making the wrong decision." The task of separating risk assessment from risk management is complex and difficult.

Risk Communication

Risk communication is the process of explaining risk. To exchange information, the risk communicator has to begin by understanding the nature, assessment, and management of a risk, but other understanding is needed. The risk communicator also has to decide whom to engage in the process, how to engage them, and at what point in the process to engage them, as well as what information to exchange and how to exchange it. Most risk communicators aim for open, two-way interactions that will enable people outside the risk analysis process to understand risk assessments and accept risk management decisions.

These definitions of the components of risk analysis are commonly, but not universally, used. For example, most risk analysts in the private sector, the states, and the federal civilian agencies accept these definitions. However, the U.S. Department of Defense reverses the meanings of risk analysis and risk assessment.

When scientists estimate the chances of getting lung cancer from radon gas in homes, they are engaged in risk assessment. When transportation experts estimate the annual chances of death from driving automobiles at different speeds, they are engaged in risk assessment. When the Environmental Protection Agency (EPA) requires a certificate of radon levels for every home sold, EPA is engaged in risk management. When Congress sets a maximum speed limit of 55 miles per hour on federal highways, Congress is engaged in risk management. If a realtors' association issues a press release explaining EPA's requirement for radon certificates, it is engaged in risk communication. When a Department of Transportation official appears on television to explain why Congress has reduced the speed limit, the official is engaged in risk communication.

Risk analysis encompasses much more than a description of a risk and an accompanying risk assessment, but the process begins with them. For example, good risk management requires an understanding of how the public views a risk, which may differ dramatically from the way that risk assessors evaluate it. *Risk perception* is a separate area of research. Understanding it is crucial for good risk communication. Perceiving a highly probable, but slight, loss expected in the near future is inherently different from thinking about a huge, but rare, loss that will only occur in the distant future and that may never occur. Our efforts to avoid a huge potential loss will color our intuitive feelings about a very low-probability event.

Procedurally, decision analysis separates probabilities from outcomes, and for this reason, many think it improves risk management. An understanding of how society will perceive a risk, and accept risk management decisions, is important. Good risk management requires a diversity of information about costs, feasibility, enforcement, legal authority, and so forth.

Risk Policy

Some add *risk policy,* a kind of meta-topic, as a fourth component of risk analysis. They have in mind issues such as guidelines for risk assessment (often different *guidelines* for different kinds of risks, such as guidelines for carcinogen risk assessment or guidelines for neurotoxicity risk assessment). We do not agree with this addition. Instead, we view the development of policies about risk analysis as something embedded throughout assessment, management, and communication, as are policies about how risk analysis should apply to different aspects of life, like environment or transportation. Much of risk policy is confined to safety assessment.

We definitely mean that the three-part structure of risk analysis applies to all forms of risk and all kinds of regulatory activities. We mostly focus on risks to health and safety, not weather, financial, social, geological, or engineering topics. Our emphasis relates to personal experience and to activities of a group of specialized federal regulatory agencies: the Consumer Product Safety Commission, the Department of Agriculture, the Department of Transportation, EPA, the Food and Drug Administration, the Nuclear Regulatory Commission, the Occupational Safety and Health Administration, and the research organizations that support these regulatory agencies.

The three-part structure also functions well over a wide range of other topics, spanning all kinds of human endeavors, including earthquakes, infectious diseases, bankruptcies, and actuarial practice in the insurance industry. While we are less familiar with, for example, the Department of Interior, the Department of Treasury, the Federal Aviation Administration, the Federal Communications Commission, the Federal Trade Commission, and the Securities and Exchanges Commission, this structure of risk analysis in this book covers their work equally well. However, the subject matter regulated by each organization differs. Risk analysis is a general concept in the abstract, but it is content specific in application.

The structure of risk analysis is helpful, just in sizing up the nature of regulatory tasks. Suppose that you become responsible for complete risk analyses of two different subjects: common colds and large meteorites striking the earth. How would you approach assessment, management, and communication of each of these risks?

Risk Characteristics

All activities, objects, and processes have risks. The idea that some activities are risky, and others are not, is illusory. Many factors motivate the idea that some process or substances can be "risk free." One factor is that our legal system seeks black and white distinctions. When people experience losses, they may resort to the legal system. Another factor is that, psychologically, people suppress their expectations of events with low probabilities as if the risks do not exist.

In our society, enforcement usually depends on *bright lines*. (*Bright lines* refers to several concepts, including the lines painted down the middles of highways.) The need for clearly understood and accepted regulatory standards is important to our sense of fairness. We allow the highway patrol to ticket people traveling faster than a certain speed and credit the troopers with the ability to make a judgment about conditions only under unusual circumstances. For example, 35 miles per hour (mph) might constitute speeding on an icy road, even if the posted limit was 55 mph per hour. For the most part, we reject the idea of unannounced speed limits or speed regulations that officials make up on the spot. To do otherwise violates our ideas of objectivity and fairness and opens up the regulatory system to abuse by government. Thus, a judge or jury may validate a fine for going 57 mph in a 55-mph zone. Such harassment may embarrass us, but it is legal. Mostly, however, we reject fines for going 53 mph in the same zone as completely unfair.

Precision and Accuracy

The differences between 53 and 57 mph also strike us as trivial. Could the police actually measure such a small difference? The scientific term for the ability to detect small differences is *precision*. Perhaps the policemen's speedometer registers 55 mph on the average when the car is traveling 55 mph, but it is not sufficiently precise to tell whether 55 mph is different from 57 mph. Government might consider equipping the police cars with more precise speedometers, perhaps with precision to one-tenth of a mph. The new speedometer also would need accuracy. Detecting a small difference in speed would not help, if the speedometer registered 35.1, when someone was traveling at 45.1 mph. The ability of the speedometer to register the true speed is its *accuracy*. Accuracy and precision commonly get confused.

Not confusing the bright lines of regulations with risks also is important, even when a standard is based on risk analysis. We all understand that driving at 35 mph in the 55-mph zone has some risk, just as driving 75 mph greatly increases the risk. Almost no one is so foolish as to believe that going 54 mph is risk free, whereas 56 mph means instant death.

Responding appropriately to a drinking water standard is equally important. If you drink water that has twice the allowable level of a substance, it does not mean that the water will kill you. You need not send for an ambulance. Just because the level of some substance in the water complies with a standard, a free society does not make you drink it. Yet, the scientific data that support a drinking water standard may change, if scientists discover some new health effect.

Safe vs. Risk Free

The notion of safety in our society generates much controversy. Dictionaries do not help. They usually define *safe* as "risk free." Nothing is risk free, however, so if this definition is operational, safety has no meaning. We might as well write about unicorns and leprechauns. People who think that a motorcycle is safe, simply because it is free of manufacturing defects, are headed for disappointment when they have an accident, especially when not wearing a protective helmet.

Science intrinsically lacks the capability of showing that some object or activity is risk free. If we repeat the circumstances that lead to an undesirable outcome, but do not observe the outcome, all we really know is that the object or activity might occur with a lower probability than we could observe in a few experiments. If we structure an experiment to observe a particular kind of outcome under a set of conditions, and the experiment is negative, we know little about either the outcome under other conditions or other kinds of undesirable outcomes. When we do observe the expected outcome in a highly controlled experimental setting, we still have to decide whether the risk will occur during more realistic conditions and whether the risk is trivial.

The insertion of personal values into our discussion of safety can generate even more confusion. Some think of motorcycle riders as intrinsically bad people and that motorcycles are not safe under any circumstances. Others think that riding a motorcycle is an expression of personal freedom guaranteed by the Constitution and that anything to do with motorcycles is good.

Risk Perception

Most people regard risks that they can control as much lower, perhaps even nonexistent, but higher, if someone else imposes the same risk on them. No one may make you ride motorcycles, but you do have to share the roadways with them. We want to regard personal apprehensions about risks as valid, particularly if we are the people who hold them. Paul Slovic and his colleagues have measured the impact of many factors, like control, on the psychological perception of risk. Methods to understand risk perceptions are available. The initial task of the risk assessor is, however, to prepare an estimate of risk that is as free as possible of human perceptions, fears, and cultural values.

Culture often shapes the perception of risk. For example, in our society we have avoided an apparently beneficial technology, hydrogen fuel. The raw material for hydrogen fuel is widely available, hydrogen is easily generated, and hydrogen produces water on burning, cleanly recycling to the environment. At least with hydrogen, we know part of the reason for our apparently irrational rejection of a clean energy source.

Before the outbreak of World War II, a German dirigible, the Hindenburg, caught fire. This incident was not the great catastrophe it seems. Only seventeen persons lost their lives out of nearly two hundred passengers. Compare this event with a modern plane wreck, say TWA Flight 800, where all of the passengers were lost. We do not react by eliminating air flight—or more extremely—petroleum fuel. We are used to these technologies and somehow accept the risks. The radio announcer at the Hindenburg crash site became hysterical, and many stations replayed his broadcast extensively. At the time, hydrogen was an unfamiliar technology. Once the idea that hydrogen was a particularly dreaded risk became imbedded in our society, developing hydrogen technology for energy purposes became more difficult.

The notion of *acceptable risk* attempts to rationalize very different, sometimes conflicting, social attitudes towards different kinds and sources of risk (Lowrance, 1974; Rowe, 1977). Its essence is that, while we are each our own risk manager, we exchange communications about our attitudes toward a particular risk collectively until society incorporates the beliefs, often informally. Thus, a group may decide to accept a risk, because of the benefits it brings. Our attitudes, however, often are cultural. We subconsciously acquire acceptance of one risk, like automobiles, and rejection of others, by observing and incorporating the attitudes of others.

Similarly, we have long avoided irradiation of food, a technology that experts believe would bring us cheaper, better-tasting foods with longer shelf lives and greater safety. The reason for this avoidance appears semantic. Some people, on seeing the term *irradiation*, somehow get the idea that the food is radioactive. The perception of experts is not like the perception of the general public.

Risk perception is an important factor in understanding how the public reacts to regulatory decisions based on risk. Any analysis of risk involves cultural values and ethics. It also involves a perception of whether the risk being considered is getting better or worse. We provide a more detailed explanation of risk perception in Chapter 11, "Risk Management: Values and Environmental Risk Decision Making."

Risk Exposure

The risk analyst has to understand a risk with physical, chemical, and biological accuracy, before comparing it with the perceived risk. Think of the task in this way: if risk analysts just recycled public perceptions, the public would not need risk analysts. The risk analyst also needs to characterize and classify risks qualitatively. Various kinds of risks exist. One approach examines the pathway from the source of risk to the object experiencing the risk. We refer to the way that risks get to humans and other living things generally as *exposure*.

The major *routes of environmental exposure* are air, water, and soil. For example, if a risk arises from releases of a substance into the air that persons breathe, inhalation exposures concern the risk analyst. A factory that releases smoke into the air may be the source of an inhalation risk when we breathe in particulate matter and it settles in our lungs. A risk of tuberculosis may arise from riding a bus because someone on the bus may have this disease. The person with the disease is the source, but the pathway still is inhalation. When the person coughs, he can spread tuberculosis microorganisms through the air, and other bus riders breathe them. The smoke particles are a chemical risk; the tuberculosis organisms are an infectious disease risk. Chemicals or infectious organisms are kinds of risks. Lung diseases, like emphysema or tuberculosis, are biological effects or outcomes.

Similarly, some pathways extend from a single source, called a *point source. Nonpoint sources* involve widespread releases. For example, automobile exhaust is a nonpoint source since it arises from many cars over widespread areas. Gasoline burned at a power plant might release exhaust up a smoke stack—a point source.

The risk analyst can describe the activity that leads to the release, for example, chemical manufacturing, shopping at the mall, petroleum refining, rock concerts, transportation, dentistry, or agriculture. But some risks arise from activities not subject to human control, such as earthquakes, hurricanes, or volcanic eruptions. We usually describe them as natural risks. We describe unintentional events as accidents.

The risk analyst can describe the agents involved. Some are manufactured substances (industrial chemicals, pesticides, food additives, or pharmaceuticals); others are not (viruses or radon). We usually describe chemical substances in units of concentration (the amount of the substance in a volume, such an milligrams per liter of water). We usually describe radiation as intensity, the energy reaching an area or volume. We usually describe infectious agents in units of concentration (the number of organisms in a volume or the minimum volume necessary to infect a host, called *infectious units*).

Risks arise from different sources. Human behaviors lead to many risks, such as accidents involving cars, guns, or bankruptcies. Geological sources of risk, such as earthquakes or volcanoes, and weather, such as hurricanes or tornadoes, cause losses of life and property. In a modern, industrial society, risks arise from sources of energy, such as electric power transmission, liquefied natural gas, motor gasoline, and from modes of transportation, such as cars, trucks, trains, and buses.

Thus, different kinds of risk exist. We have already discussed mortality as an outcome. Death may come suddenly from an acute event, like an automobile accident or a gunshot wound, but chronic diseases, such as cancer, neurotoxicities, or birth defects, also are outcomes that contribute to a loss of life. Society sometimes views the loss of function, *morbidity*, from accidents and diseases as more important than mortality. When diseases or accidents compromise personal function, society loses productivity and accrues the costs of medical care.

Components of Risk

- Sources
- Chemical and physical nature
- Environmental route of exposure
- Route of transmission
- Dissipation in the environment
- Migration from outer membranes to target organs
- Mortality and morbidity
- Biological effects in the individual
- Population effects

Communication Problems

The language used to express risks remains ambiguous. Everyday descriptions often are ambiguous about the kind of agent involved in a risk and its source. Many people get exposures and outcomes, such as symptoms, confused. For example, infection with human immunodeficiency virus (HIV) is not the same thing as acquired immune deficiency disease (AIDS). They are two different outcomes, and neither is an exposure, as is, for example, a transfusion with an HIV-contaminated blood product. Yet, the press commonly treats infection as disease. Part of this difficulty comes from the confusing language used to describe risks. Part of it comes from the dynamic nature of some risks. That is, they change with time.

Even risk assessment is inherently value-laden. It often is based on scientific data, and it has a mathematical structure that can reveal internal contradictions, but the assessor must make many choices, just to express risk, for example, as mortality or morbidity. Neither common colds nor large meteorites striking the earth are events that we accept, in the sense that we can exercise much direct control over existence of the source right now. How else would you characterize these two risks? What units would you choose for each?

Units

One outstanding characteristic of risk is the numerical expression we give it. We deal with the probabilities in daily life from weather reports to lotteries. The values range from a few percent to many tens of percent. (*See* Table 1.1) Environmental risks important to public health are much smaller quantitatively. The range of interest is usually between one in ten thousand and one in a million. In decimal notation, one in ten thousand is 0.0001, one in a hundred thousand is 0.00001 and one in a million is 0.000001. These numbers are awkward to read and write.

To simplify handling these very small numbers, we use engineering shorthand. One thousand (1,000) is 10^3, meaning 10x10x10, or ten multiplied by itself three times. One in a thousand (1/1,000) is 10^{-3}. One million is 10^6. One in a million is 10^{-6}, a rare event.

Table 1.1 Probabilities in Everday Life

Event	Probability
Weather	0.1...0.8 rain, clear, etc.
Bus or train being late	
5 min	0.5
20 min	0.05
30 min	0.01
Airplane late to take off	0.2
Medical probabilities	
Inheritable traits	
(color blindness, diabetes)	0.25
Lung cancer for a smoker	0.1
Premature baby	
living for three months	0.5
Alarm clock failure	1/365
Car failing to start	1/365
Chance of a successful event	
Cards (per draw)	1/52
Roulette (per roll)	1/38
Dice (per throw)	1/6
Typical lottery ticket	1/1,000,000

Each unit of the exponent is called an order of magnitude. One million is six orders of magnitude greater than one. One thousand and one million differ by three orders of magnitude. This shorthand allows easy comparison of risks that differ by many orders of magnitude, such as those encountered in environmental risk analysis.

The period at the end of this sentence fills about a hundred thousandth of the area of the page. (We measured it.) The chance that another dot of the same area would hit that period, at random, is 1/100,000, or 10^{-5}.

To express measurements conveniently, we also use a shorthand. You have already encountered this system, using Greek prefixes, in units like a centimeter (a hundredth of a meter) and a millimeter (one thousandth of a meter). For mass, we have the kilogram (one thousand grams). The suggested regulatory guidance for radon is a concentration of less than four picocuries per liter of air. A picocurie is a trillionth (a million millionth or 10^{-12}) of a curie. Table 1.2 lists the Greek prefixes for quantities ranging from 10^{18} to 10^{-18}.

The number of digits retained in a quantity implies its precision. Scientists call them *significant figures*. Suppose you are interested in the size of your dining table. Perhaps you want to purchase a tablecloth. You measure the table and find it is approximately 6 feet long. Your son measures it as 6.03 feet, and your daughter comes up with 6.032765 feet. You think the kids are being silly. For your purposes 6 feet long works fine. This value has one significant figure, while your daughter's measurement has seven significant figures. The number of digits counts, not the zeros. For example, both 0.000665 and 665 are correct to three significant figures. Estimates of risk in the environmental field seldom have more precision than one significant figure. To use more significant figures misrepresents the knowledge.

Table 1.2 Greek Prefixes and Numerical Factors

Factor	Prefix	Symbol
24	yotta	Y
21	zetta	Z
18	exa	E
15	penta	P
12	tera	T
9	giga	G
6	mega	M
3	kilo	k
2	hecto	h
1	deka	da
−1	deci	d
−2	centi	c
−3	milli	m
−6	micro	μ
−9	nano	n
−12	pico	p
−15	femto	f
−18	atto	a
−21	zepto	x
−24	yocto	y

Most environmental measurements have precision in the range of 10 percent at best. If a measurement is precise to 10 percent, representing it with seven significant figures makes no sense. If the value of a measurement is 167.2443, what would it be after rounding off to a precision of 10 percent? Ten percent of the value is 16.7, so the second digit of the measurement could vary some, but beyond that the significance is lost. Some people keep the third digit for completeness, but the rest is insignificant. Thus, expressing the number as either 167 or rounding off to 170 shows a precision of about 10 percent. More significant figures would be meaningless.

Most estimates of environmental risk are only correct within a range of uncertainty that spans many orders of magnitude; that is, risk often has very high uncertainty. Thus, more than one significant figure is rarely necessary to express an environmental risk.

Suppose the standard for an environmental contaminant is a concentration of 5 micrograms/liter of water. You might find it important to measure the concentration to two significant figures. For example, 5.2 micrograms/liter is over the limit, whereas 4.8 micrograms is under it. Suppose the accuracy of the measuring device is ±0.01. (Read this as "plus or minus 0.01.") 5.21 or 5.19 is still above the standard, and 4.81 or 4.79 is below. Instead, if the accuracy is ±0.4, we do not know whether the measurement of 5.2 is above or below the standard, as we might have anywhere between 4.8 to 5.6 micrograms/liter. In addition, the measurement of 4.8 ± 0.4 might indicate a violation. In this case the significant figure and range of uncertainty are important considerations. Both precision and accuracy matter.

Life Tables

The life expectancy of the average person in the U.S. has been increasing; it is now approaching 74 years. We must be doing something right. The chance that a randomly chosen U.S. resident will die during the next twelve months is approximately five in one thousand (or 0.005 or 5×10^{-3}). We expect this risk, because we know from government statistics that during 1996, 2,314,460 persons died in the U.S. Wait, you may say; that does not sound right. If we divide 2,314,460 by the number of persons living in the U.S., the number should come closer to one in one hundred (or 0.01 or 1×10^{-2}).

We could just have divided this number of deaths by the number of persons in the entire population to get a crude estimate, but we actually quoted an age-adjusted death rate to make a little bit better prediction of your risk. The current population in the U.S. has different numbers of people in different age groups. The risk of dying is not the same for all age groups. Suppose that we follow a hypothetical population of 100,000 persons over time and place the observations about them into a table. (*See* Table 1.3)

These data mimic a human population that has been studied. You might notice some interesting aspects of these data. First, the number of persons dying in each age bracket increases with age. Second, we do not start fresh with a new 100,000 persons in each bracket. The number of persons dying in each age bracket must be subtracted from the number of persons at the start of the age bracket to determine the number of persons surviving to pass into the next bracket. When we estimate risk for persons in an age bracket, the deaths increase, but we divide by a smaller number. The risk of death increases sharply with age. Still, by age 81 nearly forty percent of our population remains. Third, for infants who have not passed their first birthday, the death rate right after birth is substantially higher than for ages 1 to 9. Fourth, sometime between ages 70 and 79, only half of the original population remains, but because the risk of death increases with age, this does not mean that many people survive to age 140. Fifth, this population has not undergone some calamity like a war or epidemic that would cause a large increase in the death rate for just one age group. Instead, the risk of dying simply increases with age. Finally, the notion of starting with a group of 100,000 persons is a common convention used in population studies.

Table 1.3 Hypothetical Cohort Life Table for U.S. Citizens Born in 1894

Age Bracket (Years)	Persons at Start (N)	Deaths in Bracket (N)	Risk of Death (Per Person x 100)
0–1	100,000	1,391	1.4
1–9	98,609	463	0.5
10–19	98,146	878	0.9
20–29	97,268	1,901	2.0
30–39	95,367	3,584	3.8
40–49	91,783	5,101	5.6
50–59	86,682	7,515	8.7
60–69	79,166	15,341	19.0
70–79	63,826	25,697	40.0
80–89	38,129	26,819	70.0
90–99	11,310	10,604	94.0
100+	706		

Cohort Life Table

Population data in such formats are called *life tables*. Demographers, the experts who study populations, have several different ways to construct life tables. The two easiest ones to understand are the cohort life table and the current life table. A cohort is a group of people with some common characteristic; in Table 1.3, the common characteristic is age. Suppose that you decide to study a thousand people born in the same year in the same city, starting in 1918, approximately 81 years ago. Every ten years you determine (by making lots of phone calls, writing lots of letters, and checking with the morgue, local schools and so forth) whether each of the original thousand persons is still alive. You will have to exercise care about people who leave the city. (You may decide to remove them from the entire study, when they leave, or you might replace them with someone of the same age who moved in.) Over time, you would develop a table like the one above.

Analyzing any life table demands careful thought. For example, a common mistake in life tables that are used, like those above, to estimate risks is to divide by the average number of people in an age group, instead of by the number at the beginning of a time period. Constructing a cohort life table requires lots of patience and stable resources. A group of demographers of different ages will have to work on it, because no one old enough to start the study is likely to outlive the cohort to finish it. It explains the mortality experience of people born between 1918 and 1927, but the life table is unique to the mortality experiences of those particular persons. Every time you survey your original group, you pass up others in the population born at different times.

Current Life Table

A different way to construct a life table is to examine an entire population at one moment in time. This approach is called a current life table. (*See* Table 1.4) At the beginning of 1959, you might survey an entire population. Your table still will resemble the one above, but the persons in each age bracket will have been born at different times. It has different advantages and disadvantages from a cohort life table. In particular, an event like a war may affect persons of all ages and not stand out so prominently, but if you survey the population in the middle of an epidemic and compare it to other populations not affected by the disease, you would observe an increase in the death rate in all age brackets. Current life tables require less daunting logistics, but you have to gather data on many more people at one moment in time to have the same statistical accuracy. The advantage is that you obtain informative data quickly. (Once again, we have not tabulated persons who have not passed their first birthday.)

Table 1.4 Hypothetical Current Life Table for U.S. Citizens Alive in 1994

Birth Dates (Years)	Persons in Age Group	Deaths in Age Group	Risk of Death in Age Group
1993–1994	100,000	1,391	1.4
1985–1993	98,609	463	0.5
1975–1984	98,146	878	0.9
1965–1974	97,268	1,901	2.0
1955–1964	95,367	3,584	3.8
1945–1954	91,783	5,101	5.6
1935–1944	86,682	7,515	8.7
1925–1934	79,166	15,341	19.0
1915–1924	63,826	25,697	40.0
1905–1914	38,129	26,819	70.0
1895–1904	11,310	10,604	94.0
<1894	706		

The data in a life table need not relate to death from all causes. In this book we mostly tackle human health risks and, to a lesser extent, ecological risks. But we could, for example, record only deaths from cancer or from traffic accidents. Whatever the cause, a population under a stress that affects life span will have higher death rates, and the average member of the population will be younger. Data from either kind of life table for any kind of animal or human population will enable you to calculate age-specific death rates and predict risks.

A small, one-time risk that affects the mortality of every person in the population equally will simply reduce population size. If the risk is large, it matters whether each person has the same, additional (absolute) risk of death, or the effect of the risk is to multiply the person's baseline risk. If, however, the risk recurs each year, it will reduce the average age of a person in the population. The youngest members of the population will have many more chances of experiencing the outcome, death, and they will have less of a chance to reach old age.

Life Expectancy

Life expectancy is the number of years of life left. If a disease selectively kills old people, the population will lose fewer years of life expectancy than from a disease that primarily affects the young. As seen in Table 1.5, diseases of the elderly kill more people in our population than diseases or other factors that occur more randomly with respect to age, like accidents. The effects of these diseases on life span are different.

Table 1.5 Ten Leading Causes of Death during 1996 (U.S.)

Cause of Death	Number of Deaths
Heart Disease	733,361
Cancer	539,533
Stroke	160,431
Pulmonary Obstructive	106,027
Accidents	94,948
Pneumonia/Influenza	83,727
Diabetes	61,767
HIV/AIDS	31,130
Suicide	30,903
Liver Diseases	25,047

(From *National Vital Statistics Reports*, Volume 47, Number 9)

This book mainly addresses risks of human mortality and morbidity, the major outcomes addressed by the federal health and safety regulatory agencies. To a lesser extent, we discuss risks to the environment, particularly wildlife populations. As we explain below, many approaches to the assessment of these risks exist. Ultimately, all of them refer to life tables.

The process of analyzing risk requires us to find specific answers to many questions. If someone with a cold has a fatal car crash that they would not have had without the cold, did they die of an accident or a cold? What effects could a large meteorite striking the earth have on the world's life table? How would we go about trying to estimate these risks?

Uncertainty

The use of numerical probabilities to express risks is neither an accident of history nor merely a convention. Just because of its predictive nature, risk involves uncertainty, and probabilities indicate uncertainties in useful ways. Without getting too far into a very deep subject, probability itself is controversial. To us, a probability represents a state of belief, but measurements can strongly condition our beliefs. We usually choose to believe the available data. Yet, the probability calculus is simple. It only has four important assumptions or axioms; mathematicians can derive everything else in the probability calculus from them.

Probability Axioms

1. *A probability of one (1.0) implies that an event is certain to occur. For example, we believe that everyone will die. Thus the probability that someone will die eventually is 1.0.*

2. *A probability of zero (0.0) indicates an impossible event. For example, we believe that dogs cannot sprout wings and fly. Thus, the probability of seeing a dog flying with its own wings is 0.0.*

You may hear someone say that probabilities are "numbers between zero and one, but never exactly zero or exactly one." This statement expresses the sentiment that nothing is certain and nothing is impossible, a debatable idea. As a practical matter, we seldom, if ever, use the probability calculus with certain or impossible events.

3. *Two random events, labeled "A" and "B," with probabilities of p(A) for the first event and p(B) for the second event, such that A does not affect p(B) and B does not affect p(A), will both occur with a probability calculated as the multiplication product of p(A) and p(B).*

We represent these circumstances symbolically as p(A) x p(B). We sometimes call the condition that A and B cannot alter each other's probability *independence*. If you expect A to occur 30 percent of the time [p(A) = 0.3] and B to occur half the time [p(B) = 0.5], and p(A) and p(B) are independent probabilities, the probability of A and B is 0.3 x 0.5 = 0.15.

p(A and B) = p(A) x p(B), [p(A) independent of p(B)]

Some mathematicians describe this assumption as "and logic" or the "and condition."

You can still calculate a probability, if A depends on B or vice versa, but you need to understand the dependence. If A cannot occur when B occurs, p(A and B) = 0. For example, suppose a family has only one car. If A means "Sam drives the car," and if B means "Jane drives the car," p(A and B) = 0, because two people cannot drive a car simultaneously. (*See* Figure 1.1)

Mathematicians show dependent relationships symbolically with a vertical line. For example, p(A|B) means "the probability of A dependent on B." B must occur before A can happen. When working with *dependent probabilities,* you need to know the dependence.

4. *One or the other of two random events, one labeled AA, the other AB, with probabilities p(A), p(B), such that A does not affect p(B), B does not affect p(A), and both events have low probabilities, will occur with a probability nearly the same as the sum of p(A) and p(B).*

Suppose that you have two exclusive events with probabilities of p(A) and p(B). According to the probability calculus, the probability that either event will occur is p(A) + p(B). If you expect A to occur 3 percent of the time [p(A) = 0.03] and B to occur 10 percent of

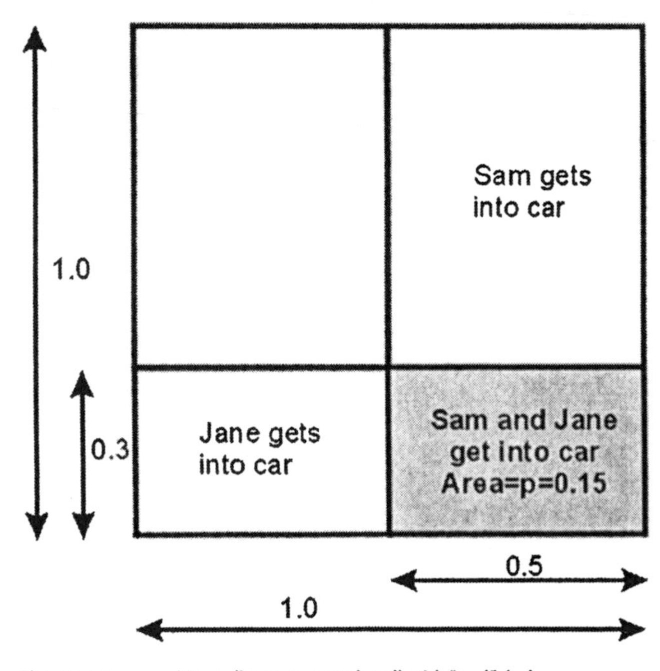

Figure 1.1 Venn diagram on unit cell with "and" dark

the time [p(B) = 0.1], and if A and B cannot occur together, the probability of A *or* B is [p(A) + p(B) = 0.13].

p(A OR B) = p(A) + p(B), [p(A) autonomous of p(B)]

If A and B occur independently, but not exclusively, sometimes A and B will occur together. Then,

p(A OR B) = [p(A) + p(B)] - p(A & B), [p(A) independent of p(B)]

Some mathematicians describe this axiom as "or logic" or the "or condition." If you expect A to occur 3 percent of the time [p(A) = 0.03] and B to occur 10 percent of the time [p(B) = 0.1], and A and B can occur simultaneously in a random way, the probability of A or B is 0.13 - 0.003 = 0.127. Notice that the probabilities of exclusive and independent events do not differ much unless the probabilities are large. As p(A) and p(B) tend toward smaller probabilities, the impact of subtracting p(A and B) from p(A) + p(B) gets less, and with very low probabilities, you can usually ignore the p(A and B) term. For example, if p(A) = 0.001 and p(B) = 0.003, p(A) + p(B) = 0.00003. (*See* Figure 1.2)

What happens if two events are neither autonomous nor independent? You can still calculate probabilities, but you need to know the dependence of A on B, or vice versa, usually obtained from direct estimation.

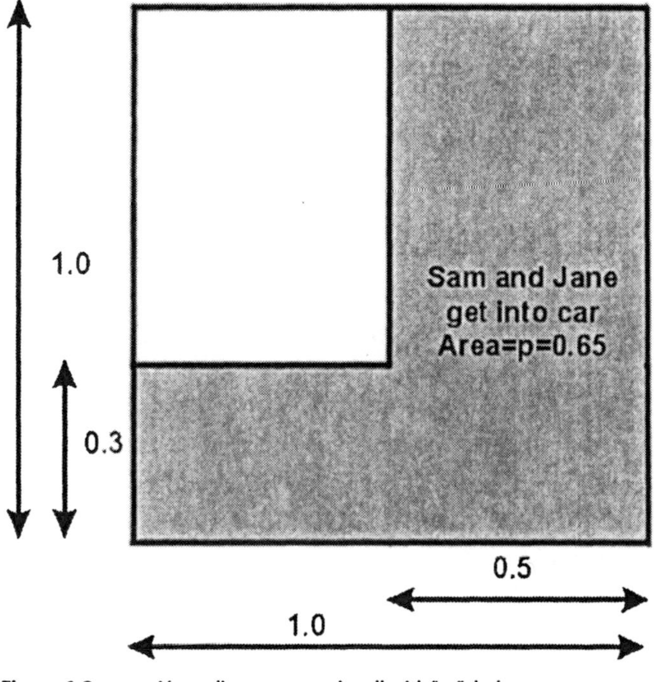

Figure 1.2 Venn diagram on unit cell with "or" dark

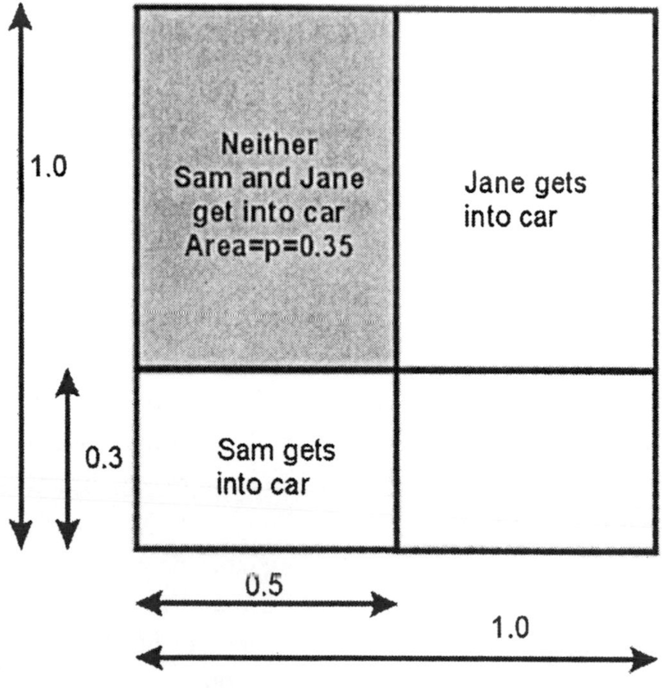

Figure 1.3 Venn diagram on unit cell with "not" dark

One application of the addition rule ("or" logic) is extremely important. The probability that an event will occur or that it will not occur seems to cover all of the possibilities; that is, it is certain that an event, A, will occur or not occur. Since the probability of a certain event is one (1),

$$p(A) + p(\text{NOT } A) = 1, \text{ or}$$

$$p(A) = 1 - p(\text{NOT } A).$$

If you expect A to occur 10 percent of the time, then you expect A not to occur 90 percent of the time. (*See* Figure 1.3)

Contrasting Views of Uncertainty as Probability

There are two contrasting views about the expression of uncertainty as probabilities. One is that maximum uncertainty occurs at a probability of 0.5. After all, you are half way between an impossible and a certain event. With risk, however, probability represents uncertainty in a different way. As the probability of some loss approaches 1.0, we become very certain of the loss. As the probability of the loss approaches 0.0, we become very uncertain of the loss. Part of the reason for this view relates to the way risk analysts gather data. If you want to base your estimates of future events on observations of past

Origin of the Venn Diagram

A famous, nineteenth century English logician, John Venn, devised a way to illustrate sets and probabilities by the areas in figures. The technique may have been a common practice simply named in his honor. The idea is to draw an area within an area to illustrate a group of objects or events that belong to some larger set. For example, people older than fifty are a subset of all persons. When two areas overlap, they illustrate a subset in common. For example, all men over fifty are a common subset of all people over fifty and all men. In probability theory, Venn diagrams have particularly useful properties. Because a certain event has a probability of 1.0, either a 1 x 1 square (a unit square) or a circle with a radius $1/(\sqrt{\pi})$ has a surface area of 1.0, which represents all possible events. You can use any convenient unit (inches, feet, meters). If you can draw accurately enough, you can illustrate the quantitative relationships between probabilities within this space and even estimate them graphically.

events, information about rare events becomes difficult to gather, and you have fewer observations.

Infrequent events are, by their very nature, unlikely to occur and difficult to observe; the more infrequent, the more difficult to gauge accurately. In areas where earthquakes occur more often, like Japan or California, geologists may gather sufficient information to estimate how often an earthquake will occur. If we can estimate that earthquakes will occur once every seventy-five years, we can predict that the risk of another earthquake is low just after a previous one, but it becomes higher, the closer we come to seventy-five years since the last one. In an area where we can only find one instance of an earthquake in the entire geological record, any prediction of earthquake risk becomes quite low and quite uncertain.

Uncertainty and the probability calculus pervade the entire field of risk analysis. The applications are not limited to risk assessment. Any good risk communicator will anticipate, for example, the consequences if, on the one hand, the audience understands a message as opposed to the consequences when an audience does not understand a message. So, the probability calculus applies to risk communication. Similarly, risk managers have to think deeply about uncertainty in selecting an option. What if the best option for managing a risk remains unknown? What if the information is so uncertain that the apparently best choice is wrong? What if the chosen approach increases, instead of decreases, risk?

Validation

Risks, like probabilities, are predictive estimates. They are not statistics, those accumulations of data about events that have already occurred, and they are not descriptions of events that happened in the past. If you estimate a risk, and the probability of the loss is high enough and occurs soon enough in the future, you can probably observe the frequency of the event you have tried to predict. Validation, as a postanalysis surveillance or an after-the-event survey, is a powerful means of assuring that risk assessment methods are working, or that a risk management decision led to risk reduction, or that a risk communication program conveyed the intended message.

Gold Standards

You need to think about the uncertainty in your data. One way is to consider categorical sources of error. Suppose that you have some known information, information about something that you want to achieve, sometimes called a "gold standard." You have a test method that you can employ that will tell you whether something meets the gold standard, but the test, like all tests, is uncertain. In risk analysis, some typical gold standards, paired with accompanying tests, are: rodent bioassays of chemical carcinogenesis and mutagenicity tests of the same chemicals in bacteria; political acceptance of a regulation to control a risk and a survey to test opinions of newspaper readers; or public understanding of the risks of radon in home and psychometric analysis of homeowners in one block of a Philadelphia suburb.

Sometimes the test will give a positive result when the outcome really should be negative (not an animal carcinogen, public resistance to a regulation, widespread misunderstanding of radon risk), falsely indicating a positive result like the gold standard. These results are called *false positives* or *Type I errors*. Based on the positive test results, you will have a false idea about the object.

The diametrical opposite is a *false negative* result. Sometimes the test will give a negative result, when the outcome really was positive (an animal carcinogen, public acceptance of a regulation, reasonably accurate public understanding of the risk of radon in homes), falsely indicating a negative result, the opposite of the gold standard. These results are called *false negatives* or *Type II errors*. Based on the negative test results, you will have a false idea about the object, when it actually behaves like the gold standard.

True positive and true negative test results accurately reveal gold standard outcomes. Obviously, we want to use tests that only give true results. Just as obviously, no test is perfect. In public health and regulatory matters, analysts usually bias their tests toward sensitivity, detecting potential effects and risks. Doing so reduces false negative results at the cost of increasing false positive results. Partially for this reason, regulators are more likely to discuss *potential effects* and risks with the implication that they view the phenomenon they protect against with some skepticism, but "better be safe than sorry." (*See* Figure 1.4)

Reproducibility

The concepts of false positive and false negative results are the heart of several important scientific ideas that also are important in risk assessment. Reproducibility is the very essence of science. If one scientist announces the observation of a new relationship during

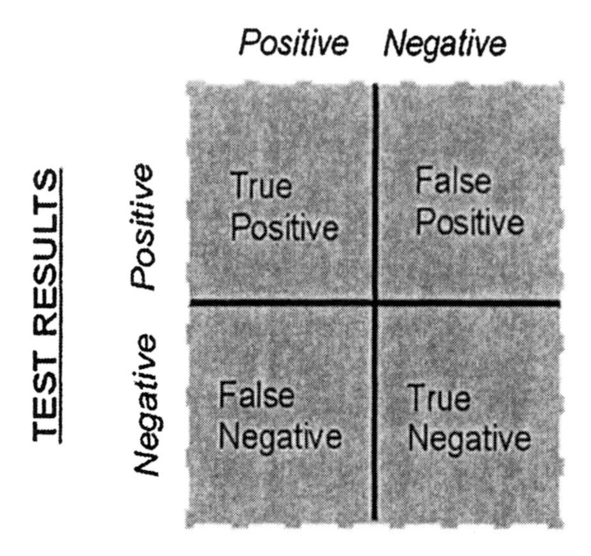

Figure 1.4 Diagram for error classification—gold standard vs. test or sample

an experiment, such as X causes Y, other scientists expect that they can repeat the experiment and get the same result. This process is the essence of validation. "If John adds acid to a solution of this dye, John gets a red color." When Cindy later adds acid to the same solution in a different lab, and no color is observed, or the color is green, John's observation cannot be validated.

In many areas of science, X does not cause Y every single time someone conducts the experiment. Instead, we have the idea that X usually causes Y, or, given X, we will see Y some of the time, more than we see Y without X. Risk assessment often involves such statistical or probabilistic expressions. When Cindy later added acid to 20 solutions of the dye, she observed no color on four occasions, but red color 16 times. John's observation was validated, but something else is going on that the experimenters do not understand. (For example, the appearance of red color may require acid and sodium, and on four occasions, the tap water in Cindy's lab lacked sufficient sodium.)

Many of the epidemiological and toxicological studies that risk assessors use to predict health effects are by nature probabilistic. If we treat mice with some dose of a dye that "causes" cancer, we do not necessarily expect cancer in every mouse. The expression of tumors requires time, and even after a long enough duration, only some of the mice will exhibit tumors, but more so than without the dye.

When a scientist states a proposition like "X causes Y," the scientist does not need to have already observed it. With or without experimental evidence, if such a conjecture closely relates to other bodies of data, we call it a hypothesis. When a hypothesis has

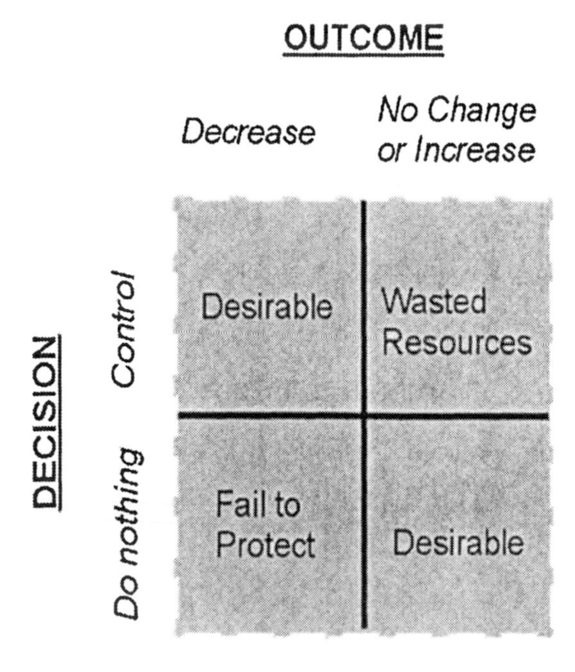

Figure 1.5 Diagram for error classification—decision vs. outcome/increase or decrease

undergone extensive validation, and it is important in a field of scientific inquiry, some describe it as an *observation* or *law*. When many validated hypotheses connect with each other and support a general proposition, we often call it a *theory*, as in Einstein's Theory of General Relativity.

Risk assessors sometimes have to work with hypotheses that they cannot validate or test experimentally. Perhaps the experiment is too expensive or will take too long. Perhaps the hypothesis is intrinsically not capable of undergoing validation. If so, the hypothesis is not scientific. For example, it might be based on a unique observation. The risk assessor has to ask whether such hypotheses or assumptions are consistent with general scientific theories. A risk assessment based on a unique observation that appears to contradict the law of gravity is simply a poor risk assessment. Instead, some assessments predict future losses based on reasonable projections, given what we know about general scientific theories. Sometimes, these supporting ideas are described as trans-science. (*See* Figure 1.5)

The Meaning of Uncertainty

Uncertainty has a concise meaning; it is not just a vague idea that we somehow know about a phenomenon in a fuzzy way or that more uncertainty means less knowledge. Knowledge of uncertainty means that we can describe a range of outcomes of some event. If we can repeat the event, the range of outcomes we observe describes the uncertainty about the event. If we only have one or a few observations, we can argue about the best way to predict this range. Part of the reasoning concerns how many repetitions we need to adequately understand the range of outcomes.

Predictions of uncertainty also depend on the state of our knowledge about a event. Part of our reasoning concerns how well we understand the event—whether we can predict unusual outcomes that are possible, but not expected, in the sample observations. Scientists are used to evaluating the quality of information. They use several criteria: reproducibility of observations, consistency with relevant theories, ability to explain the observations with plausible hypotheses, inability to disprove underlying hypotheses with experiments, capacity for quantitative measurement, and *consilience*, the reinforcing of observations through several independent kinds of evidence and ideas that interconnect. (Wilson, 1998)

Part of our concern is how closely our prediction, based on limited data, will match the observed range of outcomes. Suppose we watch the price fluctuations of the stock of a company over a period of time when nothing really changes about the company's performance. We might observe that all of the prices fell within a range from $11.00 to $23.00 per share. If we think this range is the uncertainty in the value of the stock, we would deceive ourselves. For example, we might think we got a bargain by purchasing some shares at $13.00, only to discover a month later that the company just went bankrupt, and the shares are worthless, or to find that the stock market plunges, and the shares are worth much less in a declining market.

Uncertainty vs. Variability

Imagine for a moment a randomly chosen group of one hundred men from your area. You select one man at random and measure his height. He seems fidgety, and you wonder if you measured his height correctly. You might try measuring it ten times. Instead, you might try using ten different rulers. Whatever method you chose, you will obtain slightly different heights each time. This is called *experimental variation*. Uncertainty about this man's height is exactly the range of observations, and you could attempt to predict this range after the first measurement, using various techniques. They are correct measures of uncertainty in relation to their ability to predict the variation you observed later. (*See* Figure 1.6)

What is your uncertainty about the heights of the other 99 men? If you express this uncertainty correctly, you can come back later and measure the heights of the remaining men. Your projected uncertainty should closely match these measurements. In this case the uncertainty arises, not from measurement error, but from variation in height in the population. Still, the technique you use to predict this range of variability is correct in relation to its ability to predict the variation you observed later.

Notice that, once you have a complete description of the outcomes, plus the probabilities of each outcome, you have complete description of past events. You can proceed

Figure 1.6 Measuring height

to predict outcomes you cannot observe. The simplest assumption is that future outcomes will resemble your description—heights in the future will be the same as heights in the past. The concept of uncertainty now becomes somewhat elusive. If you have a complete description, why do you have any uncertainty at all? This conundrum highlights the confusion about uncertainty and variability. If we do not know much about the heights of men, we have to rely on sampling and theory to predict the range of outcomes. If we have adequate observations, we no longer have uncertainty, but we still have variability. Variability and uncertainty affect predictions about the future differently.

Similarly, we can repeatedly observe the measurement of one person's height. The observations characterize the uncertainty in the method of measurement. That different heights, characteristic of the population, are measured differences, whereas different measurements of height characterize measurement error, not different heights, is the essential difference. It shows the difference between variability and error. It also shows why the methods to predict and evaluate with uncertainty often are identical.

In both instances we can understand the sources of uncertainties. Uncertainty about height in the population relates to factors such as human genetics and nutrition, whereas uncertainty about measuring a height relates to factors such as stretching of measuring tapes, precision of human perception, and changes in posture. When we set out to understand height in a population, however, we understand that we can progress toward a description of variability that we do not expect to change. We may come up with better methods to measure height and reduce uncertainty.

Consequences of Uncertainty

Uncertainty clearly relates closely to risk and probability. The life tables (Tables 1.3 and 1.4) we reviewed earlier tell us that mortality does not fall evenly on members of the population. We are uncertain about its causes. Conditions may change, and future events may overwhelm our estimates of risks. The very data we review are subject to measurement error. Science-based reasoning is imperfect knowledge. The risks that we do not suspect and the conditions we do not anticipate are the greatest sources of uncertainty.

Before the discovery that viruses caused common colds, how would you have estimated the risk of colds? Before the discovery that large meteorites strike the earth periodically, how would you have estimated the chances of massive species extinctions? How do you know what you do not know?

Notions of false positive and false negative pervade risk analysis, not just risk assessment. A risk manager may reduce exposure to some substance, mistakenly believing that society will experience reduced risk, only to discover later that the opposite has happened. While busy managing this risk, a different risk, believed of no significance, may have evaded the available resources, only for the risk manager to discover that an important source of mortality in the population has gone unaddressed. Similarly, a risk communicator may have the information right, but the audience walks away with irrelevant or, even worse, counterproductive ideas. However, the impact of this error may be no worse than having the audience understand exactly what the communicator intended, only to discover that the information was wrong.

Models

Models are representations of objects or processes. We will invoke models throughout our discussion of risk. Different kinds of models exist.

Physical models, like the miniature dog house in the cartoon (Figure 1.7), attempt to represent an object or process in a tangible way. Physical models are particularly useful in sales. For example, they let potential buyers understand an object, like a house, without having to visit the house in person, if it exists, and without having the sales agent lug the house around. A crash test dummy is a physical model of a person used to understand the effects of different conditions in a car accident. The physical model lets scientists improve car designs and reduce risks without putting living people at risk.

Similarly, you could create a physical model of information flow within an organizational structure, with Styrofoam blocks to represent the components and hoses of different sizes to represent information. So, moving the blocks around would let people think about the consequences of a reorganization on information flow. The model would represent a process.

Figure 1.7 The architect's doghouse

A cartoon also is a pictorial or graphic model. We are used to symbolic representations in our society. What looks like a dog but is not a dog? A picture of a dog. Similarly, a stop sign at an intersection is a symbolic command to engage in several behaviors. Diagrams, charts, graphs, and blueprints are graphic models.

Conceptual models represent objects or processes as ideas that we can explain to others. Earlier in this chapter, we described a conceptual model of risk analysis as consisting of three components: risk assessment, risk management and risk communication.

Biological models use one living organism to represent another. For example, scientists use mice as biological models of people. Much of the information in toxicology comes from studies of rodents exposed to substances under controlled conditions. Unlike a crash test dummy, a mouse has biological properties like those of a person, including development and reproduction.

Mathematical models describe some characteristics of an object or process, usually with algebraic equations. Physicists, for example, describe the motions of the planets with equations. Mathematical models confer economy of expression. Writing down a few equations is much more compact than tabulating long lists of the positions of each planet at different times. Equations are models that have everyday uses also. The equation, $2A \times B = C$, lets a seamstress conveniently calculate the area of a table cover that folds in half. The equation is a mathematical model of the tablecloth. In addition, equations let us make predictions more easily and lead to the formulation of hypotheses to test. Eventually, we gain greater understanding.

Uses of Mathematical Models

When you use a model to obtain a more precise answer within the range of data used to construct it, you *interpolate*. If, instead, you extend the model beyond the available data and theories, you *extrapolate*. Some scientists are very cautious about extrapolation with models. Extrapolation is, by nature, a highly uncertain process. Often, however, the very purpose of model construction is to make a speculation about information that we need but is outside the available data. Perhaps no one has ever made a tablecloth as large as the seamstress is about to sew, but she still expects the formula to work.

Mathematical models overlap with *statistical models*. The idea of a statistic as a collection of data about some object or process is a little restrictive, even quaint to modern statisticians. A better definition is that statistics involves ways to summarize data and make inferences. For example, we can compare one group of measurements with another group. Thus, some think of statistics as the mathematics of experiments. Statistical models help in analyzing experimental data, and in this regard, are distinctly unlike probability models. Multiplicative processes that use "and logic" and additive processes that use "or logic" both lend themselves to model approaches.

Simulation Models

Mathematical models also overlap with *simulation models*, which are usually more complex and more descriptive. For example, unlike an equation model, a simulation model may contain a table of measurements. The output of a simulation model is often a graph or a picture, instead of a numerical result. Many of these models attempt to reproduce (or simulate) processes, such as the survival of car occupants during a crash. Because the simulation model depends on computing power, an analyst can conduct many simulation exercises quickly to see how different conditions affect the occupants' mortality—less realistic but more cost-effective than a series of test crashes with real cars and crash-test dummies.

Simulation models are quantitative and computationally intensive in their operation. They have many advantages, including their reproducibility in the hands of different assessors, making them impartial in application. They can cope with large amounts of data, integrate different kinds of data, represent the knowledge of experts, and deal with complexity, yielding practical, "real world" answers. Simulation modelers can add mathematical complexity initially, to obtain a better correspondence to a situation, and test the more complex model experimentally. This dialogue between data and model is a very powerful way to advance understanding.

Simulation modelers are fond of quoting a sentence that George Box wrote in 1979, "All models are wrong, but some models are useful." It conveys the spirit of the modeling process. Simplified, often mathematical descriptions of systems also provide the basic elements of risk analysis. When a simulation model provides a framework, the model builders probably will have the most skepticism about the results. They have a sense of the limitations of the assumptions used to construct the model.

The advantages of a modeling approach are obvious. A mathematical framework quickly directs attention to missing data, wrong units, conflicts with generally accepted scientific theories, or unrealistic predictions. However, the modeling approach is not foolproof. Because models have a kind of Wizard of Oz aura, someone unfamiliar with modeling often will be the most impressed with predictions based on a model. This feature makes models dangerous.

Model Validity

A model is *valid* when it behaves the same as the object it represents in every way that is important for the analysis and under all relevant conditions. The best way to validate a model is to see whether observations under the conditions in question exactly fit the model's predictions. This requirement is something of a paradox. If you have the data to validate a model, you can just use the data. Why construct models at all? A model will point out the most important conditions to test. Generally, data gathering is much more resource intensive than model building. In addition, you can reuse the model to estimate a new situation. Having validated it once, your confidence in a model will increase. You will be less likely to obtain false positive or false negative predictions.

Even when you cannot directly obtain data about the situation you have modeled, you can do many things short of complete validation to find the shortcomings of a model. One way to think about model uncertainty is to view it hierarchically. The most understandable errors in a model are errors about variables. The seamstress trying to make the tablecloth may not have a good ruler or the cloth may be slippery. Any error in measurement will cause an error in her estimate of the area, for example, an important error, if she orders too little cloth. What if, in her frustration, she measures the length five times and the width five times? In later chapters of this book we will discuss *Error propagation* techniques that are available to understand the implications of the various measurements for her estimate of area.

Suppose, however, that the seamstress has the formula wrong. Perhaps the tablecloth really is circular, for a round table. Having the wrong formula can wreck the usefulness of a model. In the example, the seamstress is most likely to order too much material, as a circle will fit into a rectangle, and she will simply waste resources. The opposite error, too little cloth, might cause her major problems with, for example, the purchaser of the tablecloth. She needed a completely different mathematical model.

The most devastating errors are conceptual errors. Suppose that through some miscommunication the purchaser shows up at her shop expecting delivery of a table, not a tablecloth. Asking the wrong question altogether is likely to yield the largest errors. Science has a long history of this kind of error. For example, the idea that bad gases caused swamp fever caused medical scientists to invest in air conditioning to prevent malaria.

No amount of work on a model of refrigeration could yield a consistent decrease in disease, although clearing swamps to improve the climate did eliminate the breeding sites of the mosquitos that spread the disease.

Regulations and Regulatory Models

When our society experiences terrible losses, we try to reduce future losses by creating regulations. The Congress that we elect has ultimate responsibility for the direction our country takes to reduce risks to the population and the environment. When our representatives fail to address our concerns about risks in ways that have meaning for us, we can vote for someone else in the next election. The administrative branch carries out the legislation that Congress passes. The court system, the third branch of our government, adjudicates conflicts. Ultimately, however, the federal regulatory agencies that protect us (and that interest us in this book) are distinctly creatures of Congress.

Representatives and senators create these agencies for a variety of reasons, but the two most important are these: (1) to insulate themselves from direct political pressures, and (2) to insert greater technical expertise into risk reduction. In the process of creating and overseeing regulatory agencies, Congress decides whether the agencies should use risk analysis in carrying out their missions, and, if so, how. Sometimes, as under the Toxic Substances Control Act, EPA must use risk-benefit balancing in making risk management decisions. Yet under the Clean Air Act, congress prohibited EPA from engaging in risk-benefit balancing.

Politics of Regulation

Federal representatives and senators are politicians, and getting elected is sometimes their primary concern. Obviously, voters respond to a wide range of social concerns in elections, not just risk reduction. Thus, the U.S. has an imperfect system to translate social concerns into legislation. The way that you feel about a speed limit or a waste site does not translate into election of a candidate who will do exactly what you want.

We have a representative democracy. Members of Congress are not supposed to be high-minded angels engaged in philosophical discussions. They are supposed to look after interests of the people who elect them. The representative of a rural district is not elected to preside over the future of American cities. The representative of an urban district is not elected to pay much attention to agriculture. The interactions between these representatives balance different interests.

This system does surprisingly well at creating institutions that prevent losses, but it has two major deficiencies. A representative democracy tends toward episodic outbursts of attention to a subject. Some incident calls public attention to the need for regulation of a substance or process. Congress responds, sometimes in overwrought or impetuous ways. The idea of passing responsibility onto an agency that can sustain focused consideration about the situation seems a wise approach. Some have a more cynical view, however, of politicians taking credit for responding to public demands, then giving responsibility to an agency with more technical resources to resolve problems, and ultimately blaming the agency for the problem.

A representative democracy also is vulnerable to demands from special interests, particularly from interests that can command public attention through the media. At the time we write this book, concerns about children have become a dominant political pressure. How does an elected politician resist accusations of being against kids? It is a nearly impossible situation. Thus, any time a special interest can align legislation favorable to its cause with the interests of children, and then spend sufficient funds on publicity about how its cause benefits children, it can virtually guarantee the passage of legislation.

Regulatory agencies are where the action is found. Once Congress creates and funds them, they can create and enforce regulations, which have the force of law. Regulatory enforcement follows regulatory development. Representatives write the basis into statutes, but the agencies do extensive work to translate them into actionable requirements. Most regulations pass through a notice-and-comment process before they become final., which can take many years. Then, after an agency achieves a final regulation, contending parties can sue, and the lawsuits may take many years to come to resolution. Yet, once the regulations are in place in the U.S., most compliance is voluntary. The judiciary recognizes that many decisions are forced, and a large body of case law protects the officials responsible for these decisions.

Legislation creating regulatory agencies or giving existing agencies like EPA new authority has forced attention on regulatory risk analysis. No one is quite sure why the U.S. Congress chose to write "risk" or equivalent language into so many statutes over the past thirty years. Sometimes, as with the Clean Air Act, it has changed its mind. The Consumer Product Safety Commission, the Department of Agriculture, the Department of Transportation, EPA, the Food and Drug Administration, the Nuclear Regulatory Commission, the Occupational Safety and Health Administration, and the federal research institutes that support these agencies have been forced to pay attention to risk analysis to discharge their responsibilities and obligations.

Responding to Technological Change

The regulatory system changes with time as social concerns change. Many advances in technology, such as new biologically engineered crops or cloning, have raised concerns. Over the past few thousand years, we must be doing something right with technology. Average life span has nearly tripled. Yet, many remain uncomfortable with the rapid pace of modern advances. Lifestyle changes seem to come nearly annually with the introduction of new technologies. Some skepticism about these large scale experiments is in order, particularly when we feel that participation is involuntary. In addition, the successes of regulation have raised expectations. As we have mitigated larger risks, smaller risks have become more important.

In addition, the technological fixes of the risks imposed by other technologies have their own risks. Some social reaction to a new technology is important, particularly since its developers usually underestimate its risks. Typically, some of it also is nonsense, a wistful embrace of a past that never existed. At times, some reaction even serves bizarre agendas. The debate often has emotional content, and sometimes the debate is irrational. For example, agriculture is a technological invention that is continually changing. This technology has supported a huge growth in the human population, and higher population densities also create new problems.

Risk analysis provides a very useful set of tools for reducing technological risks. Regulatory models have protection as their focus, not accuracy. The notion is one of boundaries. If a substance or process has some undesirable property, we can simply prohibit it. For example, nickel refining is associated with cancer in the work force. We do not need risk analysis to eliminate nickel refining. Risk analysis comes into play when we try to decide how to control nickel refining to reduce as much risk as we reasonably can and still enjoy the products this industry provides. The idea is to generate a mathematically coherent framework of scientific fact to support more complicated policy decisions.

Process of Risk Analysis

Suppose that you have to decide how high to build levies to protect a city against floods. If you build them too high, you will waste resources that the city could employ elsewhere for schools and hospitals. If you build them too low, floods will occur, and education and medical care would become difficult, if not impossible. The resources you would save would go to waste. To get started, conduct a risk assessment. Gather facts. Ask scientists to look at river banks for historical information about flood levels. Find out whether activities upstream might change these patterns. Do climatologists expect a change in rainfall in the river basin?

Having found and integrated this information, you now start on the risk management portion of your task. You are not likely to set a goal of protecting against the average flood of the past one hundred years. If you do, slightly less often than every other year, the city will flood. Instead, you want to construct levies that will protect most of the time. The basic concepts of probability and uncertainty are crucial to this task. Once you decide, for example, to protect against all but the highest flood you expect every fifty years, how do you explain your decision to the residents. How will you speak with them in a few years, when by chance the river floods to the highest level seen in the past two hundred years?

We usually estimate an average first, then try to evaluate uncertainty. Trying to estimate a boundary condition first, without worrying about the average, seems strange, but it is often more efficient for regulators to aim directly for the boundary condition. Sometimes, it is all they can reasonably achieve. For example, the geologists may not discover the average height of the water, but they can show that the river never rose above a specific height in the past one hundred years.

Regulatory models, which we will discuss in the following chapters, require approval by scientists, regulators, affected stakeholders, and the public. Some call these models *conservative*, but this term is confusing, because sometimes *liberal* politicians prefer *conservative* models. To avoid this problem, we will call them *protective* models or boundary models. All kinds of models can have regulatory versions, and they occur in all phases of risk analysis.

From the perspective of risk analysis, a small loss that occurs frequently is inherently different from a potentially huge loss that is a rare event, like a change in the climate or an earthquake. We sometimes call the latter high-consequence-low-probability (HCLP) risks. Regulatory models for HCLP risks will differ from regulatory models for risks that, from a societal perspective, occur frequently. For example, with approximately 20 percent of the population dying from cancer, small predicted changes in the risk of cancer have immediacy, but not much impact. We will analyze them in distinctly different ways.

The Rest of the Book

In this chapter we have introduced you to some information about risk, the probability of a future loss. We showed that risk analysis, the development, study, and use of risk-based information, has a structure consisting of risk assessment, risk management, and risk communication. Finally, we saw that risk analysis has important applications in the regulatory efforts of federal agencies.

Risk analysis is a general concept in the abstract, but it is content specific in application. Congress wrote the law creating the regulatory agencies, which interest us in this book, in response to specific situations. Thus, the Food Safety Inspection Service of the Department of Agriculture primarily inspects meats. EPA really is a collection of little agencies, with the most prominent ones concerned with environmental routes of exposure: air, water, waste sites, chemical substances, and pesticides. Similarly, the Food and Drug Administration looks after a variety of risks, drugs, food contaminants, food additives, medical devices, and radiological devices. We will start in the abstract, but cover some of the specific applications in case studies and examples.

A life table of the U.S. population helps in comprehending human health and ecological risks, the major emphasis in this book. Uncertainty is intimately connected to risk and probability. Models are very useful in appraising risk, or components of risk, such as exposure. Quantitative aspects of risk analysis are unavoidable, and regulatory applications of risk analysis require some background knowledge of their organizational context. In the next two chapters, we expand on these two themes.

The middle of the book follows the structure of risk analysis. Initially, we provide a more detailed treatment of risk assessment, then expand our views about risk management and risk communication. Toward the end, we have included some specialized topics that range over all three components.

Our approach throughout this book is to provide you with introductory information, essentially to set the stage. Without producing a multivolume encyclopedia, we cannot provide all of the information about risk analysis you might need. Instead, we point to additional information, through literature citations, useful texts, Internet addresses, and listserver addresses. Obviously, we cannot guarantee the availability of these reference materials, and Internet addresses change frequently.

Two professional societies maintain useful Internet sites that will lead you to other risk-related topics. The Society for Risk Analysis home page is located at www.sra.org, and the Risk Assessment & Policy Association is located at www.fplc.edu/tfield/rapa.htm. Similarly, several mailing lists allow participants to discuss recent topics about risk among themselves. The most active one is RiskAnal. If your name is John Doe, send the message "subscribe riskanal John Doe" to lyris@lyris.pnl.gov in order to subscribe. RISKNet, another mailing list devoted to risk and insurance issues, has its own home page at www.riskweb.com/riskmsg.html.

Several private sector and nonprofit think tanks have Internet sites that focus on risk analysis. One of the most useful is RiskWorld, with a home page at www.riskworld.com. The Risk Sciences Institute is a component of the International Life Sciences Institute at www.ilsi.org/rsi.html, and the Center for Risk Analysis is part of Resources for the Future. It has published a short but excellent primer, *Understanding Risk Analysis*, at www.rff.org/misc_docs/risk_book.htm.

Many academic institutions support research institutes focused on risk. We will mention only the few that we know well: the Center for Technology Risk Studies of the University of Maryland at www.enre.umd.edu/ctrs, the Institute for Reliability and Risk Analysis of George Washington University at www.seas.gwu.edu/seas/institutes/irra, and the International Institute of Forecasters at weatherhead.cwru.edu/forecasting. For some interesting discussion of risk assessment models, see Guelph University's site, tdg.uoguelph.ca/omafra/risk.html.

Several federal sites have a wealth of risk-related information. For example, see the Department of Energy's "What is Risk?—Principles, Policy, Papers,and Primers" at www.em.doe.gov/irm/question.html. Several of DOE's National Laboratories have extensive information available about risk and risk-related subjects. Similarly, see the National Center for Environmental Assessment in the Environmental Protection Agency at www.epa.gov/ncea/mcpolicy.htm or the Office of Risk Assessment and Cost-Benefit Analysis in the Department of Agriculture at www.usda.gov/agency/oce/oracba/oracba.htm. Books written by Committees of the National Research Council have influence in federal institutions. Many are still available through the National Academy Press at www.nap.edu, and you can order them over the Internet. One of their publications, the "red book," still is required reading for anyone undertaking regulatory risk analysis projects. [*See* Committee on Institutional Means, *Risk Assessment in the Federal Government: Managing the Process*. National Academy Press, Washington, DC pp. 191 (1983).]

We recommend the following three books to students as the best information about risk analysis. The last one, in particular, is not directly about risk analysis but more in the nature of a provocative and insightful inquiry into the nature of probability. Unfortunately, the publishers of the last two books have not kept them in print. So, you will more likely find them in a library than in a book store.

Glickman, T. S. and M. Gough. 1993. *Readings in Risk Resources for the Future*, Washington, DC.

Morgan, M. G. and M. Henrion. 1992. *Uncertainty: A Guide to Dealing with Uncertainty in Quantitative Risk and Policy Analysis*. New York, NY: Cambridge University Press.

Von Mises, R. 1981. *Probability, Statistics and Truth*. Reprinted by Dover Publications, New York, NY.

References

Lowrance, W. W. 1976. *Of Acceptable Risk: Science and the Determination of Safety*. Los Altos, CA: William Kaufmann, Inc.

Larry Laudan. 1997. *Danger Ahead*. New York: John Wiley & Sons.

Kaplan, S. and B. J. Garrick. 1981. On the Quantitative Definition of Risk. *Risk Analysis 1(1)*: 10–27.

FAO/WHO. 1995. Expert Consultation Report on the Application of Risk Analysis to Food Standards Issues.

North, D. W. 1995. Limitations, Definitions, Principles and Methods of Risk Analysis. *Review of Science and Technology 14(4) 915*.

Committee on Institutional Means. 1993. *Risk Assessment in the Federal Government: Managing the Process*. Washington, DC: National Academy Press.

Rowe, W. D. 1997. *An Anatomy of Risk*. New York: John Wiley & Sons.

Palmore, James A. and Gardner, Robert W. 1994. *Measuring Mortality, Fertility, and Natural Increase: A Self-Teaching Guide to Elementary Measures.* Honolulu, Hawaii: East-west Center.

Starr, Chauncy and Whipple, Chris. 1969. Social Benefit versus Technological Risk, *Science* vol. 165 899: 1232–8.

Morgan, M. G. and Henrion, M. 1992. *Uncertainty: A Guide to Dealing with Uncertainty in Quantitative Risk and Policy Analysis.* New York: Cambridge Univ. Press.

Wilson, E. O. 1998. Scientists, Scholars, Knaves and Fools. *American Scientist 86:* 6–7.

Ash, C. 1993. *The Probability Tutoring Book.* Piscataway, NJ: IEEE Press.

Taylor, J. R. 1997. *An Introduction to Error Analysis.* (2nd Ed.) Sausalito, CA: Univ. Science Books.

Box, G. E. P. 1979. Robustness is the Strategy of Scientific Model Building. (In) R. L Launer and G. N. Wilkinson (Eds.), *Robustness in Statistics.* New York, NY: Academic Press.

Byrd, D. M. and Barfield, E. T. 1989. Uncertainty in the Estimation of Benzene Risks: Application of an Uncertainty Taxonomy to Risk Assessments Based on an Epidemiology Study of Rubber Hydrochloride Workers. *Environmental Health Perspectives 82*: 283–287.

Chapter Two ....

Functions, Models, and Uncertainties

Qualitative Approaches to Risk

As we have seen in the introductory chapter, using conceptual models of risk, such as qualitative or quantitative ones, allows us to sort out the many details. Qualitative notions, in the form of models, sometimes have advantages of speed and simplicity. But, in a society lacking math skills, learning to express yourself in language, through analogies and ideas, is essential to good risk communication. In this chapter we will discuss the different ways to sort out risk, both qualitative and quantitative.

Techniques exist to work with *qualitative data*. For example, the *median* of a population is its middle. Think about the heights of some randomly chosen men. Suppose you need to find the man who has the median height. You could measure the height of each man. Then you could sort the measurements, and count until you reached the man halfway between the tallest and shortest. You do not need to make any measurements, however. You could just rank the men from tallest to shortest by lining them up against a wall and comparing their heights by eye. Then, you could count from the shortest man until you reached the middle one in line. Unless you are having trouble with your measuring rod or your eyesight, you should select the same man with either method.

The techniques for analyzing qualitative data are ranking methods (Gibbons, Olkin, and Sobel, 1977). Using them, you can accomplish many tasks. You can estimate how different in height two populations of one hundred people might be. You might discover, for example, that men are significantly taller than women.

You do not need to measure the women for this task. You can rank them, as you did the men. Then, you can compare the shortest man with the shortest woman, the next shortest man, and so forth. When you finish, you will discover that the male to female comparisons mostly go in one direction: the shortest man is taller than the shortest woman, and so forth. This male-female difference is an instance of *sexual dimorphism* in humans. You can even calculate the statistical significance of the difference between the heights of men and women, without knowing the difference in inches or meters.

Frank Wilcoxon devised a rank sum test to compare two populations, and statisticians named the test for him. It is a good example of a ranking approach and its relation to statistics (Wilcoxon, 1945). The procedure is simple. You mix the two populations together and rank all the members in the mixed group. Assign a numerical rank to each person in order of height, using first =1, second =2 and so forth. You use the average rank for ties. Then, add up the ranks for the men and the women separately.

If ten men and ten women, randomly chosen, have the same average rank, the rank sum for either population would be 105. (Try making this calculation yourself.) If the men and women alternated in height, with each man just taller than the woman beside him, the rank sum of the women would be 100, and the rank sum of the men would be 110. If all of the men are taller than any of the women, the rank sum of the women would be 55, and the rank sum of the men would be 155. We usually will accept the hypothesis that men are taller than women, if the sum of the ranks for the men is greater than a value we can obtain from tables in statistical reference books (or from a large sample approximation). Here, a rank sum of 110 is consistent with a probability of 0.170. Seventeen percent of the time seems like a common occurrence. However, any rank sum for men of 128 or more will happen with a probability of less than .05, or less than once in twenty at random, a probability many people think is significant.

Other techniques exist to conduct risk assessments with qualitative data (Chicken and Hayns, 1989). Component data, such as greater exposure at site A than site B, are ranked and processed, enabling a rapid comparison of the two sites. The net result of the operation of a qualitative risk assessment will be that some risk (A) is greater than another risk (B). Qualitative approaches are useful, and they illustrate the minimum data necessary to understand risk. They have the advantages of speed, simplicity, and transparency.

Ranking methods depend on simple comparisons. Is A greater than B ($A > B$)? Is A less than B ($A < B$)? Are A and B approximately the same ($A = B$)? These methods do not incorporate the magnitude of the difference. If A and B are our only interest, the size of the difference between them may not matter. When we want to compare the difference between A and B to the difference between two other entities, C and D, it may matter a great deal. Similarly, we will encounter sharp limits in our ability to conduct accurate mathematical operations on ranked data.

Converting quantitative data to qualitative data is usually a one-way street. If we convert some measured heights to ranks and throw away the measurements, we cannot recover some potentially important information. Perhaps the tallest people in a small population are more than seven feet tall, but with ranks, we cannot tell that the population differs from a more typical one. Until we mix two populations and compare their ranks, all we know is that each population has some taller members. A basketball coach would notice the difference immediately.

If we keep our quantitative measurements, we can always go back to them. If we substitute qualitative expressions, such as high and low, or likely and unlikely, and get rid of the original data, we lose much information. Very different entities may get lumped into the same category. So, if we never have quantitative measures, but only ranks, we cannot know some important things. Like other crude approaches, however, ranking methods also are robust. Robust means reliable, consistent, dependable. Sometimes qualitative data are all that will be available to you. If so, you still have some tools and notions of how to analyze the data.

Nonparametric, a term you will encounter, literally means without parameters. Currently, risk analysts use this term in ambiguous ways. Some analysts will characterize a population of quantitative measurements, like the heights of the men, as nonparametric data. They mean that no one has created a mathematical model that will describe the population. If someone had created a model, the model would have parameters. Others use *nonparametric* to mean a population without quantitative measures, similar to the ranked populations that we just discussed.

Besides qualitative attributes that indicate a relative amount, some qualitative categories indicate the possession or lack of a property, such as substances that are carcinogens and noncarcinogens. Often, these properties reflect a one-way approach. For example, an analyst might ask what data about a substance reflects the properties of a carcinogen. Lacking such data, the substance may be conditionally called a noncarcinogen; that is, it is a noncarcinogen until someone gets some data suggesting that it now is a carcinogen.

This treatment of a property or quality reflects the *null hypothesis*, the original idea you set out to understand when you undertook information gathering. Examples of a null hypothesis are: $A > B$, $A < B$, and $A = B$. Again, false negative and false positive notions govern how you go about testing a hypothesis, depending on how you include them.

You can express your views about the quality of data available to make an inference, and the inference itself. Qualitative *descriptors* often lead, however, to misperceptions. For example, one way to approach carcinogens is to define a class of all known human carcinogens.

The alternative descriptor is a suspect human carcinogen. If you lack data to support classification of a substance as a known human carcinogen, all you can do is keep an open mind about new data that might arise in the future. Thus, the category of suspect human carcinogen includes all substances that lack test data and, possibly, some with negative test data. Referring to substances with negative test data as suspect human carcinogens is particularly confusing for many people. The basic problem is that they perceive the term, suspect human carcinogen, as a kind of qualitative risk estimate, instead of a category defined by a lack of dispositive data. A suspect human carcinogen sounds like something that we would like to avoid.

When an assessor characterizes a risk with a qualitative descriptor, such as "likely," or "trivial," a quantitative impression still supports the portrayal. The problem is that these descriptors may mean different things to different people. Ultimately, risk implies a probability, and we best state probabilities numerically. We believe that assessors often could express risk numerically, when instead they too often use qualitative descriptors. If they provided numerical expressions, at least we would not have confusion about the probabilities intended by likely or trivial risks.

Unfortunately, the common human tendency is to attach some universal meaning to descriptors, outside (or beyond) the circumstances used to develop the test data. Scientifically, for example, we may not mean to imply that human carcinogenic effects are a property of a substance like a molecular weight, that remains a characteristic of the substance no matter the circumstances. Instead, the substance may only cause cancer in human under a narrow range of conditions. *Conditional descriptors* should express conditional probabilities.

A better approach would use standard test conditions. You could decide which substances have carcinogenic properties, which substances have anticarcinogenic properties (a two-way approach), which substances have neither property, and which substances have not been tested adequately under the standard conditions. You could express your uncertainty about these categories separately. So, collections of uninterpreted data obtained under standard conditions, such as the bioassay data maintained by the National Toxicology Program, are particularly valuable resources for risk analysts.

With standard test conditions, you also can state what properties the test methods generally imply. We could then attach conditional interpretations to your categories. We probably cannot generalize much beyond the test conditions, but at least we would have coherence and consistency in our descriptors. This alternative approach is, however, seldom employed by regulatory agencies of interest in this book or the research institutions that support them.

Quantitative Approaches to Risk

Another of the founders of modern risk assessment, Professor Richard Wilson, often observes that a scenario concerned with risk does not begin to make sense until you try to attach numerical values to your ideas. You try to manipulate these values, looking for coherence and consistency in your estimates. We agree. Wilson's idea resembles the one in language that emphasizes the need to define a word before using it.

Risks imply quantitative probabilities, even if assessors describe the estimates with qualitative descriptors, such as likely or trivial. When a descriptor is qualitative, we expect that the assessor could substitute a numerical expression of the risk, given sufficient time and resources. For example, the probability that you will die of cancer during your lifetime is approximately 0.2 $lifetimes^{-1}$. (Read these units as "per lifetime.")

If you experience an exposure to radon of approximately four (4) picocuries/liter of air throughout your seventy (70) year expected lifetime, the probability that you will die of lung cancer adds to your existing risk, and your risk of dying from cancer increases to 0.22 $lifetimes^{-1}$ (0.2 for the background rate and 0.02 for the extra risk due to radon exposure). To us, this is a large risk. What to you think?

Numerical expressions of risk lack ambiguity and potential misperception. They have potential advantages of both accuracy and precision. You might think that an increase in lifetime risk of death from 0.2 to 0.22 is small. Whatever our conflicting descriptors, if we state the risk numerically, we cannot have confusion about the risk. Quantitative risk descriptors have an advantage: they do not incorporate subjective terms in their expression. Unfortunately, we live in a world that is largely enumerate, that is, it lacks mathematical skills and understanding. Of necessity, perceptions about numerical expressions still include value judgments, and in our society the perceptions vary widely. A story, possibly an apocryphal one, will help you grasp what we mean.

An electric utility sought a permit to build a new plant. The manager of the plant-to-be brought blueprints for the design to a local citizen's council for approval. He spent long hours preparing for the meeting. When asked about the total lifetime risk to local residents from the operation of the plant, he proudly declared a lifetime risk of premature death of only 7×10^{-5} *per average resident per lifetime. The council members did not understand his notation. So, he had to explain orders of magnitude. To his dismay, the council was unwilling to approve the permit. They requested a lower risk design.*

Back to the drawing boards went the manager. At significant cost and with more expensive components elements, he obtained a design with a risk of only 7×10^{-6} *per person. The council also did not approve this permit. Desperate, the manager designed a plant with overprotection everywhere, triple barriers, and such great expense that he doubted his company would ever construct it. Still, the council turned down a plant with a lifetime risk of premature death of only* 7×10^{-7} *per average resident. "What do you want?" cried the poor manager. "I have reduced the risk of an already safe plant by two orders of magnitude." "We like what you have done with the orders of magnitude," the president of the council declared, "but we want to see a little progress with the seven as well."*

Risks have units. Using a quantitative descriptor does not relieve the assessor of the obligation to clearly state the units. For example, a 0.02 additional probability of premature death from cancer during a 70-year lifetime of exposure to 4 picocuries of radon/liter of air.

Table 2.1 lists some estimates of mortality risk. For each source, we divided the number of persons who died during a one-year period from each cause by the size of the population at risk. Thus, we have the risk for an average U.S. resident, not for a specific person.

Table 2.1 Some Common Risks of Accidental Death in the U.S.

Risk of death Qualitative source	Average risk to one individual during one year
motor vehicle accident	2.4×10^{-4} (person-year)$^{-1}$
home accident	1.1×10^{-4} (person-year)$^{-1}$
fall	6.2×10^{-5} (person-year)$^{-1}$
drowning	3.6×10^{-5} (person-year)$^{-1}$
firearms	1.0×10^{-5} (person-year)$^{-1}$
electrocution	5.3×10^{-6} (person-year)$^{-1}$
tornadoes	6.0×10^{-7} (person-year)$^{-1}$
lightning	5.0×10^{-7} (person-year)$^{-1}$
air pollution	2.4×10^{-8} (person-year)$^{-1}$

Another way to express risk is to describe the time it takes for one-half of the exposed population to die. Table 2.2 also lists some sources of risk, but with the alternative units. These estimates only apply to the persons engaged in the activity. So, they answer the question, how long you take to die if you start this activity. If the activity is climbing mountains, and if you behave like a typical mountain climber, you have a 50 percent chance of dying after two years. Clearly, the shorter the time, the higher the risk.

The comparison between motor vehicle risks in the two tables is particularly interesting. How, you may wonder, can we have a risk of death per lifetime that seems high, whereas the length of time for someone to accumulate half their risk from motor vehicles seems so long? The answer is that it takes a long time for a risk near 10^{-4} to affect lifespan. When the source of the risk is familiar, like riding in a car, we tend to dismiss it. The entire population is exposed to automobiles, a commonplace technology.

The dot at the end of this sentence that represents a period occupies approximately 2×10^{-2} (0.02 or $1/50$) of the surface area of one space on this page in a typical 14 point font.

We can treat a dot on the page like a Venn diagram. You can ask what is the chance that another dot of the same size would fall at *random* on top of the period.

Table 2.2 Amount of Time It Takes to Die for Half of Those Exposed to Some Common Risks of Death in the U.S.

Risk of death Activity	Average time for half of the exposed individuals to die
Mountain climbing	2 years
Riding motorcycles	12 years
Flying scheduled airlines	35 years
Smoking cigarettes	63 years
Disease and old age	79 years
Riding in private cars	84 years
Riding railroads and buses	94 years
Ambient radiation (5 rems/year)	1600 years

Notice that this estimate differs from the risk of typing a period on top of another period. A typical 8.5 x 11-inch page has one-inch margins, and within them, we have approximately 4.7×10^3 spaces in 14 point font. A typed period lines up within a space the same way each time. So, the risk of a period collision (one next to the succeeding one), while typing periods into spaces at random, is approximately $1/4{,}700$ or 2×10^{-4}. When you visualize the risk of two periods colliding, the risk does not seem so small, but for many people, risks of approximately 10^{-4} do seem low. But, this risk level is the highest allowed for any standard promulgated by the U.S. Environmental Protection Agency.

The random chance of a dot landing on another dot, as if we threw tiny darts at the space within the margins, is, however, less than the risk of a period collision. It is approximately 4×10^{-6} [$=(2 \times 10^{-4})(2 \times 10^{-2})$]. The value 4×10^{-6} is close to, if not less than, a risk that some assessors consider trivial.

The quantitative aspects of risks are essential to understanding and managing them. Only a few tools are essential to understand risk quantitatively. This chapter reviews them by focusing on population distributions and on exposure-response relationships. The math skills required are essentially those in a high school curriculum for college preparation.

Gaining an understanding of risk does require quantitative analysis. You need to understand how the assessors obtained the numerical values and units of risk in risk estimates. Otherwise, at best you will mislead, confuse, and bewilder yourself—at worst mislead others. Some people have math phobias. We hope these conditions are treatable. Many persons seem to believe that reworking and reviewing calculations is boring. We agree. It is tedious. Calculators and spreadsheets help some. Far worse, however, is that many analysts with good skills behave as if displaying calculations, asking others to refigure simple arithmetic, or even to reveal the underlying algebra, is somehow humiliating. We disagree.

We do these kinds of calculations routinely. We make errors easily and often. Preserved miscalculations can create disastrous situations. For example, twenty years ago, the substitution of a single character in a computer program turned a very expensive space exploration rocket into a fireworks display. Similarly, we recently saw some assessors mistakenly enter a 10 percent decrease in litter size into a calculation of a safe level by subtracting 0.1 from the values. If a litter had twelve pups, instead of using 10.8 (12 – 1.2) pups/litter, the assessors used 10.8 (12 – 0.1) pups/litter to estimate the safe level. Fortunately, peer reviewers found the error before the wrong level, 0.15 ppm, became a final regulation, instead of the correct value, 4.2 ppm.

Ask others to check your calculations! Your best ally is someone who is thorough and not afraid to criticize you. The cost of public embarrassment from an error far outweighs either the cost of the tedium involved in checking calculations internally or the pride of authorship. Arithmetic is an essential component of risk assessment, as is algebra, as is the necessity of expressing values numerically. Trying to avoid them is like trying to cook without pots and pans. However, we view the necessary skills as simple, even if their application lacks excitement. The quantitative aspects of risk are essential to managing and communicating risk.

To set the stage for further exploration, suppose that you have two numerical parameters, x and y, and you observe that y changes when x changes. Mathematicians call x the independent variable and y the dependent variable. They say that y is a function, f, of x. Usually, working with a function is not so abstract. It implies that you have a specific mathematical expression or formula, which relates x to y. For example, $y = 3x$, or $y = x^2$. However, $y = f(x)$ can be any mathematical expression.

Many risks are products of two or more other factors. You calculate the risk by multiplying these factors. Now, suppose you have some function, $y = f(x)$, and f is composed of other functions, for example, E and P. If $f(x) = E(x) \times P(x)$, f a *multiplicative function* of E and P. In the abstract, a multiplicative function, f, lets you extract a second function, E, from it and have another function, P, left over: $P(x) = f(x)/E(x)$. You can recover f, by multiplying E times P.

The Chain Rule

The chain rule is indispensable to the practice of risk assessment because it depends on so many multiplicative models. When you break a risk model into its components, you need it to be sure that the reassembled components will behave, and you will not inadvertently leave out some part. Usually, working with a function implies that you are interested in the behavior of *y*. For example, you might want to know about the rate of change of *y* with *x*, a process of differentiation, or you might want to calculate the area under a curve described by a function, when you examine a graph of *x* compared with *y*, a process of integration.

The chain rule is a mathematical manipulation that is useful in differentiating complex functions. Instead of differentiating the function directly, you can define a new function, for example, *u*, which equals a more familiar function. Then,

$y = f(u)$ and $u = g(x)$,

$y = f[g(x)]$, and

$dy/dx = dy/du . du/dx$

If you want to differentiate a function like $\ln(x^2)$, you set $u = (x^2)$, then

$dy/dx = dy/du \; du/dx = 1/u \; 2x = 1/x^2 \; 2x = 2/x$.

A multiplicative model of risk will behave properly in a mathematical sense, if it obeys the chain rule.

The most important factors in the assessment of risks from substances are *populations, exposures* and *potencies*. Exposure is the amount of a substance available to cross the membranes of an animal and become an internal dose. Because we usually measure the concentration of a substance (or the intensity of an energy) in the environment, we must do some work to convert monitoring data into exposures. Potency is the biological effect we expect from some amount of a substance. Some statisticians use a different term, the damage function. Thus, in the assessment of an environmental risk, we can break down risk into two functions called potency and exposure, such that for an average member of the population, risk = potency x exposure. For example, given a typical U.S. resident, if the potency of a substance is 10^{-3} (deaths/mg.year) and the continuous exposure to the substance is 10^{-3} (mg/person), the average risk will be 10^{-6} (deaths/person year). If we have functions to calculate exposure and potency, and values to insert into the functions, we can calculate risk. In essence, we have broken the problem of estimating risk into two smaller problems, estimating exposure and estimating potency.

Imagine that you plan to drive from Chicago, IL, to Washington, DC, and you become curious about your risk of death during the drive. Your exposure is the number of miles you will drive. What is the potency, or damage function, for driving? Essentially, it is the probability of death per mile driven. You can obtain data about both exposure and potency from many sources. To save your time, we have done this work for you. We have provided the average death rate per mile driven and the driving distance.

According to the National Highway Transportation Safety Administration, as of 1996 approximately 1.7 fatalities occurred per hundred million vehicle miles traveled (1.7×10^{-8} deaths/miles). This value is the potency you need. According to the

Rand McNally Travel Atlas, the distance from Chicago to Washington, D.C., is approximately 715 miles. This value is the exposure you need. Multiplying, we get:

$$(1.7 \times 10^{-8} \text{ deaths/miles})(715 \text{ miles/trip}) = 1.2 \times 10^{-5} \text{ deaths/trip}.$$

The potency of car travel is a useful perspective about risks. Given the current fatality rate, you could drive approximately 60 miles to accumulate a risk of 10^{-6} or one-in-a million, a value often cited as a trivial risk in risk management contexts. A risk of death per lifetime less than 10^{-6} means that you would confine yourself to no more than sixty miles of automobile travel during your lifetime. A risk of death per year less than 10^{-6} means that you would not travel more than sixty miles by car during a one year time. These values may seem absurdly risk averse to you; however, automobile transportation has a high potency.

If you did your own research and looked into the risk of automobile transportation in more detail, you would find that neither the potency nor the exposure are precise values. The distance between Chicago and Washington, DC, is a legal fiction. It depends on where you think Chicago begins and where you think you arrive in Washington, DC. It depends on the highways you travel and how efficiently you drive. Even changing lanes often will add to the distance traveled.

Even more surprising is that the fatality rate has very high variation. It depends on the kind of car you drive, when you drive, and your driving habits. Speeding or consuming alcohol while driving increases your risk. So does the time of day when you drive and wearing seat belts. Even the fatality rate and the driving distance are dependent probabilities. You will behave differently when driveing a long distance than you do when driving short distances in your neighborhood.

How much would you pay to insure your life during this trip?

Graphs and Mathematical Models

How do we discover the functions for risk, potency and exposure? Consider an environmental contaminant at two different exposure levels (E) yielding the following risk of death per organism (P). Eventually, we will show you how to extract these estimates from observational or experimental data. For purposes of discussion, we will provide data in the examples in this section.

E(mg)	P (deaths/animal)
1.0	0.159
3.16	0.309

Technically, P in this example stands for expected *prevalence*, the number of observed events that you believe will occur in a population divided by the population size. For example, you might read a line in the table as "at an exposure of 1 milligram, the probability is that 15.9 percent of the mice will die." (We have left the duration of exposure out of the units. Depending on how you conducted the experiments or obtained the observations, the duration could vary widely.)

Some confusion may result from correctly describing these data as *exposure-response relationships*, whereas many risk assessors commonly describe them as *dose-response relationships*. The difference is simple. If you have data about exposures, you can develop an exposure-response relationship. You probably will not have data about doses very often. (*See* Chapter 5, "Dosimetry.") If you have data about exposures but want to develop a dose-response relationship, you will need to convert the exposures into doses.

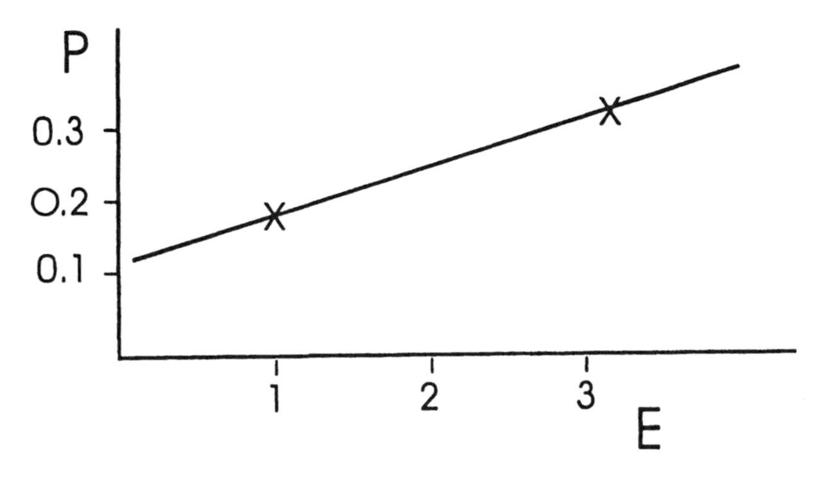

Figure 2.1 Two data points from experimental data plotted on a linear graph showing the resulting straight line.

Figure 2.1 shows these points and the straight line (linear fit) that joins these points. You can lay a ruler on the two points and draw the line. Instead, you can derive an equation for the line, as follows:

$$P = a + bE$$

where a and b are parameters determined by the actual data. We can determine what a and b are by substituting the values above as follows:

$$0.159 = a + b(1)$$

$$0.309 = a + b(3.16)$$

Subtracting yields:

$$0.15 = 2.16b$$

Thus,

$b = 0.0694$ (Note that we carry out the calculation keeping three significant figures, although we doubt that our data merit more than one.)

From the first substitution equation above,

$$a = 0.159 - 0.0694 = 0.0896$$

and thus the equation for a straight line that fits these data is:

$$P = 0.0896 + 0.0694E$$

Notice that the parameter a is the value of P when E = 0 or the intercept of the straight line and the vertical axis on the graph and b is the slope of the line (ratio of P to E). This equation represents a "fit" of a linear model to the data. For environmental exposures you usually will need to know the risk at much lower exposures than shown here. You can determine these values by substitution in the equation above yielding the following results:

E(mg)	P(deaths/animal)
10	0.784
0.1	0.159
0.01	0.0965
0.001	0.0897
0.0001	0.0866
0.00001	0.0896
0.000001	0.0896

Notice that, as you move to lower values of exposure, the probability levels off at a constant value. Since the probability of a risk is likely to reach zero at E = 0, this is an unsatisfactory fit. You might want to look for a more complex model that will fit better.

To define the slope of the exposure-response curve at lower exposure levels, we need to include an experimental point at a lower level. It is financially prohibitive to expose enough animals at the very low levels found in the environment since the numbers required to establish statistical significance are beyond the total available.

For purposes of discussion here we will add one more data point. For consistency the same set of exposure-response data will be used throughout the rest of this chapter. Also the symbols will be constant. Here E is the exposure, and P is the probability of response.

E(mg)	P(deaths/animal)
0.1	0.023
1.0	0.159
3.16	0.309

Thus at an exposure of 0.1 milligrams, the expected prevalence (or risk) of death is 2.3 percent, at an exposure of 1 mg the expected prevalence is 15.9 percent, and at an exposure of 3.16 mg the expected prevalence is 30.9 percent.

To fit the above data (the three data points) with a straight line, you might try the equation:

$$P = a + bE$$

In this case you can write three equations;

$$0.023 = a + 0.1E$$

$$0.159 = a + 1.0\,E$$

$$.309 = a + 3.16\,E$$

You have three equations and only two parameters. Thus, your problem is over-determined, meaning that you have more equations than parameters. In this case you use a least squares fit. A least squares fit is part of any standard statistics package, and you can use a key to perform this operation on most hand-held calculators. The idea is to find the line that minimizes the perpendicular lengths from the three data points to the fitted line. Figure 2.2 illustrates this concept. For your three equations, the best fit is

$$P = 0.0372 + 0.089E$$

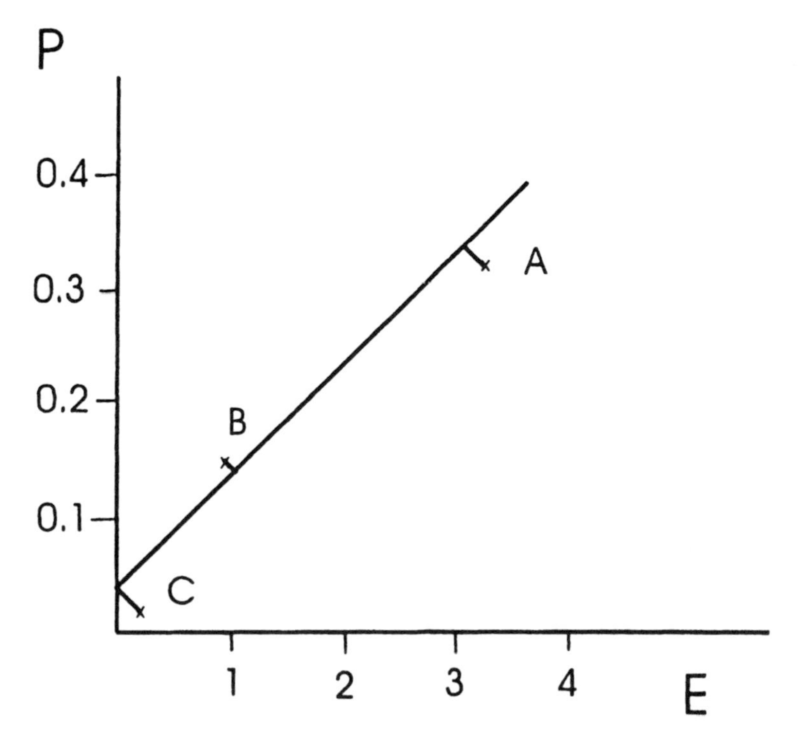

Figure 2.2 Least squares fit procedure; to minimize the sums of the squares of the lengths A, B, and C.

Now lets try for a more sophisticated mathematical expression or model. Consider adding a term, E^2 to the linear equation we used before making a new one that is a quadratic equation;

$$P = a + bE + cE^2$$

The quadratic equation fitting these three data points is (*see* Appendix A)

$$P = 0.00570 + 0.18E - 0.0267E^2$$

Substituting into this equation we can provide values in the region of the data and for the lower environmental exposure levels.

E	P
1	0.186
0.1	0.0234
0.01	0.0059
0.001	0.0057
0.0001	0.057
0.00001	0.057

Here again the curve levels off at lower exposures and does not go through the origin. We will have to seek a more sophisticated model.

In order to better understand the curve fitting process we will use different kinds of graph paper. Here the exposure-response data are plotted on two types of graph paper. Figure 2.3 shows a plot on standard linear graph paper and Figure 2.4 shows the same

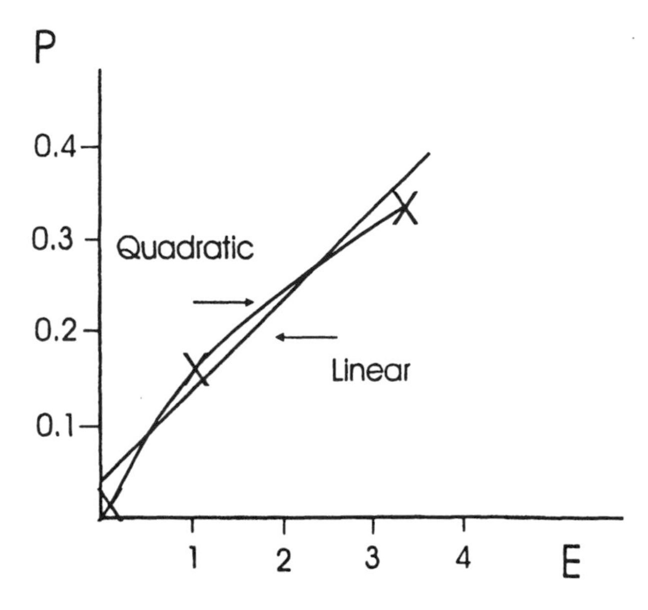

Figure 2.3 Three data points from the experimental data showing a least squares fit to a straight line and a fit to a quadratic function.

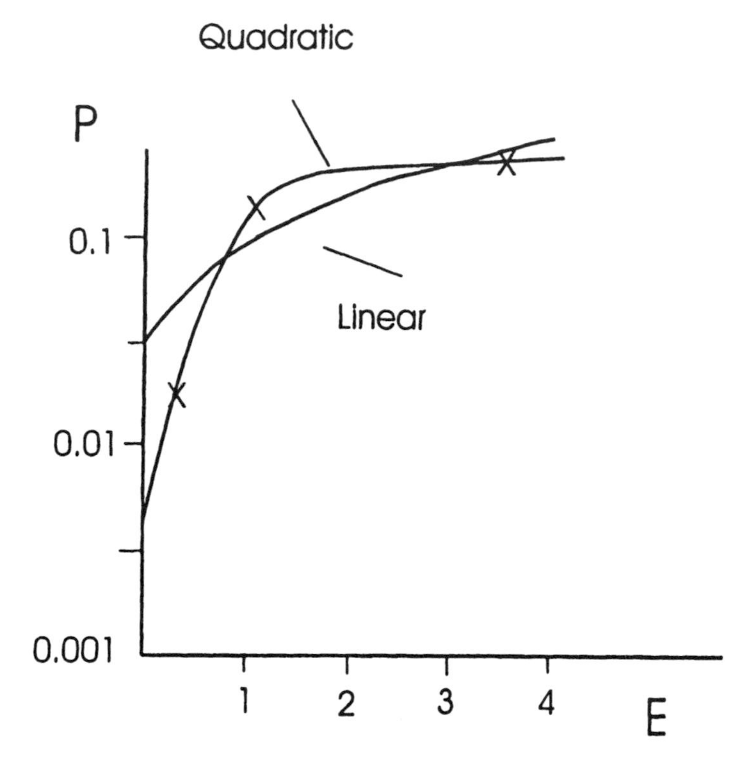

Figure 2.4 Semi-logarithmic plot of the three experimental dose-response values showing the least squares fit to a straight line and the fit to the quadratic function.

data plotted on semi-logarithm (semi-log) graph paper. The vertical scale in Figure 2.4 is the logarithm of P. Environmental exposures such as those you are exposed to would be much lower than 0.1 mg. You need to take a magnifying glass to see the part of the graph between 0 and 0.1 mg and several orders of magnitude below the lower horizontal axis.

Notice that the quadratic fit is perfect for all three data points. That is because there are three parameters and exactly three data points. If more data exist (they seldom do), you will need a procedure like the least squares procedure. The fit for the least squares for a straight line (linear) does not appear to be a straight line on the semilog paper. The use of different kinds of graph paper allows us some flexibility in how we view the data and how we fit it with a line. We can generalize and call these line fits models. Thus we have fit a linear model and a quadratic model to the data.

Notice that the shape of the curve connecting the three points is different on each different plot. Generally the risk or probability is known only at high exposures. Environmental risk and standards set to protect us from environmental exposures of interest generally are in the range of 1 in 10,000 (10^{-4}) to one in a million or 1 in 1,000,000 (10^{-6}). This range is shown in the "bricked" area in Figure 2.5. Clearly the linear and quadratic models do not pass through this box. You could simply draw a straight line from the data points towards

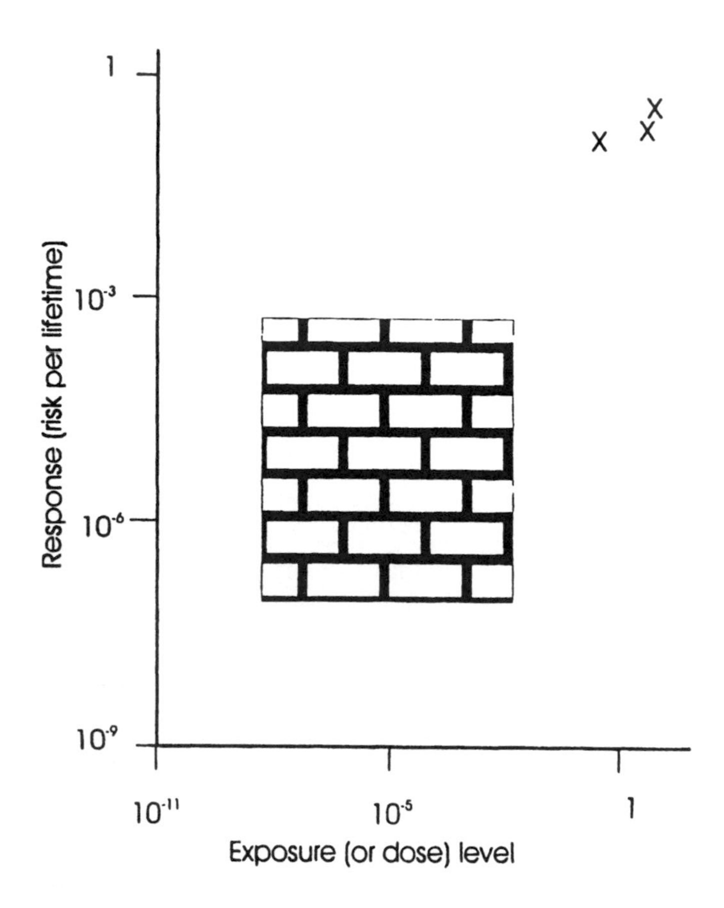

Figure 2.5 Plot showing the relationship of the experimental data and the region of doses more likely found in environmental situations.

the origin. However, you could also develop a more sophisticated model that will go through the origin, as we will describe later in this chapter.

Although scientists prefer not to extrapolate a line from the data points into unknown territory, risk assessors do so regularly. The reason it is improper is that the curve in fact could actually not pass through the box at all. For example, there could be a threshold at 0.01 mg and the curve would drop down to the horizontal axis at that point. Or there could be some kind of hermetic effect that could cause the curve to go up and actually be above the box. In any case when the exposure gets to zero, the probability of risk should be zero also. (You cannot show an origin of zero on semi-log or log-log plots.)

In Figures 2.3 and 2.4, you can with some confidence predict the probability or risk at exposures between the data points. Which fit is better—the linear or quadratic (with the squared term added)? Or do you need a more complex equation? You have developed a mathematical model (or function) to fit these curves to the data. Actually, you want to know the exposure for probabilities in the environmental range, which are in the range 10^{-4} to 10^{-6} per lifetime shown by the box ("bricked" area) in Figure 2.5. The log-log plot magnifies the region that will interest you, and thus, it is the most useful plot.

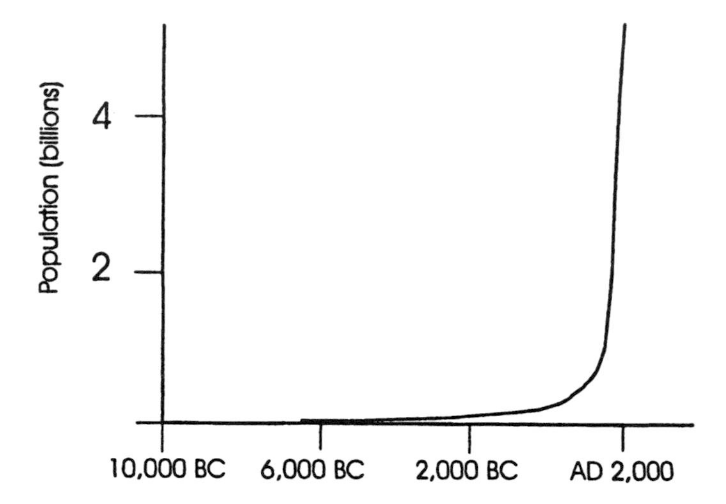

Figure 2.6 Estimated human population from the last ice age to the present

The kind of graph paper used to plot data can have a strong influence on the interpretation. Consider the data in Table 2.3 showing population growth as a function of time. Figure 2.6 shows this data plotted on linear graph paper. The conclusion here is that population is growing at an alarming rate. We might be headed for a catastrophe. Now, if you plot the same data on semi-logarithmic graph paper, you will get Figure 2.7. Here, you could interpret the increase at 10,000 years ago as a consequence of the discovery of

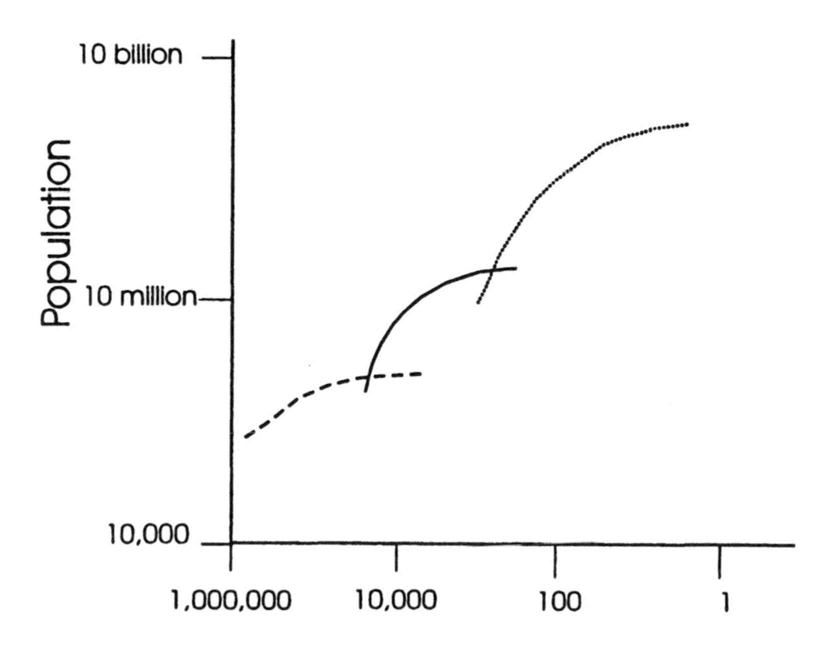

Figure 2.7 Plot of the same data as used in Figure 2.5 on a log-log plot

agriculture. The increase a few hundred years ago might relate to the industrial revolution. These two technological discoveries allowed the planet to support more people. You can see that the population is leveling off. Perhaps a new technological advance will allow the population to grow even more.

Growth Rate, Doubling Time, and Logarithms

To develop better models to fit exposure-response data, we will use the concepts of natural logarithms and the normal distribution curve. We will first examine the relationship between growth rate and doubling time to see how natural logarithms can enter. In the following section we will develop the notion of the normal distribution.

Consider the example of compound interest. Let D_0 stand for the initial sum (principal) at $t = 0$, with t in years. Let D_t be the amount at time t, r the rate of interest and x the number of times per year that the interest is computed. For example, at 6 percent per year, twice yearly, then after one year:

$$D_1 = D_0 (1 + 0.03)(1 + 0.03)$$

and if $D_0 = \$100$

$$D_1 = \$100\ (1.03)^2 = \$106.09$$

and after two years,

$$D_2 = \$100\ (1.03)^4 = \$112.55$$

or in general,

$$D_t = D_0 (1 + r/x)^{xt} = D_0 [(1 + r/x)^{x/r}]^{rt} \qquad (1)$$

Consider the quantity

$$(1 + r/x)^{x/r},$$

as x gets very large and approaches infinity.

For example for r = 2

x	$(1 + r/x)^{x/r}$
2	2
4	2.25
8	2.44
16	2.56

If you repeat this calculation with different values for r, you will find that the quantity approaches 2.7183. This is a "natural" number, and it is quite useful in mathematics, sta-

Table 2.3 Estimate of Growth in Human Population (Millions)

Year	Population (millions)	Years Ago	ln (years ago)	ln (population)
1,000,000 BC	0.125	1,002,000	13.8	11.7
300,000 BC	1	302,000	12.6	13.8
10,000 BC	4	12,000	9.39	15.2
5,000 BC	5	7000	8.85	15.4
4,000 BC	7	6000	8.70	15.8
3,000 BC	14	5000	8.52	16.5
2,000 BC	27	4000	8.29	17.1
1,000 BC	50	3000	8.01	17.7
500 BC	100	2500	7.82	18.4
200 BC	150	2200	7.70	18.8
1	170	2001	7.60	19.0
200	190	1800	7.50	19.1
600	200	1400	7.24	19.1
800	220	1200	7.09	19.2
1000	265	1000	6.91	19.4
1200	360	800	6.68	19.7
1500	425	500	6.21	19.9
1600	545	400	5.99	20.1
1700	610	300	5.70	20.2
1750	720	250	5.52	20.4
1800	900	200	5.30	20.6
1850	1200	150	5.01	20.9
1875	1325	125	4.83	21.0
1900	1625	100	4.60	21.2
1920	1823	80	4.38	21.3
1930	1987	70	4.25	21.4
1940	2213	60	4.09	21.5
1950	2516	50	3.91	21.6
1960	3019	40	3.69	21.8
1970	3693	30	3.40	22.0
1980	4450	20	3.00	22.2
1990	5333	10	2.30	22.4

tistics, theoretical physics, chemistry, and risk analysis. It is so common that it is universally given the symbol "*e*".

Using *e*, the above expression (equation 1) can be written

$$D_t = D_o e^{rt}$$

Now ask how long it will take to double your money, or when does

$$D_t = 2\ D_o$$

$$= D_o e^{rt}$$

or

$$e^{rt} = 2$$

Using the rules of logarithms (logarithm of a base to a power is just the power) or taking the natural logarithm of each side of this equation yields

$$rt = \ln 2 = 0.693$$

or

$$t = 0.693/r$$

Therefore for a 7 percent rate,

$$t = 0.693/0.07 = 10 \text{ years}$$

Thus if the interest rate is 7 percent per year, you will double your amount invested in 10 years.

The Logarithm

The logarithm is a handy notation to describe and manipulate the wide ranges of numbers that descriptions of risk often involve. Consider the inverse function $x = a^y$. Formally, we write $y = log_a x$. This reads as the logarithm of *x* to the base *a* of *x*. The common logarithms using the common base of 10 are:

x	$y = \log_{10} x$
1	0
10	1
100	2

and so on. An entire well known logarithmetic set is based on this called natural (or Naperian) logarithms (base *e*), and is abbreviated, ln *x*, whereas common logarithms (base 10) are written log *x*.

Just as an exercise to refresh your knowledge of logarithms, calculate *e* squared, cubed, to the fourth power and so on. (The answers are 7.3891, 20.0855, 54.5982 and so on). Then the logarithm of 20.0855 to the base e (natural logarithm) would be 3 and the logarithm of 54.5982 would be 4, and so on.

Statistics about Pedestrians

Statistics are not probabilities. The difference is not subtle. If you have a favorite baseball team, you may want to keep track of its batting averages. You will use statistical tools to analyze the performance of a player on the team. Statistics are descriptive! Probabilities are predictions. If you are predicting the player's batting average for the coming year, you should use the probability calculus. You would not make the mistake of assuming that this player will perform exactly the same next year as he did last year. You might, for example, plot his batting average for each of the past six years versus this year and try to understand whether he is getting better or worse in batting skills. If his contract will

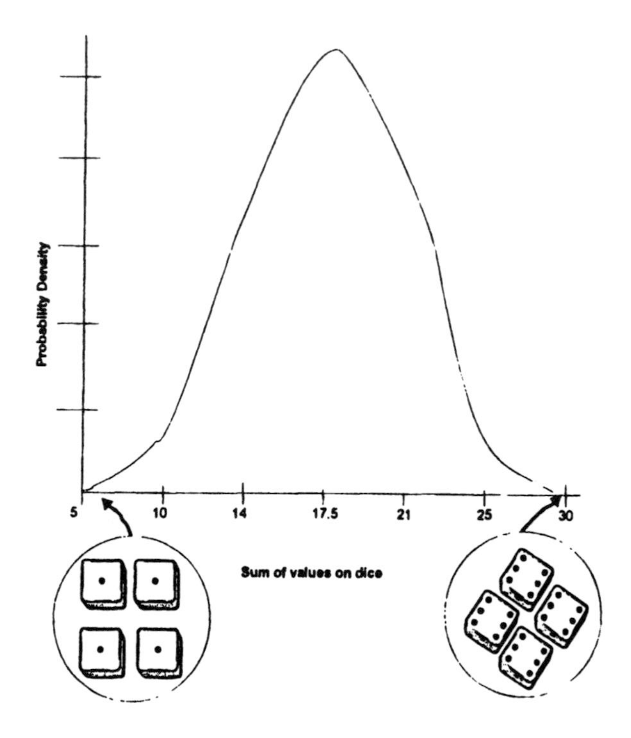

Figure 2.8 Normal curve generated by five dice in a Yatzee game.

come up for renewal or a new player from the minor leagues will try to take his position in the coming season, you might expect him to perform better.

Suppose that you have five dice from a Yatzee game and throw them onto a table. Then, you add the numbers on the dice together. The lowest number you can get will be 5 and the highest 30. When you throw one die many times, its average value will be 3.5, since [(1+2+3+4+5+6)/6 = 3.5]. Suppose that you carry out the process of throwing five dice over and over, recording the numbers each time. The average will be halfway between 17 and 18 (5 x 3.5 = 17.5). After awhile, if you plot your data, you will get a chart like Figure 2.8. We could draw a smooth curve connecting the points on Figure 2.8. This curve is called a normal curve. It results when the underlying processes that generated the data are additive.

Suppose you collected data on the body weights of 100 students, and plotted the numbers that fall within intervals, as a function of their weight. Such a plot might look like the bar graph in Figure 2.9. The more data you collect the closer it will come to the curve shown in Figure 2.9. This curve has a natural character and results from plotting many different kinds of data. The curve is called the normal curve.

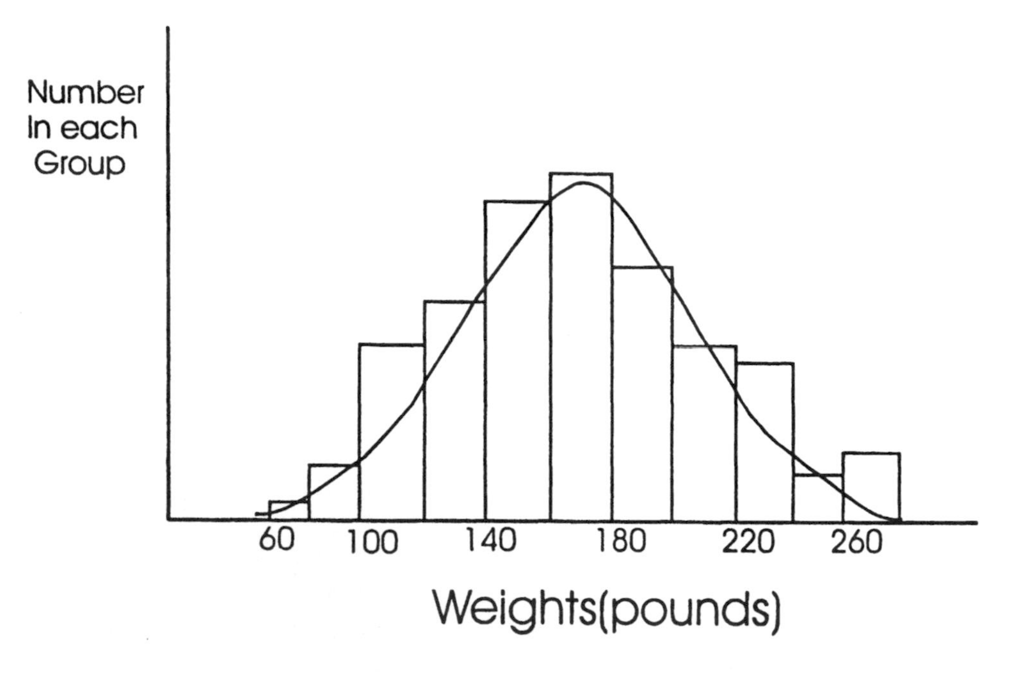

Figure 2.9 Data from sampling weights and the continuous normal curve.

Statisticians often describe the width of this curve in units of standard deviation, which are units relative to the shape of the curve. *Standard deviation* relates to the probability that a value will occur. The distance on the horizontal scale from its maximum, or peak, to the inflection point, where the curvature changes, is one standard deviation. Figure 2.10 shows a normal curve and the locations of one, two, and three standard deviations from the mean. Statisticians often use the small Greek letter sigma (σ) to indicate standard deviations. Thus, the standard deviation is a measure of the variation of a distribution, and it is a way to compare different distributions. A smaller standard deviation indicates less variation and thus a narrower range of values.

Normal distributions are found almost everywhere: weights and heights of people or animals, yearly temperature averages, stock market fluctuations, and achievement test scores. In general, these data approximate normal curves when plotted. They sometimes do not fit exactly, because the distributions involve measurements, but the normal distribution is continuous. Sometimes the data may not represent a random sample of the total population. (Human height varies with socioeconomic class, for example. If your sample overrepresents wealthy people, or if you have a disproportionate number of men, you will not achieve a normal distribution.) Often enough, however, real world data closely approximate a normal distribution. So, it is worth using the function for a normal distribution to describe the data.

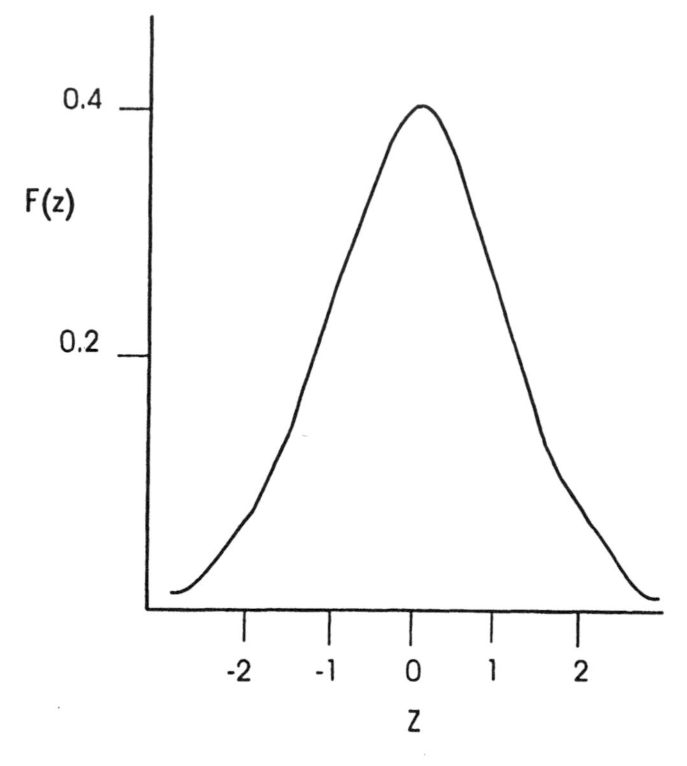

Figure 2.10 The normal curve plotted as a function of the z score.

To find the average (μ) of a data set of values of *x*,

$$\mu = \frac{1}{n}\sum_{i=1}^{n} x_i$$

where there are *n* values of *x*. The large Greek sigma (Σ) reads as "a sum over the x_i values from 1 to *n*. Thus, if $n = 5$; $\mu = (1/5)[x_1 + x_2 + x_3 + x_4 + x_5]$.

The width of any distribution is an important characteristic, which may vary widely from data set to data set and distribution to distribution. For a discrete distribution, the sum of the differences between the data values and the mean would indicate the variation. If the distribution is symmetrical, 1/2 of the values will be larger, and 1/2 smaller than the mean. To avoid getting zero variance for every symmetrical distribution, whether it varies widely or little, you could take the squares of the differences, add them up, and divide by the number of measurements. Technically, this approach yields a value called a *variance* and the standard deviation (σ) is the square root of variance. (Also for technical reasons, we divide by n – 1, not n).

$$\sigma = \sqrt{\frac{1}{(n - 1)} \sum_{i=1}^{n} (x_i - \mu)^2}$$

The normal curve is symmetrical and it is convenient to express it in a form that represents the symmetry—such as,

$$f(z) = \frac{1}{\sqrt{2\pi}} e^{-\frac{z^2}{2}}$$

where,

$$z = \frac{x - \mu}{\sigma}$$

This simpler function relates to the continuous curve in Figure 2.10 through the concept of a z score where x is the weight in Figure 2.9 and $f(x)$ is the number of students within each weight group. You can express this idea as: $x = \mu + \sigma z$. The square root included in the function for this prototypical curve generates a total area under the curve of 1.0, the probability of the entire population.

Figure 2.10 shows a normal curve. Approximately 68 percent of the curve is between plus and minus one standard deviation, 95 percent between plus and minus two standard deviations, and 99.7 percent between plus and minus three standard deviations. Figure 2.11 shows normal curves for a range of means (μ) and standard deviations (σ).

We provide a partial table of values for z, $f(z)$ and the area under the normal curve in Table 2.4. Notice that

$$\frac{1}{\sqrt{2\pi}} = 0.399$$

For more details refer to a standard statistical text or to mathematical tables in reference books.

The concept of a z score involves the mean, μ, and the standard deviation, σ, where,

$$z = \frac{x - \mu}{\sigma}$$

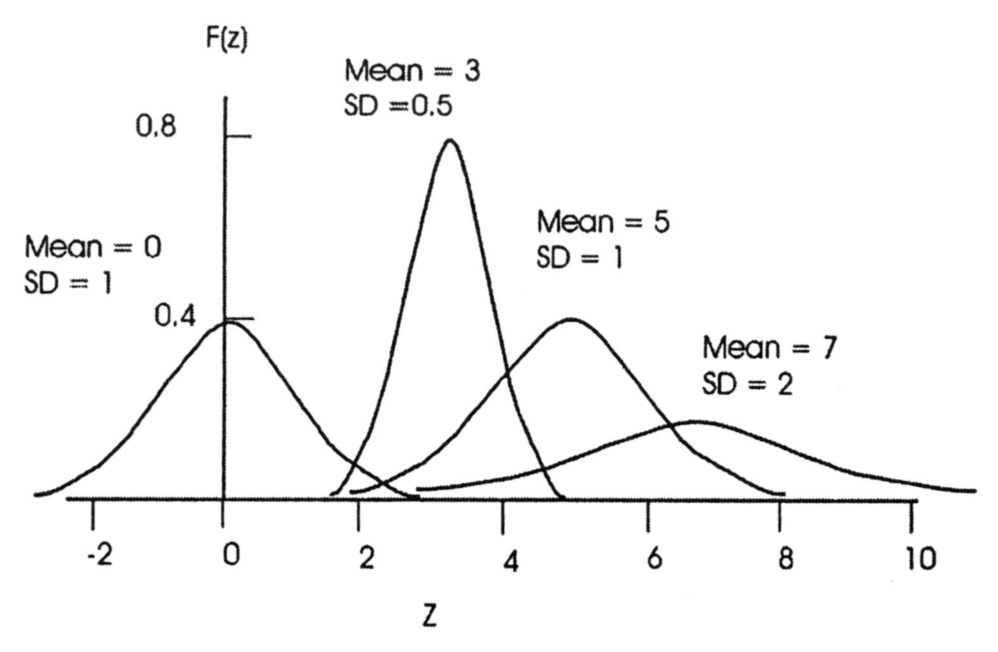

Figure 2.11 Plots of four different normal curves with different means (μ) and standard deviations (σ).

For the lognormal distribution,

$$z = \frac{\log x - \mu}{\sigma}$$

and for μ = 0 and σ = 0, z = x. The probability of an event is represented by the area under the normal curve as,

$$P = \frac{1}{\sqrt{2\pi}} \int_{-\infty}^{z} e^{-\frac{z^2}{2}}\, dz \qquad (2)$$

For those of you without a calculus background, this represents the area under the curve from the left hand infinity up to the value of *z* of interest. From Table 2.4, you can see that the area under the curve from the peak (z = 0) to one σ is 0.3413, and the area under the curve from the peak to two σ is 0.4773 and from the peak to the far right is 1/2.

Probability Distributions

We have already looked at one probability distribution, the normal distribution. Some of these probability functions are very simple. The *uniform distribution* captures all of your information, when you know that the population exists between two values, and nothing more. If all of your data lie within a range, but you do not know where, you have a uniform distribution, which some analysts also call a square wave distribution or a rectangular distribution. (*See* Figure 2.12a.)

Similarly, a *triangular distribution* captures all of your information, when you expect none of the population to exist below one value, the most members to have a second, higher value (the *mode*), and none of the population to have a value more than a third, highest value. (*See* Figure 2.12b.) These simple functions are very useful in developing risk models of poorly understood processes.

The *Poisson distribution*, named after the French mathematician, Simeon D. Poisson, is another probability distribution, very useful for work with populations. One way to think about it is to imagine that you have many cells or receptacles. You can throw ping pong balls at random into the cells until you have some average number of balls per cell, λ (the small Greek letter lambda). Think back to our example of period collisions. If you type periods at random on a page, when you type 10 percent of the number of spaces on the page, you have $\lambda = 0.1$. If you have 1,000 cells, and you throw 500

Table 2.4 The Normal Curve

z	area under one half (or one tail)
0.0	0.500
0.1	0.460
0.2	0.421
0.3	0.382
0.4	0.345
0.5	0.309
0.6	0.274
0.7	0.242
0.8	0.212
0.9	0.184
1.0	0.159
1.1	0.136
1.2	0.115
1.3	0.097
1.4	0.081
1.5	0.067
1.6	0.055
1.7	0.045
1.8	0.036
1.9	0.029
2.0	0.023
2.1	0.018
2.2	0.014
2.3	0.011
2.4	0.008
2.5	0.006
2.6	0.005
2.7	0.003
2.8	0.003
2.9	0.002
3.0	0.001

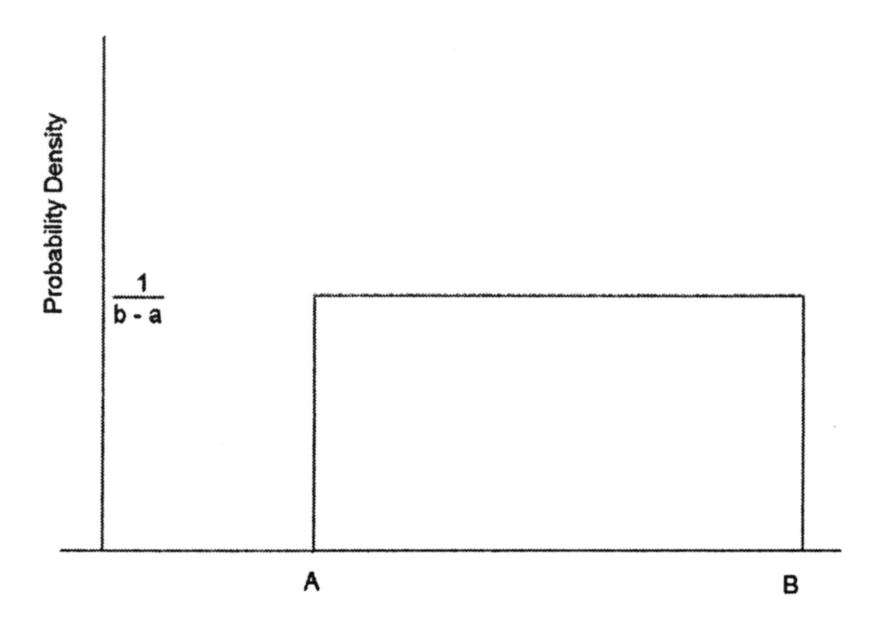

Figure 2.12a Example of square wave distribution (uniform distribution). (a.. x.. b), mean + (a+b)/2

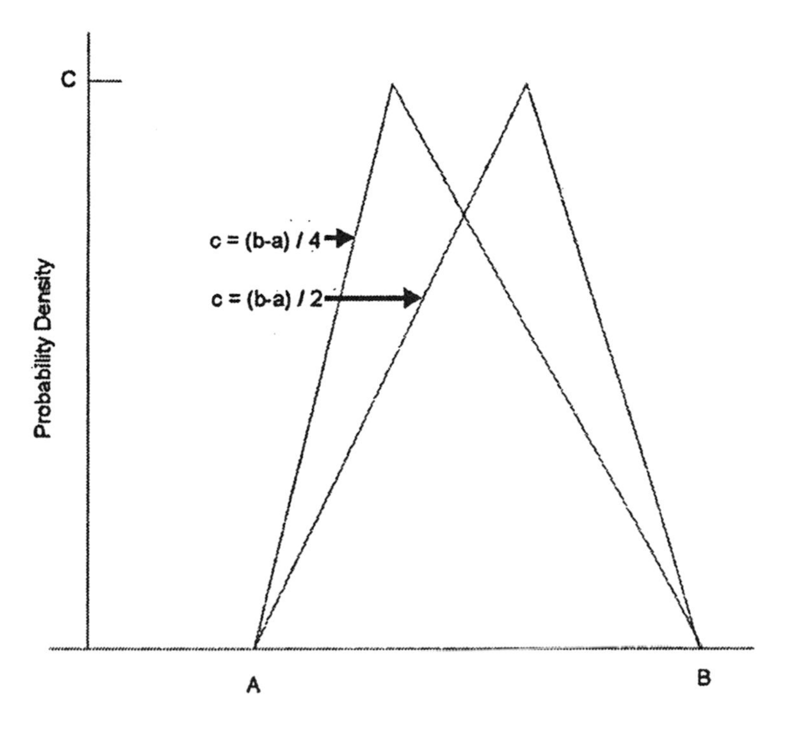

Figure 2.12b Example of triangular distribution. (a.. x.. b, mode + c), left curve mean + (a+b+c)/3

balls into them, you have $\lambda = 0.5$ (500/1,000). The *probability function* for a Poisson distribution is:

$$p(x) = (\lambda^x \cdot e^{-\lambda})/x!$$

where x is some integer, 0, 1, 2, 3 and where $x!$ is the factorial of x, *e.g.* $5! = 5 \times 4 \times 3 \times 2 \times 1$.

For example, if we are interested in the number of empty cells, or $x = 0$, this formula becomes:

$$p(x) = e^{-\lambda}$$

So, the number of empty cells for $\lambda = 0.5$ above is $e^{-0.5}$, or 0.6065. You can expect approximately 61 percent of the cells to have no balls in them. Some cells will have only one ball, some two, and so forth. You can calculate these numbers. According to the probability calculus, the number on nonempty cells will be:

$$1 - p(x) = 1 - e^{-\lambda} = 1 - 0.6065 = 0.3935$$ or approximately 39 percent.

As λ gets very large, the Poisson distribution resembles a normal distribution.

Statisticians know the standard deviation (usually symbolized by the Greek small sigma) of the Poisson distribution mathematically. It is easy to calculate:

$$\sigma = N^{\frac{1}{2}}$$

This formula is very useful. If you are counting objects, events, or anything, you can expect the error in your count to be in proportion to the square root of the number you counted. Thus, if 100 data points are in a category (say 100 people voting for the candidate) the standard deviation is ±10, and you can expect that approximately 70 percent of the time you will have between 90 and 110 data points.

The way that we have thought about probability functions so far is to plot the number of persons (or events or objects) that occur each time we increase x by some amount, that is, how many fall into an interval. We call the distribution that we get a *probability density function* or *pdf*. The pdf for the Poisson function that we just analyzed looks like the bottom of Figure 2.13

Instead, suppose that you started with a low value of x, and as you increased it, you added the number in the next increment to your total. We call this result the *distribution function* (or sometimes the cumulative density function), as illustrated in the top of Figure 2.13. The pdf is the first derivative of the distribution function. It shows the rate of change of the distribution function, as x changes. Moving in the other direction, the distribution function shows the area under the curve of the pdf.

Think again about our earlier example of the five dice. Instead of adding the numbers on the five dice, now suppose that you multiply them, for example, $1 \times 5 \times 2 \times 4 \times 2 = 80$.

Now, you repeatedly throw the dice several thousand times, plotting the number of times you get each multiplication product. We call this distribution a log-normal distribution. When plotted as multiplication products, it looks skewed to the right (*See* Figure 2.15). Because the antilog of the log of a number plus the log of another number is the same as the log of the multiplication product of the two numbers, you could have added ln(1) + ln(5) + ln (2) + ln(4) + ln(2) to get this result. So, a plot of the logarithms of the numbers you generate in your experiment will be a normal distribution. This property of multiplicative models is important, and it generates *lognormal distributions.*

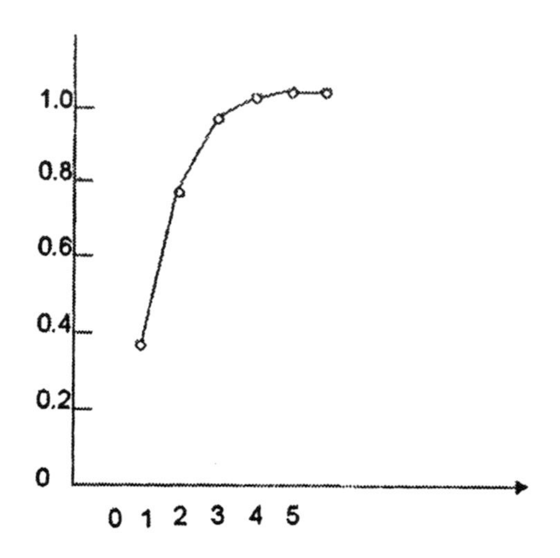

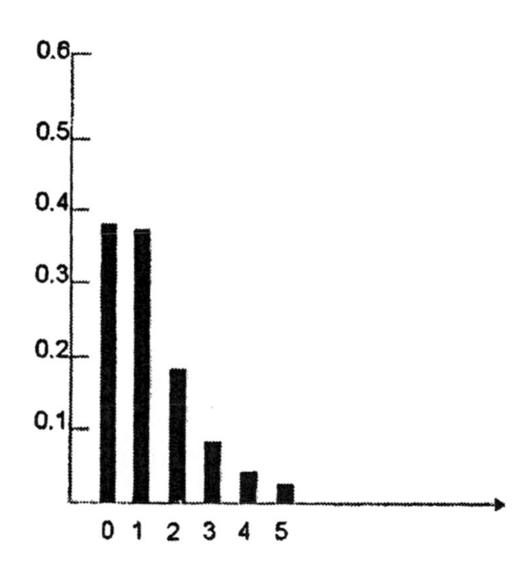

Figure 2.13 Plots of the probability distribution function and the cumulative probability distribution function.

We have three general ways of carrying out mathematical operations, such as addition or multiplication, on two or more probability distributions. (1) If we know the mathematical formulas for the probability functions, we may be able to obtain an algebraic formula for the result of operations on two (or more) distributions. In practice, the algebra may prove difficult, if not impossible, to solve. (2) If we have data for each of two (or more) populations, we can carry out the desired mathematical operation with all of the data. This approach is combinatorial, and combinatorial models are useful in risk analysis, particularly with small population sizes or data sets, and complex mathematical operations. (3) If we have appropriate distribution functions and parameters for one or more populations or data sets referenced in a model, we can apply a Monte Carlo approach.

The term "Monte Carlo" comes from gambling casinos in the south of France. In this approach, an analyst selects a probability distribution of some variable, such as body weight, instead of a single value in a mathematical model that requires body weight, such as an exposure model. The analyst may have generated the underlying exposure model independently, without reference to any population measurement. We can change the variables of the distribution function and compare them with observational or experimental measurements of a population.

A *Monte Carlo method* will select a value for body weight from the probability distribution at random and calculate exposure. The analyst (actually the analyst's computer) repeats this calculation often (usually many thousands of times), producing a result that fully expresses the implications of varying body weight and generating a prediction about the population distribution of exposures. Because most environmental models are multiplicative, the outputs of most Monte Carlo models are log normal distributions.

Because the Monte Carlo methodology can use more data and use data that relate more closely to the factors that contribute to exposure, they can produce more accurate predictions. Thus, Monte Carlo models critically rely on data and on assumptions about data. Obtaining the data can prove expensive and time consuming. Yet, assumptions about population distributions also can mislead. Imagine, for example, substituting a triangular distribution for the actual distribution of body weights. After passing this distribution through a Monte Carlo model, the exposure predictions will become distorted, more so in the tails of the distribution than near the mean.

Because the analyst can replace the distribution of body weights of randomly chosen people with, for example, the body weights of aging, overweight men, Monte Carlo models can predict the exposure of a subpopulation, a useful characteristic. Unfortunately, Monte Carlo models also can predict some wildly improbable, if not impossible exposures. We might obtain one prediction for a 120-year-old, twelve-pound man. Thus, Monte Carlo models require the insertion of practical constraints to prevent nearly impossible predictions, a difficult task not yet accomplished by environmental models. The midpoint of an exposure distribution produced by a Monte Carlo model is usually accurate, but the tails of the predicted distributions are suspect because of the input data.

Risk analysts find a few other operations on distributions useful. The average of two uniform distributions with the same mean is a symmetrical triangular distribution. The sum of many triangular distributions with the same mean is a normal distribution. The sum of two or more Poisson distributions (adding every point in a Poisson distribution to every point in another Poisson distribution) is another Poisson distribution. As λ in a Poisson distribution becomes large, the Poisson distribution becomes a normal distribution.

Do you recall our earlier example of the risk of driving from Chicago to Washington, DC? Both the exposure (distance) and the potency (fatality rate) values were uncertain. Suppose that we look at pdfs for estimates of the distance and the fatality rate. To multiply one pdf by another, we need to multiply each value in one pdf by each value in the other pdf. Earlier, we multiplied single values, either the average or best value we could obtain, to generate a point estimate. Figure 2.14 shows two pdfs, and the pdf that results from multiplying them.

Because we used a multiplicative model for the risk of a fatal accident, the pdf is skewed, tending toward a lognormal distribution. The distance from Chicago to Washington, DC is not an exact value. The potency of car travel for death is not a single value, but depends on many factors. Thus, we have an average prediction of the chance of death during a car trip from Chicago to Washington, DC, but the prediction is uncertain. The distribution in Figure 2.14 illustrates our uncertainty.

Also, recall that a z score involves a mean, μ, and a standard deviation, σ, where,

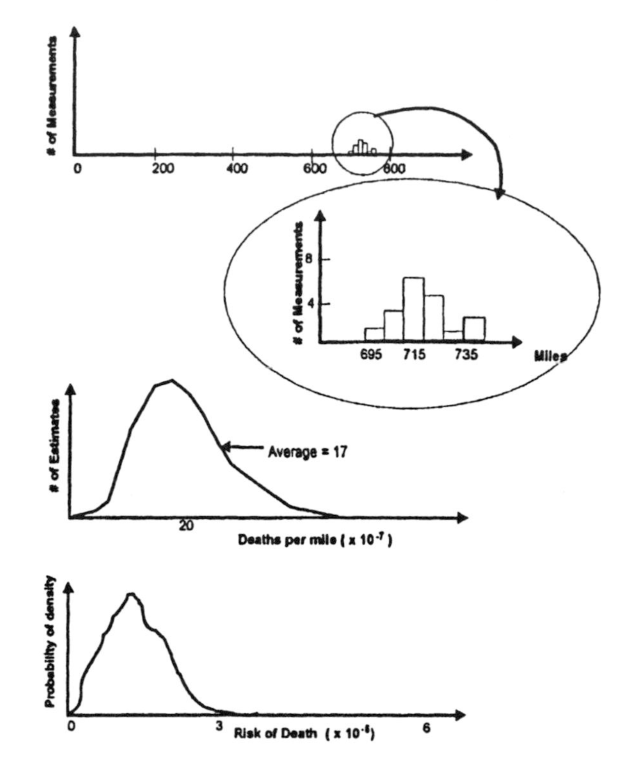

Figure 2.14 Plot of two probability functions and their product. (Multiplication of two probability densities to estimate the risk of a car fatality while driving to Chicago, IL.)

$$z = \frac{x - \mu}{\sigma}$$

For a lognormal distribution,

$$z = \frac{\log x - \mu}{\sigma}$$

and for μ = 0 and σ = 0, z = log x. (*See* Figure 2.15)

The probability of an event is represented by the area under the normal curve by:

$$P = \frac{1}{\sqrt{2\pi}} \int_{-\infty}^{z} e^{-\frac{z^2}{2}} dz \qquad (2)$$

This formula describes the area under the curve from negative infinity up to the value of z of interest. From Table 2.4, you can see that the area under the curve from the peak (z = 0) to one σ is 0.3413, and the area under the curve from the peak to two σ is 0.4773, and from the peak to the far right is 1/2.

So, we have the idea that the distribution could represent the variability or uncertainty in the data that go into the distribution. Most environmental exposure measurements are lognormal in their distribution. (*See* Figure 2.15) That is, the logarithm of x, the quantity of interest, distributes normally. Environmental measurements have distributions skewed by more values above the highest point on the curve (called the *mode*) than below.

Once you understand a probability distribution in the abstract, you can apply it to any data. A *variate* is a set of values generated by a random process. All variates have associated probability functions. However, given a variate, you may find it difficult to discover the density function or pdf. To gain an initial idea of the fit, you can *transform* or *map* a known model in another coordinate system into dimensions that apply to your data.

For example, in the Yatzee example discussed earlier, we initially could take a normal distribution with a mean of zero and a range from –0.5 to +0.5 (a unit normal distribution), and transform it to approximate our data from our experiment with five dice. We know the mean, 17.5, so we can add 17.5 to any μ and get a new μ' that is the approximate mean of the data. Similarly, the unit normal distribution extends to a maximum σ of one in the positive direction, and we know that the maximum value in the data is 30. So, we can set σ' at 12.5 (30 – 17.5 = 12.5) by multiplying σ by 25. A normal distribution transformed from the unit distribution to $\mu' = 17.5$ and $\sigma' = 12.5$ will fit our data well.

Transformation of data or a probability model is a simple operation mathematically. It directly relates to the idea in risk analysis of a *threshold*. If we transform a unit normal distribution to $\mu' = 17.5$ and $\sigma' = 12.5$, the lowest possible value will be 5 (17.5 – 12.5 = 5). You

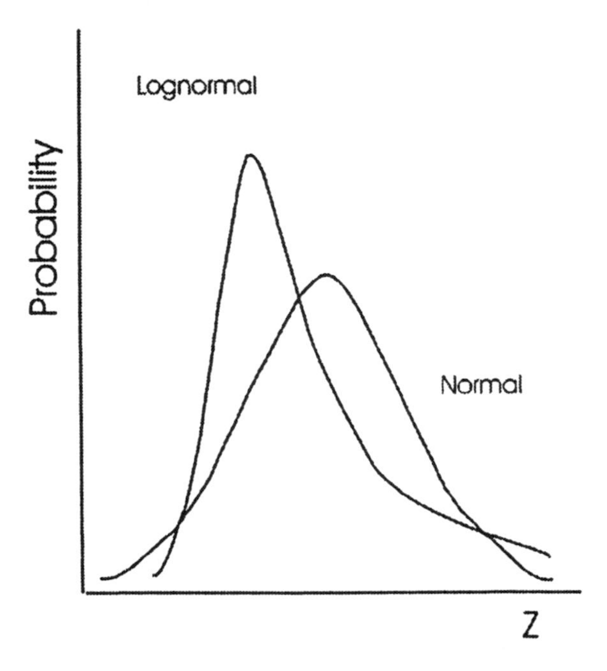

Figure 2.15 Comparison of normal and lognormal curves

can never throw five dice such that the sum of the numbers on the dice is zero. Therefore, five is the threshold value.

As you might imagine, thresholds are important in risk analysis. When measured values are lower than the threshold of a risk model, they suggest that no risk exists. This circumstance is the true meaning of safety. But a nonthreshold model has a lowest possible value of zero or less, such as some negative value. When this circumstance occurs in an exposure-response relationship, the model suggests that no safe exposure exists.

Much confusion exists in the risk analysis literature between nonlinear models and thresholds. Theoretically, these notions are independent of each other, and we have four kinds of models, as noted in Table 2.5.

Linear, nonthreshold models are particularly important in regulatory risk analysis, both in the history of the field and in practice. Certain kinds of biological processes that have regulatory prominence, including carcinogenesis, may generate linear, nonthreshold models. In addition, linear, nonthreshold models are easy to use. With a linear, nonthreshold model of potency, you will get the convenient feature of potency that does not change with risk. For the other three categories of models, this feature does not apply. Instead, potency will change depending on the exposure, making your calculations harder.

No matter how bad the fit, you can always draw a straight line through your data to the origin and use it as a first approximation of risk. Whatever the exposure, you need only engage in simple multiplication with the same potency. Doing so may distort values, but often these simple models will support regulatory decision-making. Some would say they are good enough for government work.

Suppose, for example, you draw a straight line to the origin above all of your data points. If this exposure-response model and the highest exposure you expect only gener-

Table 2.5 Categories of Some Exposure-Response Models

	Threshold	Nonthreshold
Linear	*Segmented line models, segmented regression*	*Straight line to the origin, linear nonthreshold models, single hit models, linearized multistage models*
Nonlinear	*Benchmark dose models, log-linear models*	*Power relationships, models, polynomial models, probit models, Weibull models*

ates a trivial risk estimate, you do not need to worry about regulation. However, if you draw a straight line below all of your data points, and the lowest known exposure shown by this model generates a risk that seems high, you have a risk that suggests regulatory attention. So, if nothing else, these simple models of boundary conditions are useful in setting priorities.

Tolerance-Distribution Models

If we have three or more data points, we wish to seek mathematical expressions that will fit our data better than the linear and quadratic equations that we fitted to our exposure-response curve. The mathematical expressions are models. In the language of risk analysis, we fitted linear and quadratic models to our test data.

Risk analysts use two categories of mathematical functions with exposure-response data: tolerance distribution models and mechanistic models. Tolerance distribution models often are confused with threshold models, but you can transform the coordinates of a tolerance distribution model so that it describes either a threshold or nonthreshold process. You can also transform the coordinates of a mechanistic model to obtain a threshold model.

First, we will discuss three tolerance distribution models in sufficient detail for you to fit data to them. We will use the same exposure-response data from earlier in this chapter. These models are the log-logistic (logit), Weibull, and log-normal (probit) functions. We use a standard nomenclature to allow comparison of these models.

P = probability of risk

E = exposure

σ = standard deviation

μ = mean

The approach we use changes these complex mathematical expressions into linear forms. So, we can use the linear or straight line fit or a least squares fit.

Log-logistic (logit) models are useful in evaluating epidemiological data. They have two parameters, a location parameter or mean, *a*, and a scale parameter, *b*. For these models probability is distributed according to the following expression:

$$P = \frac{1}{1 + e^{-(a + b\log E)}} \qquad (3)$$

This is another variation of an exponential mathematical expression, where a and b are curve-fitting parameters. Rearranging Equation 3:

$$P + P\, e^{-(a + b\log E)} = 1$$

$$P\, e^{-(a + b\log E)} = 1 - P$$

$$e^{-(a + b\log E)} = \frac{1 - P}{P}$$

and taking the natural logarithm, ln, of both sides, we get the general linear equation:

$$y = (a + b\log E) = -\ln\left(\frac{1 - P}{P}\right) \quad (4)$$

We will use the same test case data to fit this equation. Substituting P in Equation 4 and taking log(E) yields the values for y and log(E) shown in Table 2.6.

Table 2.6 Experimental Exposure-Response Data Fitted with a Logit Model

E (mg)	P (deaths/organism)	y = a + blog(E)	log E	y(eqn)
0.1	0.023	−3.749	−1	−3.73
1	0.159	−1.666	0	−1.75
3.16	0.309	−0.805	0.5	-0.76

Now plot the data points on a graph of y vs. log(E) and determine the best straight line fit. (*See* Figure 2.16) This can be done using a see-through ruler to get the best visual straight line fit or by using the least square procedure. Using a least squares fit, a = −1.75 and b = 1.98. Plot the data points and the straight line y = −1.75 + 1.98log(E) to convince yourself that this a good fit. Thus the probability is:

$$P = \frac{1}{1 + e^{(1.75 - 1.98\log E)}}$$

Figure 2.17 shows how the curve of Equation 3 fits the data points. The curve for the logit model "fits" the data points in that it is the best representation. The difference can be seen in the values presented in Table 2.6. Compare the values in the *y* column to those generated from the straight line fit in the *y(eqn)* column.

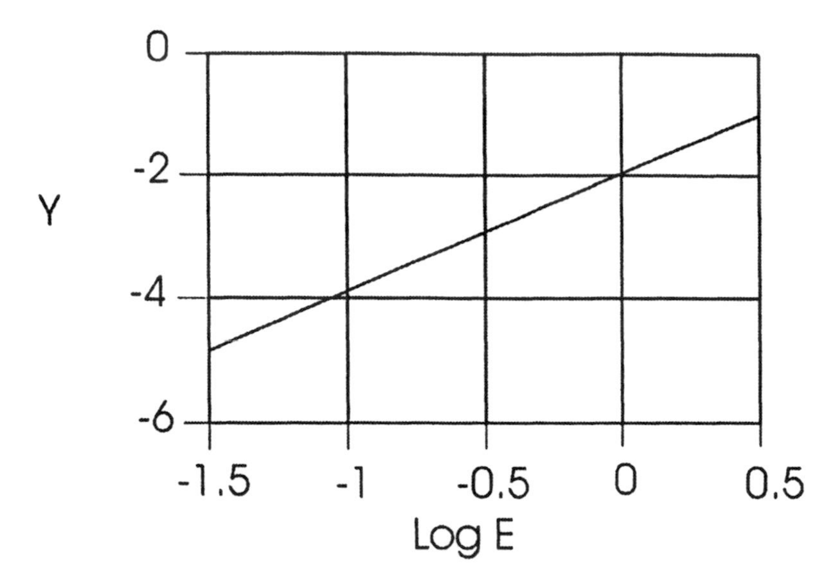

Figure 2.16 Plot of the functions used in logit model fits.

To find the probabilities for environmental exposures (E), substitute the values shown in Table 2.7 in this equation.

Table 2.7 Probabilities for Environmental Exposure Levels for the Logit Model

E		P	
0.0001	(or 10^{-4})	0.0000631	(or 6.31×10^{-5})
0.00001	(or 10^{-5})	0.00000872	(or 8.72×10^{-6})
0.000001	(or 10^{-6})	0.000000956	(or 9.56×10^{-7})

The *Weibull model* has two shape parameters. As such, it is more flexible and can fit a wider variety of distributions. Following the idea of a Poisson distribution, Weibull models involve mathematical expressions that are exponential in form; that is, they involve *e* raised to some power. The distribution function for a Weibull model is:

$$P = 1 - e^{-aE^b} \qquad (5)$$

where a and b are curve-fitting parameters. From Equation 5:

$$e^{-aE^b} = 1 - P$$

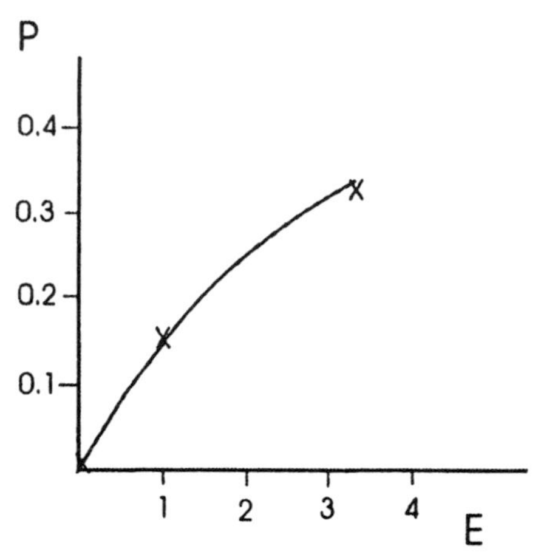

Figure 2.17 Plot of logit model with the data

Taking the ln of both sides yields:

$$-a(E)^b = \ln(1 - P)$$

Taking the ln again of both sides yields:

$$\ln a + b\ \ln(E) = \ln[-\ln(1 - P)] = y$$

Substituting the data values for P in this equation and evaluating ln(E) produces the values shown in Table 2.8.

Table 2.8 Experimental Values for Exposure-Response Fit for Weibull Model

E (mg)	P	y	ln E	y(eqn)
0.1	0.023	−3.77	−2.3	−3.70
1	0.159	−1.75	0	−1.85
3.16	0.309	−0.995	1.15	−0.923

Now plot y vs ln(E), and determine the best straight line fit. The least squares fit yields a = 0.157 and b = .806. (*See* Figure 2.18) This yields the values shown in Table 2.8 and the curve in Figure 2.19.

Substituting from Table 2.8:

$$\ln a + b(-2.3) = -3.75$$

$$\ln a + b(0) = -1.75$$

and subtracting yields:

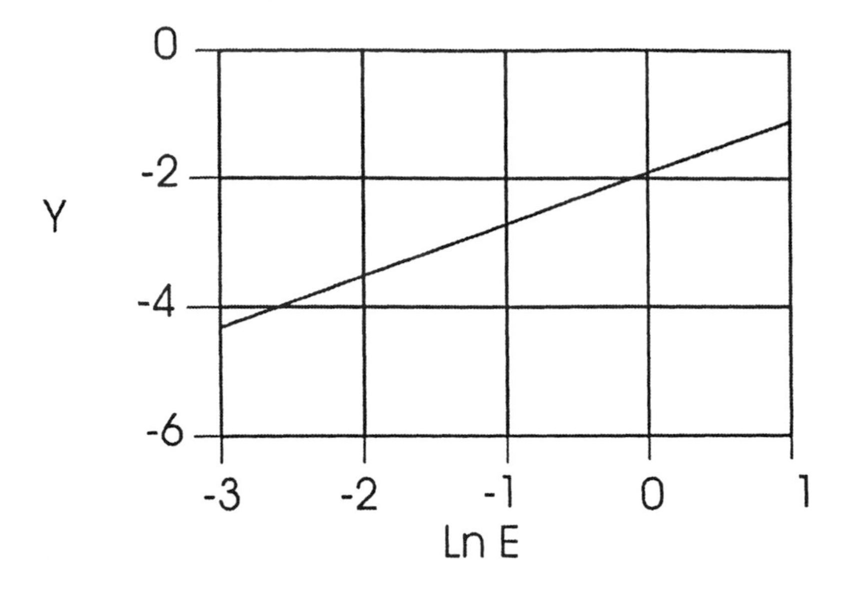

Figure 2.18 Plot of the functions used in Weibull model fits

$-2.3\ b = -2$, and thus $b = 0.8696$

$\ln a = -1.75$, and thus $a = 0.174$

$\ln a + b \ln(E) = \ln[-\ln(1 - P)]$, and thus

$y = -1.75 + 0.8696 \ln(E) = \ln[-\ln(1 - P)]$ as shown in Figure 2.19.

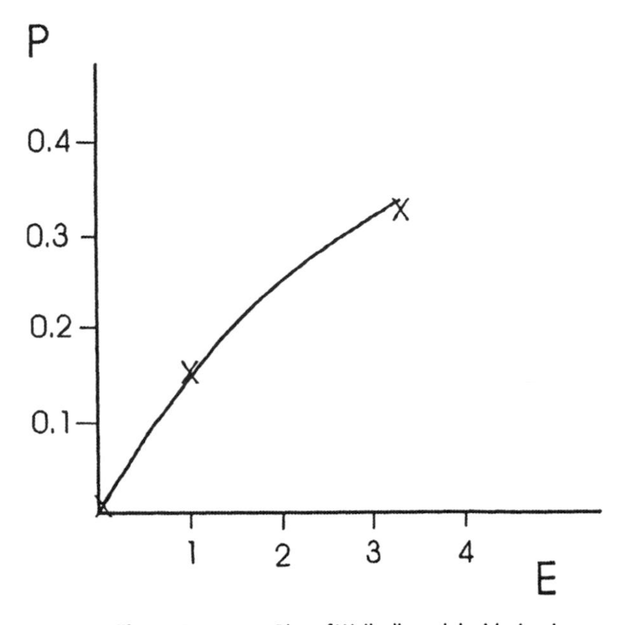

Figure 2.19 Plot of Weibull model with the data

We can now complete Table 2.8. To generate P values for environmental exposures, substitute in the general equation for P = 0.0001:

$-1.75 + 0.8696 \ln(E) = -9.21$

$\ln(E) = -8.58,$

$E = 0.000636.$

And for P = 0.00001,

$\ln(E) = -11.5,$

$E = 0.0000636.$

And for P = 0.000001

$E = 0.00000636.$

We have already studied the *lognormal (probit) model* above. Most environmental exposure data sets, as well as most potency models for environmental pollutants, fit lognormal models. Since a lognormal model multiplied by a lognormal model yields a lognormal model, most risk distributions generated by environmental exposures to environmental pollutants have lognormal distributions. This is a very convenient property.

We can now use the concept of probability being the area under a normal or lognormal curve to develop a more sophisticated model to fit the exposure-response data. Following the idea of a normal curve, these models involve different mathematical expressions that are exponential in form. That is, they involve *e* raised to some power.

This model uses the normal curve as the mathematical expression. From experience with the distribution of environmental quantities, we assume that the exposures involved distribute normally on a logarithmetic scale. Thus the z axis is log(E). The mean or highest point on the curve is $\mu = \log(E)$. To make the scale look like that of the normal curve in Figure 2.10 where the peak is at z = 0, the variable z is replaced by $\log(E - \mu)$. To create a scale with standard deviations of 1, 2, 3,... (as in Figure 2.10), divide by σ. Thus, $z = [\log(E - \mu)]/\sigma$. Using Equation 6 for the normal curve, this approach yields a mathematical expression for the probability of risk using the probit model as:

$$P = \frac{1}{\sqrt{2\pi}} \int_{-\infty}^{\log E} e^{-\frac{1}{2}\left(\frac{\log E - \mu}{\sigma}\right)^2} d\frac{\log E}{\sigma} \qquad (6)$$

Again the integral represents the area from the left hand side of the distribution to the exposure, or log(E) in this case. The relationship is:

$$z = \frac{\log E - \mu}{\sigma} = \frac{\log E}{\sigma} - \frac{\mu}{\sigma}$$

This is a parametric expression where the general form is z = a log(E) + b, where a = $1/\sigma$ and b = $-m/\mu$. The parameters a and b are curve-fitting parameters to be determined by the specific data to be fit. We determine the parameters a and b by substituting the values of E and P from Table 2.6.

A probability of P = 0.023 is the area under the curve from –4 to the z value we seek. The total area from $-\infty$ up to the peak of the normal curve is ½ the total area since the curve is symmetrical. The area or probability is 0.023, which corresponds to z = –2. (Check this out in Table 2.4). This value along with the log(E) is put in Table 2.9. For the probability P = 0.159, z = –1. For P = 0.309, z = –0.5.

Table 2.9 Experimental Values and the Corresponding z Values

E (mg)	P (deaths/animal)	z	log E
0.1	0.023	–2	–1
1.0	0.159	–1	0
3.16	0.309	–0.5	0.5

Now plot these points on a z vs. log(E) graph. We have reduced the complex integral for probability shown in Equation 6 to a linear plot of the form: z = alog(E) + b. This straight line can be determined in one of two ways. We could plot the three points and use a ruler to find the line that best fits these points. For most risk analysis work this is accurate enough. Alternatively, we could use the least squares fit procedure.

The parameters a and b that best fit the data are a = 1 and b = –1. Thus the equation is z = log(E) –1 and this is plotted in Figure 2.20. We can determine z values for other exposures, E, and knowing the z value, we can look up the probability in Table 2.4 (or a more complete table).

Now to generate the probabilities for the environmental exposures of 10^{-4}, 10^{-5}, and 10^{-6}. For example for E = 10^{-4}, z = logD – 1 = –4 –1 = –5, and using a more complete table of the normal curve values (or using a calculator), we can complete Table 2.10. These values are plotted in Figure 2.21. Fortuitously, these points lie on the curve represented by Equation 6.

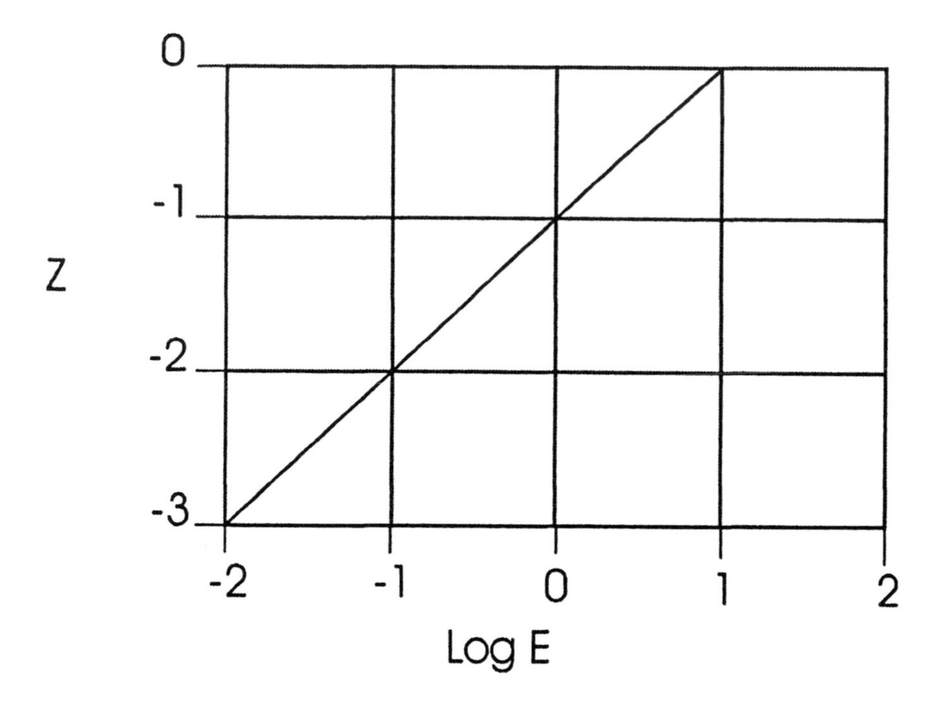

Figure 2.20 Plot of the functions used in probit model

Figure 2.22 shows the plots of the three models we have generated values for in the experimental and environmental exposure regions. Note that the models fit the three data points and have very different values at low exposures such as those commonly found in environmental situations.

Table 2.10 Probabilities for Environmental Exposure Levels for the Probit Model

E (mg)	P (deaths/animal)	z
10^{-4}	2.9×10^{-7}	–5
10^{-5}	9.9×10^{-10}	–6
10^{-6}	1.3×10^{-12}	–7

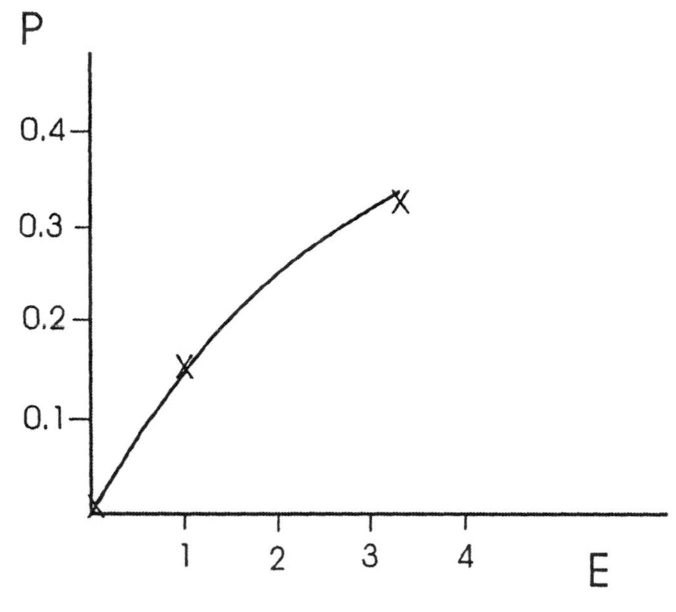

Figure 2.21 Plot of the probit model with data

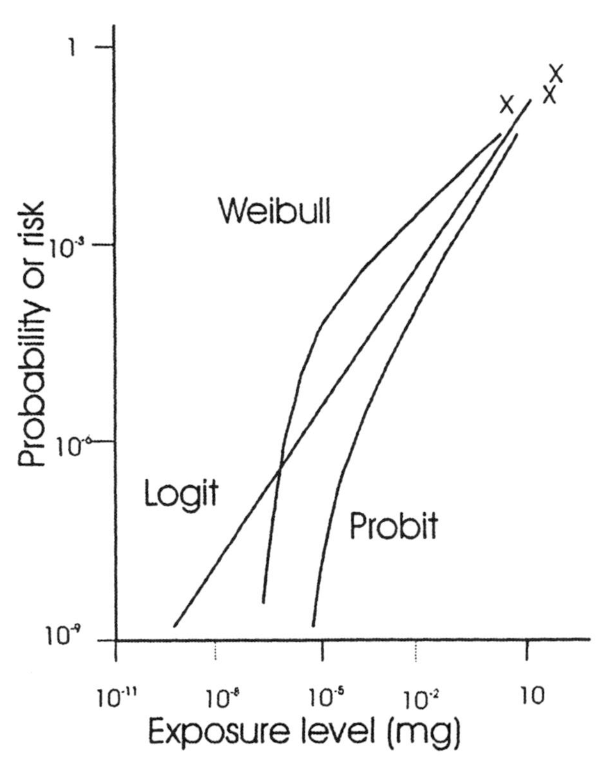

Figure 2.22 Log-log plot of curves fitted to the experimental data using probit, logit, and Weibull models

Mechanistic Models

Some biological processes may operate the same way that our example of the Poisson distribution worked. If you throw a small number of ping pong balls at random into a group of receptacles, you fill a small fraction of the receptacles with balls, and you fill very few with more than one ball. Suppose that the receptacles are organisms (mice or people) and the ping pong balls are "hits," toxic lesions of a particular kind. Hits have an all-or-none property; that is, when an organism receives a hit, it gets the effect in the same way that a light switch turns on or off. No gradation of effect occurs, and the effect of two hits is the same as the effect of one hit. (Either the light is on, or it is off. You can not turn it on more than once.)

Toxic lesions of DNA, the genetic material, might operate in this way. Once the genetic code changes, all of the progeny will inherit the change. The code change might occur in a germ line cell, and the next generation of organisms will inherit the change. These lesions are heritable mutations. Suppose that the genetic code changes in a cell that is not part of the germ line, but the code change causes a cell to lose its control program for growth. This cell now proliferates out of control. Uncontrolled cellular replication is the essence of cancer. Thus, at least two toxic effects, carcinogenesis and mutagenesis, might resemble hits in their mechanisms of toxicity because they might operate in all-or-none ways.

We could use the same general procedures to generate probabilities for environmental exposures using these models. Since you already have a general idea of how to do this, from the tolerance distribution examples in the previous section, we will just list the equations for two mechanistic models.

The *one-hit model* has exactly the same form as the Poisson distribution function for the proportion of events with no hits:

$$P = 1 - e^{-aE}$$

The proportion of cells or organisms without one or more hits is exactly the value we seek. At low exposures (or a low numbers of hits), this model becomes approximately the same as

$$P(x) = a(E).$$

Thus, we expect that hits will occur in proportion to some exposure, perhaps an environmental concentration or a radiation intensity. You may notice that at low exposures, this model has linear, nonthreshold properties. We do not expect every molecule or energy wave to produce a carcinogenic or mutagenic lesion. We expect that a fraction, *a*, of the molecules or energy the organism experiences can lead to a hit and that *a* does not change when exposure changes. Suppose that a did change with exposure. Most experts think that, if a changes, it will become smaller as exposure gets lower. If so, when you fit a one-hit model to data at high exposures, it might overestimate, but would not underestimate, risk. In this sense, the one-hit model predicts an *upper bound* of risk.

The one-hit model is not very flexible in fitting data. Efforts to derive more information led to the Armitage-Doll, or *multistage model* of carcinogenesis. Mathematically, it is a filtered poisson model. Assume that a cell passes through k > 1 stages before its response is irreversible and it becomes a cancer cell. Only one cell needs to survive each stage to go on to the next. Then,

$$P = 1 - e^{-\sum_{i=1}^{k} a_i E^i}$$

Thus, for k = 2

$$P = 1 - e^{-(a_1 E + a_2 E^2)}$$

If the biological process that leads to the initiation of a cancer cell occurs at the second stage of a multistage process, the squared term will dominate the exposure-response relationship at high doses, where you fit the model to data. If only a little of the additional risk of cancer comes from hits on the first stage, however, the squared term will contribute little to risk at very low exposures. Again, the equation becomes approximately:

$$P = 1 - e^{-a_1 E}$$

Thus, the multistage model also is an upper-bound model at low doses.

The multistage model as presented here is not the same as the linearized multistage (LMT) model that some regulatory agencies use. In the Armitage-Doll model, the constants in the polynomial part of the equation relate to the time between stages. As the LMT is applied to animal data (carcinogenesis bioassays), these constants relate to the number of exposures used in an experiment. In addition, the LMT has other constraints. For example, an LMT model cannot have any negative constants.

A full explanation of the details of these models lies beyond the coverage of this text. In fact, we have barely scratched the surface of cancer modeling. For example, much recent energy has gone into Moolgavkar-Knudson-Venzon (MVK) models. They are two-stage models, similar to the multistage model depicted above, but a cell that survives the first stage is allowed to replicate. Thus, the number of initiated cells at risk for a second hit can vary in a MVK model.

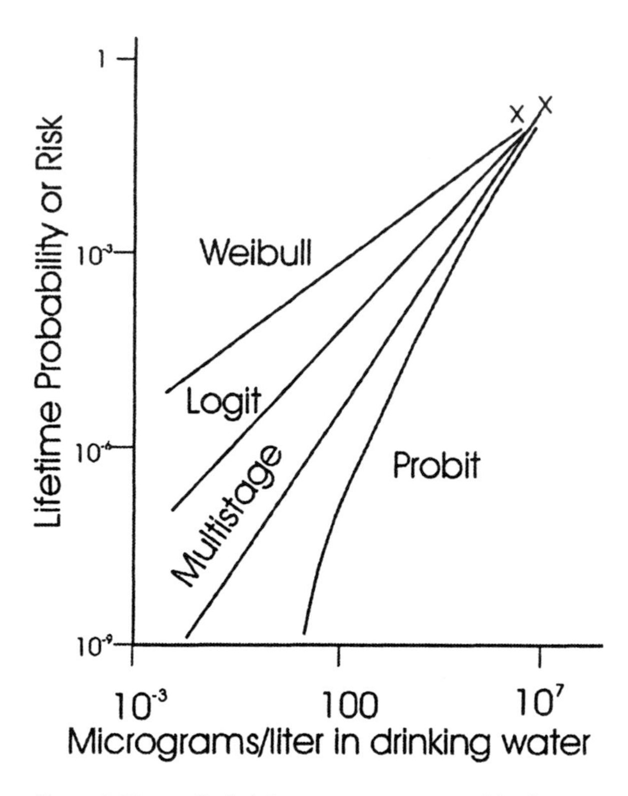

Figure 2.23 Typical dose-response curve resulting from analysis of animal data. This one is for drinking water exposure of mice to trichloroethylene.

Figures 2.23 and 2.24 show some typical fits to experimental exposure-response data. Figure 2.23 uses exposure-response data from exposure of mice to trichloroethylene in drinking water to show fits of the multistage model with the probit, logit, and Weibull models for comparison. Note that the plot is on log-log axes. There are two data points shown by the stars in the upper right hand corner. Generally, animal exposure-response data includes a control group and two, or sometimes three, exposure levels. Shown on the plot are the curves that were fitted and the upper 95 percent confidence limits. Typically these upper limits are used in setting environmental standards. From a public health perspective, the regulatory argument is that a false positive prediction is better than a false negative prediction. The upper limits represent a worst-case scenario.

The area of interest for environmental exposures in Figure 2.23 is in the range between 10^{-4} and 10^{-6}, and this region is far from the actual data points. Consider the range responses or probabilities for the exposure of about 80 mg/L. The range of responses for the four models at this concentration extend from a low of about 8×10^{-9} for the probit model to about 10^{-2} for the Weibull model. This range is just over 6 orders of magnitude of uncertainty. The problem of setting standards with these fits should be clear. The range of responses are from 8×10^{-9}, which is clearly below any concern level, to a value of 10^{-2}, which is clearly unacceptable. In this instance, the process of developing models

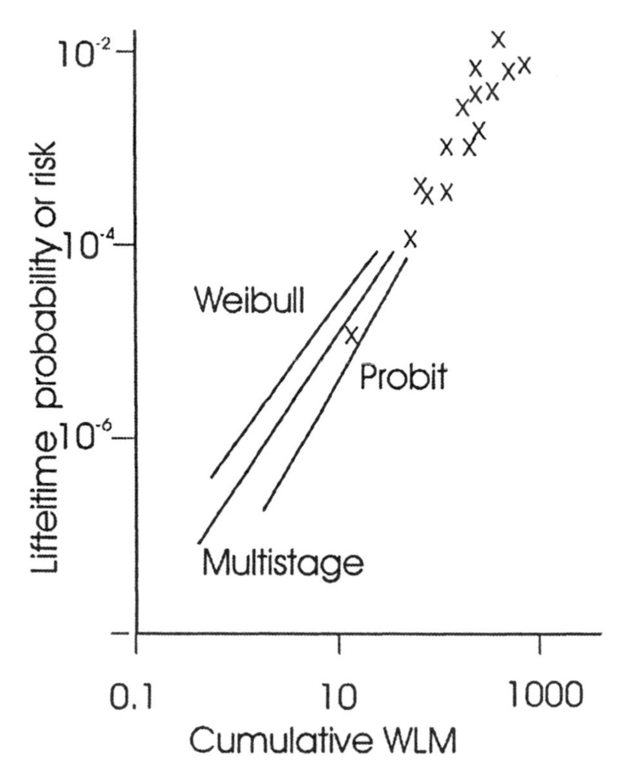

Figure 2.24 Dose-response data and model fits to data from exposures of miners to radon

does not contribute to regulatory decision-making because the risk is below regulatory concern.

A similar plot is shown in Figure 2.24 for exposure of humans (mainly miners) to radon. The exposures are in working level months. At equilibrium of the radioactive progeny of radon, one working level corresponds to 100 pCi/L. For lower levels, like those found in homes (a few tenths to a few pCi/L), the corresponding lifetime risk in percent is about ½ the pCi/L value. Thus, for the EPA suggested guidance value of 4 pCi/L, the risk of that exposure for a 70-year lifetime is about 2 percent of the risk of getting lung cancer. There are many differences between Figures 2.23 and 2.24:

- the data in Figure 2.24 are for humans while the data in Figure 2.23 are for animals and must be converted to apply to humans;
- the data in Figure 2.23 are two closely spaced points while the data in Figure 2.24 span two orders of magnitude;
- the data in Figure 2.23 are for one study while the data in Figure 2.24 represent four independent studies;
- the range of variation is much less in Figure 2.24, and
- the environmental area is far from the data in Figure 2.23, while the environmental and actual data overlap (for the lowest point) in Figure 2.24.

We conclude that the risk for exposure to radon is known far better than that for trichloroethylene.

Uncertainty

Uncertainty is perhaps the most misused and misunderstood concept in the field of risk analysis. It means different things to different people. It can indicate measurement error, a feature of the measuring instrument, vague language, or just plain ignorance. Yet, it is an essential idea.

To make matters more confusing, risk analysts use many different terms as crude synonyms for uncertainty when they mean something quite technical, including standard deviation, variability, random error (chance), or lack of knowledge. They use these terms interchangeably. We have already discussed standard deviation, which relates to the normal distribution. Variability refers to the standard deviation, but may be more complex since many distributions are not normal. Determining the standard deviation of a non-normal curve can be complicated and is beyond the scope of this book.

We have arbitrarily grouped different sources of uncertainty as follows:

1. subjective judgment
2. linguistic imprecision
3. statistical variation
4. sampling
5. inherent randomness
6. mathematical modeling
7. causality
8. lack of data or information
9. problem formulation

Subjective Judgment

In very few cases is the environmental measurement known well. Usually an estimate of risk contains some component that is either partly known or unknown. In some cases a poll is taken of the judgment of experts—called the Delphi approach or method. The results are then fed back to those making the judgments and the process is repeated. Experts may have a better sense of the likelihood of a given value, but this is still subjective and subject to error or uncertainty.

All of us have biases. Some think that the environment is getting worse and others think it is doing all right. Some think cancer is more important than neurological, immunological, or developmental maladies. These biases color our thinking when judgment is required and introduce subjective uncertainties. For example, the decision about which pollutant to test and which endpoints to test for depends on our biases, and this introduces subjective judgmental uncertainties. Also, uncertainty is introduced by studying selected groups, such as children and the elderly. The choice of model for analysis of exposure-response data or exposure plot introduces further uncertainty.

Linguistic Imprecision

Beauty is in the eye of the beholder. A 25-year-old person may seem ancient to a 5-year-old child and quite young to an octogenarian. Thirty miles per hour is speeding in a school zone and recklessly slow on a superhighway. *Ancient/young, fast/slow* are words that have different meanings in different situations. Other words that have multiple meanings are *large, huge, small, very small, negligible, high, low, minimal,* and *highly probable* to mention only a few.

Another form of linguistic imprecision is found in expressing error or uncertainty. The terms *overestimate* or *underestimate* are not usually helpful. The real question here, as with any other relative words, is what is the numerical value? How slow or fast? How short or tall? By how much is it underestimated?

Thus we need to be careful when we encounter or use words that have relative meanings. The solution is to be as quantitative as possible. Generally speaking, the more perspective we can give the reader or listener, the less uncertainty there will be.

Statistical Variation

As an experiment, measure the length of your table ten times. You will get a variety of values, but all will be very close. A typical ruler has marks every $1/16$ inch, so you cannot do much better than measure to the nearest $1/32$ inch. There is an inherent range of values likely within a range of $1/16$ inch above the actual value to $1/16$ inch below the actual value. This range represents an uncertainty and is present because there is a minimal unit on our measuring device. There is an unpredictability about what the next measured value may be. Thus we are dealing with a random uncertainty or error. If the average measurement of the table is $6\frac{1}{2}$ feet, we could express the measured value and error or uncertainty as:

$$6\tfrac{1}{2}\text{ feet} \pm 1/16\text{ inch}$$

if we are conservative in estimating error, or as:

$$6\tfrac{1}{2}\text{ feet} \pm 1/32\text{ inch}$$

if we estimate error liberally.

It may be that our ruler has been worn at the end and the last 1⁄32 inch has been worn off. In that case all our measurements will be 1⁄32 inch low. This is called systematic error or bias.

Another kind of statistical uncertainty results from incomplete or uneven mixing of a contaminant in the atmosphere, water, or soil. Thus there can be "hot spots" or locations where the concentration is quite high or very low, and these can occur in a random manner.

The random scatter about a control or average point due to measurement limitation or imperfect mixing is called precision. (*See* Figure 2.25) If a measurement is biased it is called inaccurate.

The standard deviation is a way to express the statistical variation. It is expressed as the mean ± one standard deviation:

$$\mu \pm \sigma$$

For a normal distribution, we have roughly two-thirds confidence that the actual value lies within this range.

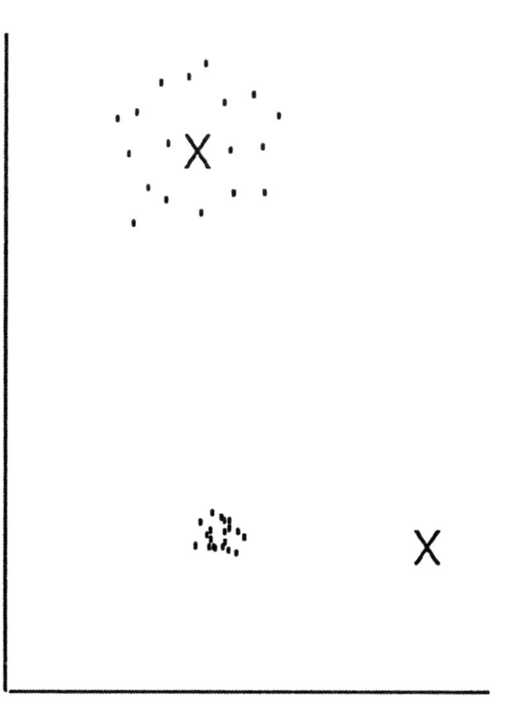

Figure 2.25 Plot showing the relationship of data scatter to a random scatter for precision and a biased or accurate representation

Sampling

Consider a dump or what we now call a municipal waste site. Try to develop a plan to sample the site to describe its contents in a representative way. How many samples would you take? In what areas and at what depths would you sample? And remember that you have a limited budget. Can you develop a plan that will produce results that are representative of the site's contents, short of sorting and weighing everything in it? The more samples you take, the better your description becomes. Making assumptions, like wanting to be confident the estimate was within 50, 5, or 1 percent of the actual value, provides a basis for developing a plan that is statistically representative of the site being sampled.

Most environmental measurements are anecdotal. That is, someone takes a few samples without any plan and ignores such factors as the flow of the river in the middle and edge, the distance from the smoke stack, and the uneven character of the soil. Too often we are only looking for hot spots and points of interest.

Inherent Randomness

On the quantum level Heisenberg discovered the principle that both the momentum and location of an entity, such as an electron, cannot be known with absolute precision. If we know the location of an electron, we have no idea what its momentum is; if we know its momentum, we have no idea where it is. Thus the concept of an electron cloud surrounding the atomic nucleus was formed.

There is an inherent variability in the way our world is constructed, and there is a limit to how precisely we can measure quantities. Although it is interesting philosophically, this limitation does not interfere with environmental measurements—even for the quantity in the femto or atto range.

Model Uncertainty

In this chapter, we described how risk analysts fit mathematical models or equations to environmental data, such as exposure-response curves. The underlying theory of causation usually is not known. Thus the appropriate mathematical expression is not clear. Different equations or models yield different estimates of risk. An uncertainty or range of values can evolve from a range of possible equations or models.

Another source of modeling uncertainty is the complexity of the values used to estimate risk. Consider the equation:

$$\text{concentration in drinking water} \;\; X \;\; \text{ingestion rate} \;\; X \;\; \text{body weight}$$

$$\text{Risk} = \frac{(\text{mg/L})(\text{L/day})(\text{kg}\backslash\text{person})}{\text{potency}(\text{mg/kg/person/day})}$$

Each of these quantities has a range of values or a description. (*See* Figure 2.26) The mathematical problem is how to multiply and divide these distributions. We could multiply the averages or propagate the errors with, for example, a Monte Carlo approach. A curve, such as the one shown for risk in Figure 2.26, results. This repetitive process is possible because of the speed of modern computers. A drawback of this method is that it requires a knowledge of the individual distributions. We seldom know these individual distributions with much accuracy.

Causality

Epidemiology data can show a correlation between a possible cause and an effect. For example, a pack-a-day cigarette smoker has a 10 percent lifetime chance of developing lung cancer. This correlation does not prove cause.

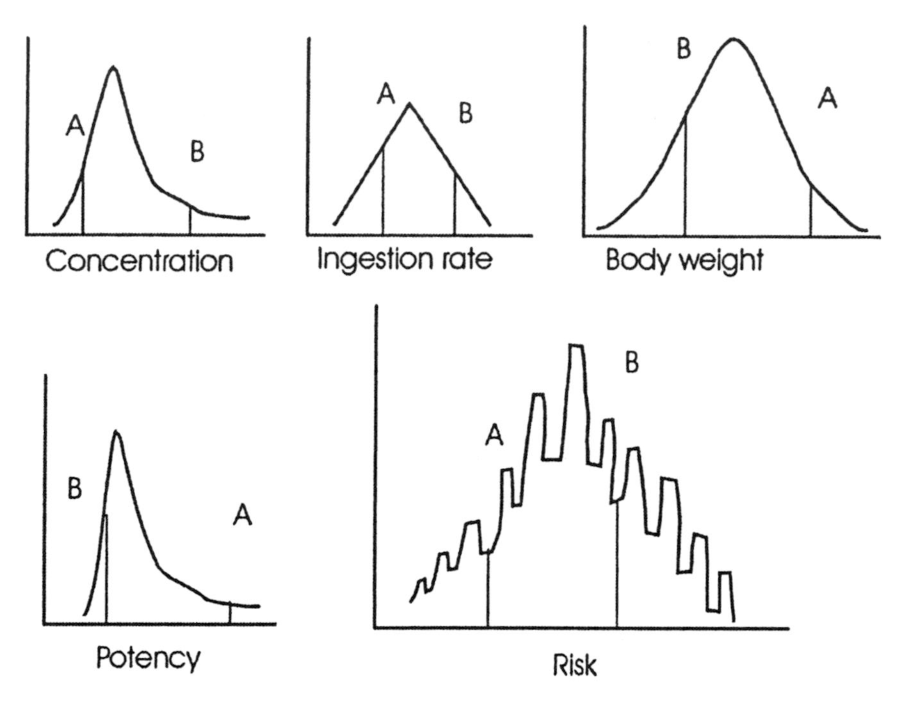

Figure 2.26 Plot showing how individual distributions are multiplied to get the overall risk distribution

To show or prove cause is very scientifically demanding, complex, and difficult. For example, to determine the cause of cancer (if we ever do) will require knowledge of how molecules interact in terms of the kind of reactions, the timing, and the intensity of the reaction as well as other properties not known at this time. The complexity of the biological world often precludes knowing the actual cause of a disease, toxic effect, or negative outcome.

Lack of Data and Information

For most risk estimates there is data and information missing. It may be known that a given concentration of a contaminant causes x number of fatalities due to liver cancer. But how many people are exposed (the epidemiologist calls this the cohort) may be poorly known or not known at all. In this case, the default assumption can be used. For example, it can be assumed that the concentration for everyone is the same as that for the one location sampled. Or it can be assumed that none of the contaminant was metabolized or excreted. All risk assessment procedures have built-in assumptions. Not being aware of these assumptions can cause uncertainty. Even knowing them does not eliminate the uncertainty about what actually happened.

Problem Formulation

Even if you have a perfect model, accurate data, a good understanding of the sources of random and sampling error, notions of causality, and precise terminology, you can still misunderstand a risk. For several centuries, scientists pursued ways to eliminate swamp gas to decrease malaria. Because clearing swamps eliminated mosquitoes, the scientists reached some correlation. Their efforts led to advances in refrigeration and an improved understanding of gases. Even so, they attempted to solve the wrong problem. Well past the American Civil War, well-intentioned physicians continued to bleed their patients, causing many deaths and exacerbating injuries. Lest you pass this off as ancient history, only in the past few years has the medical community realized that ulcers are an infectious disease. Most previous therapies and theories of causation of ulcers were misguided.

Making Estimates

Because we know that the number of people killed in automobiles is approximately 50,000 a year, we can estimate how many will be killed over a weekend in the Washington, D.C., metropolitan area. We may not be exactly sure of the local population, but 4 million may not be far off, and we all know that the population of the United States is about 250 million people. Thus, we can estimate the number of automobile fatalities in a given area on any weekend:

$$(4/250) \times (2/365) \times 50{,}000 = 4.38 \text{ or about } 5$$

Other problems you might like to try for homework are these:

- estimate the number of barbers in the United States
- estimate the number of drug stores in the United States
- estimate the number of piano tuners and cobblers in the Washington, D.C., metropolitan area

In many cases we only have limited information, but can supply the missing pieces to make an estimate that may allow us to make a decision. An important element of this process is to know how good our estimates are. For example, consider the estimate of how high the water would get in the flood of the Red River in Grand Forks, North Dakota, in the spring of 1997. The estimate from the weather forecasters was widely quoted as 49 feet. Most people understood this to be accurate to the nearest foot. The message from the weather forecasters that this was an estimate based on crude information about the local geography and that it could be off by five to ten feet did not get through. The sand bags were laid up to 49 feet (or perhaps to 50 feet). But the river crested at 54 feet and a disaster ensued. The lesson is that estimates are great, but we also need to know how precise these estimates are.

Example 1: Bladder Cancer from Saccharin

In 1977 the Food and Drug Administration proposed a ban on saccharin. This resulted from several experiments with animals that showed that saccharin can cause bladder cancer. This caused a furor among the public because millions had come to rely on saccharin to sweeten coffee, soft drinks, and more. We can use the available data to estimate how much of a problem use of saccharin could be.

We start with the dose-response evaluation. A daily dose of 2500 mg/kg of body weight in animals produces 24 percent excess incidence of bladder cancer; that is, in a lifetime, 12 of 50 animals developed bladder cancer. But there is an uncertainty and the upper 95 percent confidence limit for 18 of 50 animals or a probability of 0.36, which we round off to 0.4. We will have to make some assumptions. For example, we will assume that the effect in the lifetime of the animal is the same for humans with a lifetime of 70 years. We will also assume that the dose-response curve is linear. Thus 250 mg/kg would produce no more than 4 percent tumors, and 2.5 mg/kg would produce no more than 0.04 percent.

By reading the label, we can find out that one bottle of soda contains about 140 mg of saccharin. Thus for a 70 kg adult this is about 2 mg/kg. The individual lifetime risk of a tumor after consuming one bottle would then be:

(2/2.5) x 0.04% or about 0.03% of 0.0003

If everyone consumed 140 mg/kg a day for a lifetime, the annual cancer incidence for saccharin consumption would be:

Annual incidence = 0.0003 x 250,000,000 x 1/70 = 1071, or about a thousand cases per year

But we know that not everyone in the United States consumed 140 mg/kg per day. The actual consumption was about five million pounds/year. If everyone consumed 140 mg/kg/day, the yearly total would be:

(140 mg/kg/day)(1 kg/1,000,000 mg)(250 million)(365 days) = 12/775 million = approximately 13 million (or about three times the five million quoted above)

Thus, the estimated annual incidence would be 1000/3 = 300 cases of bladder cancer per year.

Example 2: Penetration of a Liquid Natural Gas (LNG) Facility by a Meteorite

We know that to facilitate the transportation of natural gas it is liquefied. It is stored and transported in ocean-going vessels in a liquid state. Liquid natural gas (LNG) does not burn, but it vaporizes quickly and it is flammable when it is sufficiently dilute.

Our concern is that a meteorite could penetrate a pipeline or a storage tank, releasing the liquid which would vaporize and then drift towards densely populated areas and ignite. We could learn from an encyclopedia that small meteors are constantly entering the earth's atmosphere at velocities ranging from 40,000 ft/sec to 300,000 ft/sec, but most of these completely disintegrate during their encounter with the earth's atmosphere. However, a few still manage to strike the earth's surface. These are the ones we are concerned about. When we work out the physics, it turns out that a meteor weighing more than a pound will have an impact greater than one million foot-pounds.

Now, let us consider how to estimate the risk to humans if a meteorite hits an LNG facility. If N meteorites fall randomly on a given reference area per year, the probability of one striking a target area per year is given by:

$$F = (\text{area of target})/(\text{reference area}).$$

This is like a Russian roulette problem where the probability of no hits is:

$(1 - F)^N$, where N is the number of hits

Then, the probability of at least one meteorite striking the target area per year is one minus this or:

$$P = 1 - (1 - F)^N$$

If $F << 1$ then we can approximate $(1 - F)^N$ by $1 - NF$ and thus

$$P = NF$$

The average number of meteorites weighing more than one pound which fall on the United States annually is 65. The area of the United States is 1.05×10^{14} square feet. Thus, the annual probability of striking the area A is:

$$P = 65A/1.05 \times 10^{14} = A \times 6.2 \times 10^{-13} \text{ per year}$$

Let us consider the following table of information which shows the area of LNG facilities and the probability of their being struck according to the above equation:

Facility	Area (sq. ft)	Probability (P)
LNG tanks	8×10^4	5×10^{-8}
Pipelines	1.7×10^4	1.1×10^{-8}
Tankers*	7.8×10^4	5×10^{-8}

*(0.16 ships docked on average)

However, we realize that not all meteorites will have sufficient energy to penetrate the atmosphere. We can estimate that 5.5×10^{-2} will do so. Then the probabilities above become 2.8×10^{-9} for tanks, 6.1×10^{-10} for pipelines, and 2.8×10^{-9} for tankers.

For us to be able to estimate the actual risk of death, injury, or property loss we would need to know the local meteorology, heat conduction from the ground, ignition potential, and so on.

References

Armitage, P. and Doll, R. 1954. The Age Distribution of Cancer and a Multistage Theory of Carcinogenesis. *British Journal of Cancer 8*:1–12.

Armitage, P. and Doll, R. 1961. Stochastic Models for Carcinogenesis. IN *Proceedings of the Fourth Berkeley Symposium on Mathematical Statistics and Probability* (Ed.) Lecam and Neyman.

Bernstein, P. 1996. *Against the Gods: The Remarkable Story of Risk.* New York: Wiley & Sons.

Chemical Rubber Company. 1995. Standard Mathematical Tables. Boca Raton, FL: CRC Press.

Chicken, J. C. and Haynes, M. R. 1989. *The Risk Ranking Technique in Decision Making.* Pergamon Press, New York.

Evans, M., Hastings, N., and Peacock, B. 1993. *Statistical Distributions.* (2nd Ed) New York: John Wiley & Sons.

Food Safety Council. 1980. *Proposed System for Food Safety Assessment.* Final Report of the Scientific Committee of the Food Safety Council, Washington, DC.

Gibbons, J. D., Olkin, I., and Sobel, M. 1977. *Selecting and Ordering Populations; A New Statistical Methodology.* New York: John Wiley & Sons.

Hallenbeck, W. H. 1993. *Quantitative Risk Assessment for Environmental and Occupational Health,* (2nd Ed). Boca Raton: Lewis Publishers.

Hollander, M. and Wolfe, D. A. 1973. *Nonparametric Statistical Methods.* New York: John Wiley & Sons.

Morgan, M. G. and Henrion, M. 1992. *Uncertainty: A Guide to Dealing with Uncertainty in Quantitative Risk and Policy Analysis.* New York: Cambridge University Press.

Poulter, S. R. 1998. Monte Carlo Simulation in Environmental Risk Assessment—Science, Policy and Legal Issues. *Risk: Health, Safety and Environment* 9: 7–26.

Walker, V. R. 1998. Risk Regulation and the Faces of Uncertainty. *Risk: Health, Safety and Environment* 9: 27–38.

Wilcoxon, F. 1945. Individual Comparisons by Ranking Methods. *Biometrics 1*: 8083 (1945).

Chapter Appendix

Solving three equations for unknowns is an excercise you may remember from high school algebra. Since it is an excercise that may bring back memories and sharpen your mathematical skills, we include this exercise in this appendix.

We can write three equations for the three data points, as follows:

$$0.023 = a + 0.1b + 0.01c \qquad (7)$$

$$0.159 = a + b + c \qquad (8)$$

$$0.309 = a + 3.16b + 9.99c \qquad (9)$$

To determine the values of the three parameters a, b, and c, subtract Equation 7 from Equation 8:

$$0.136 = 0.9b + 0.99c \qquad (10)$$

Subtract Equation 8 from Equation 9:

$$0.15 = 2.16b + 8.99c \qquad (11)$$

Divide Equation 10 by 0.9:

$$0.151 = b + 1.1c \qquad (12)$$

Divide Equation 11 by 2.16:

$$0.0694 = b + 4.16c \qquad (13)$$

Subtract Equation 13 from Equation 12:

$$0.0816 = -3.06c$$

Thus $c = -0.0267$. Substituting this into Equation 13:

$$b = 0.0694 - (4.16)(-0.0267) = 0.180$$

Substituting this into Equation 8:

$$a = 0.159 - b - c = 0.159 - 0.180 + 0.0267 = 0.00570$$

Chapter Three

Regulation

The Regulatory Process

Why have regulations at all? If you took a car trip recently, you probably found the travel so commonplace that you did not notice some interesting aspects of the ride. In this country everyone drives on the right side of the road. This convention seems arbitrary. Indeed, in some countries such as England, the convention reverses and everyone drives on the left side. Yet failure to observe the local conventions can lead to disaster, as can driving on the wrong side of the road. Governments generally provide statements of how people should behave through laws.

In this chapter we describe the U.S. regulatory process in sufficient detail for you to understand the use of risk assessment in regulation. In this section we briefly sketch a linear process that begins with the creation of a regulatory authority and ends with socially desirable changes in behavior. We do not mean to imply that regulation is a dominant activity. Much socially appropriate behavior stems from forces outside the regulatory system, such as culture, product liability, contracts, insurance, self-regulation, and ethical principles.

In a representative democracy like ours, Congress has the authority to create, modify, fund, and evaluate regulatory authorities, which are not quite the same as regulatory agencies. Sometimes one agency will have responsibility for many statutes. In contrast, sometimes a single statute contains several authorities. Congress passes such legislation out of a variety of motives, but the most important is to respond to perceived public needs. In other words, we do not necessarily need that new regulatory statute. What matters is that most people believe they need it.

Congress is ill-equipped to create regulations and set standards, much less enforce them. The public elects the members primarily because of their political skills, not their technical knowledge. Further, most federal representatives consider these kinds of tasks a dilution of their primary responsibility of responding to the legislative desires of the public.

Instead, Congress gives the Administrative Branch of our government the responsibility and authority to carry out the intent of a statute through an agency. The implicit basis of this delegation is that the agency will provide scientific data, technical expertise, and policy-making skills that the legislature lacks. However, most modern regulatory statutes go well beyond this primary delegation and include many proscriptions. The extent of these details, the "shoulds" and "musts" of the laws, primarily reflect a distrust of regulatory agencies by Congress. Thus, our regulatory legislation has become extensive and complex.

Try stopping by a local law library sometime and looking at the many volumes of the *United States Code* (USC), which contains our nation's laws. Then, consider that this huge amount of legislation is not adequate to explain what is going on in the legislative process. In addition, for example, you must understand how statutes interact with each other and what the legislative history says. The *Congressional Record* captures much of the floor debate about pending legislation, but Congress holds some hearings without publishing them and publishes some hearings separately. In some situations, regulatory agencies and the courts have to resort to the legislative history to understand what the statutes mean.

Congress leaves it to the Administrative Branch to manage these agencies in creating regulations, which are sometimes generic, sometimes specific. The regulatory agencies do not do this work in isolation, however. They work under general oversight, guidance, and constraints, such as the budgets that Congress authorizes. Some legal restraints are quite mundane. The employees have to abide by the strictures of the Office of Personnel Management. The General Services Administration provides buildings and furnishings for agency personnel and establishes conditions for the use of these facilities. If the agencies need outside assistance, they must go through complex federal contracting procedures.

We are interested in the general ways of developing regulations and the direct constraints on this process. Several are important. The Freedom of Information Act (FOIA), the Administrative Procedures Act (APA), and the Federal Advisory Committee Act (FACA) control the overall public and stakeholder access to (and involvement in) the regulatory process.

When an administrative agency wants to make binding regulations, it usually goes through a process called informal rulemaking, as defined by the APA. In the rulemaking process, the agency first proposes a regulation, then takes public comments and/or holds public hearings. After reviewing these comments to see if the proposal has any flaws that merit alteration in the proposal, the agency issues a final regulation. This process sometimes takes a long time. The stakeholders and affected parties have the right to comment

on the proposal and to have their ideas taken seriously. Often, litigation occurs at steps in the process, bringing the courts into the dialogue. In the United States, however, the benefit of this laborious and intense process is that most parties abide by the final regulations. While enforcement does occur, penalizing those entities that fail to comply with regulations, most compliance is voluntary. Enforcement serves more to ensure a level playing field for all parties affected by regulations.

For example, three periods of environmental activism have influenced the development of environmental laws or other efforts to regulate environmental risks. First, President Theodore Roosevelt initiated a progressive era of natural resource conservation. This addressed what we now call sustainable development. Second came the New Deal era of President Franklin Roosevelt with conservation projects such as public works, including the Tennessee Valley Authority, the Civilian Conservation Corps, the Public Works Administration, and the Soil Conservation Service. Third, President Richard Nixon launched the contemporary era by forming the Environmental Protection Agency. A series of environmental laws based on command-and-control approaches has dominated this period from 1970 to the present. Regulations based on these laws have reduced air and water pollution. Unfortunately, the volume of these regulations now exceeds human comprehension. Again, try stopping by a local law library and looking at the extensive *Code of Federal Regulations* (CFRs) or the *Federal Register* (Fed. Reg., also sometimes abbreviated as FR), which covers notices about hearings and proposed or new regulations.

In a simplistic sense, the United States has controlled most major and visible sources of pollution. We are now trying to address more difficult environmental problems, including genetic engineering products, radon, indoor air, stratospheric ozone, nuclear power, acid rain, cryptosporidium in drinking water, and global climate change. These problems require different regulatory approaches, such as pollution prevention, pollution trading, and economic motivation such as pollution bubbles, pollution credits, and tax incentives. Perhaps we will soon begin a fourth period of environmental activism.

We try to achieve environmental protection through legal mechanisms. However, public awareness and spontaneous public actions supplement regulations. The United States also tries to achieve a balance between federal, state, and local regulation. Overall, the system has become so complex that it has spawned entirely new professions of environmental law, environmental engineering, and environmental auditing in order to keep track of the effort.

Regulations about occupational, pharmaceutical, nutritional, consumer-based, and other risks have their own histories and complexities. Obviously, regulations in each area have their own politics. Usually, advocacy or public interest groups apply pressure to Congress and the federal agencies. Similarly, trade organizations representing industry sectors look after the interests of their members. Think tanks are concerned with the more intellectual aspects and impacts of regulation.

Why Have Regulatory Agencies?

Our system of government consists of three branches: executive, legislative, and judicial. All three are involved in environmental laws and regulations. An oversimplified picture is that the legislative branch, the Congress, delegates to the executive branch through laws the responsibility for environmental regulation. The judicial branch acts as a kind of referee between the political and policy conflicts that inevitably emerge.

Universal observation of an appropriate convention is vital to our health. The carnage that would result if we each decided separately which side of the road to drive on is unimaginable. Periodically, a driver does drive on the left side of the road, failing to observe the driving regulation to keep to the right. Serious accidents result. Yet, driving on the right side of the road is a convention, and as we have said, the driving convention in some other countries is to stay to the left side of the road.

English Common Law vs. Napoleonic Code

All states in the United States, except Louisiana, follow the principles of English common law. Under this system of law, the courts rely on precedent—the past history of relevant judicial decisions. The Napoleonic Code applies in Louisiana and in many other parts of the world, including most of the countries of the European Union, except for the United Kingdom. Under Napoleonic Code, the judicial system relies on the primary statutes; a judge need worry less about contradicting previous decisions by other courts.

As a society, we abhor accidents so much that we have traffic police, and to prevent future accidents we authorize them to arrest and punish anyone driving on the wrong side of the road. When an accident does occur, the driver who fails to stay on the right side of the road is liable for any damages that result. If the liability is contested, our system of law usually finds the driver on the wrong side at fault for damages. Often, the legal system adds criminal penalties to the damages. Yet, complying with an appropriate convention may inconvenience an individual.

Similarly, governments post stop signs in places where a failure to observe traffic while motionless or to yield the road to traffic on a different road has a high probability of resulting in an accident. Stop signs have a conventional shape, color, and design. The appearance of a stop sign is universally observed. Governments try to insure that all drivers recognize stop signs through licensing. As a deterrent measure, police try to catch drivers who do not observe stop signs. Courts will assess damages to a driver who causes an accident by failing to stop at a stop sign and additionally may send the driver to jail and/or assess fines. These examples from car safety contain most of the elements of the regulatory process.

Through legislation, government develops ideas about how we should behave to have a safer or better-functioning society. Regulatory agencies sometimes function like a legis-

lature and further develop the ideas in the legislation through regulations. Regulatory agencies also function like police when enforcing the regulations.

Ultimately, the effectiveness of regulations depends on public attitudes and beliefs. When government or its regulatory agencies get too far from the truth or too far from the capacity of people to comply, lawlessness results. Why, you may wonder, should we even have regulatory bodies? Could we not have the very legislators who pass laws do the job themselves? That very circumstance occurs when the courts decide that the legislators have delegated too much authority to the regulators or that the regulators have usurped too much legislative power.

Licenses and Permits

Government can issue a permit to do something in the future, or a license to engage in some activity, to persons with appropriate qualifications. Most people are familiar with the idea of a driver's license. The possession of a permit or license is not a right, only the ability to apply for one. Once in possession of a license or permit, it becomes a property value. For example, a taxi driver's license enables the holder to engage in activities that have economic compensation. Licensing is one way that government can control social behavior. EPA grants a registration to manufacture a pesticide to firms that meet stringent requirements, including the submission of data about the pesticide. This is how the government ensures that pesticide manufacturers carry on their business in a socially responsible manner.

Food Safety: An Example

The United States has the safest food supply in the world, thanks in large part to an interlocking monitoring system that watches over food production and distribution at every level: local, state, national, and international. Food inspectors, microbiologists, epidemiologists, and other food scientists working for local health departments, state public health departments, and federal agencies monitor the food supply and foodborne diseases continually. Laws, guidelines, customs, contracts, and other directives dictate their duties. Some agencies monitor only one kind of food, such as milk or seafood. Others work strictly within a specified geographic area. For example, customs agents are concerned only with exports and imports. Others are responsible for only one type of food establishment, such as restaurants, seafood processors, or meat-packing plants. Together they make up the U.S. food safety system.

Food safety has a longer regulatory history in the United States than environmental protection. Historically, under tort law anyone could obtain compensation for injuries from bad food by suing the producer for damages. The injured parties had to prove to a jury that, more likely than not, the food was adulterated or contaminated, the adulteration or contamination generally caused the kind of injury experienced, the injured persons consumed the food, the producer allowed the adulteration or contamination, and the injured persons experienced the injury. These requirements made successful litigation a difficult matter, and the costs of litigation often outweighed any damages recovered. Even so, fear of product liability is an important component of food safety. The adverse publicity that accompanies litigation may prove very costly to a food producer, adding to the motivation to produce high quality food.

Food safety first received close government attention during the 1930s, when Dr. Harvey Wiley and a group of volunteers, called the "poison squad," began testing food additives on themselves. Congress passed the 1938 version of the *Federal Food Drug and Cosmetics Act* (FFDCA) to recognize this work and provide it better legal authority. However, the early food safety provisions were weak. For example, the 1938 statute had no provisions for premarket approval of added chemical substances. FDA had to demonstrate consumer risk to remove an additive from the market. Manufacturers did not have to provide the information needed by FDA to establish a tolerance.

Congress began amending FFDCA in 1954 to create tolerances for pesticides used on raw agricultural commodities. Manufacturers now had to provide data to FDA. FDA did not have to prove a risk and could weigh benefits against the risk. In 1958, the Food Additive Amendments applied similar premarketing provisions to food additives and required that FDA show, through a petition process involving scientific judgment, that food additives would be safe.

Tort Law and Criminal Law

Under criminal law, society decides what actions to prohibit and penalizes parties that do not abide by these restrictions. A criminal offense is a violation of someone's obligations to society. Tort law, instead, relates to the civil obligations of one party to another party. Often, contracts rule these interactions. The courts become involved when one party sues another. Under common law doctrine, the obligations of one party towards another do not depend upon laws written by the legislature or King. Instead, juries decide what social obligations one party has toward another (or toward the property of another). Most regulatory laws allow for two different kinds of enforcement: criminal penalties and civil penalties. Under English common law, the losing party pays the costs of litigation, providing an incentive to settle cases. In the United States, each party usually pays its own portion of the costs of litigation.

Congress did not define *safe*, but an accompanying report suggested "a reasonable certainty of no harm" as a standard. Scientifically, FDA could not assure an absence of risk but could reasonably establish a relative standard through studies of potency and exposure. These definitions no longer called for consideration of benefits. Instead, the notion that food additives are a societal benefit was intrinsic to the amendments. Why establish procedures to allow the use of food additives if no benefits accrued with their use? More recently, the *Food Quality Protection Act of 1996* explicitly recognized "reasonable certainty of no harm" as a definition of safety for food additives and extended this standard to pesticides.

FDA is far from the only federal agency with responsibility for food safety. While FDA has much of the enforcement authority for pesticides, EPA registers pesticides and establishes their food tolerances. FDA also regulates bottled water, wine-based beverages with less than 7 percent alcohol, residues of animal drugs in meats, animal feeds, milk, shellfish, retail foods (both domestic and imported), and eggs still in their shells. FDA does not monitor meat and poultry. The Bureau of Alcohol, Tobacco and Firearms, part of the U.S. Department of the Treasury, has responsibility for the safety of beverages containing alcohol, except wine beverages containing less than 7 percent alcohol, and EPA's Office of Water regulates unbottled drinking water. The Food Safety and Inspection Service (FSIS), part of the U.S. Department of Agriculture (USDA), has responsibility for products containing meat or poultry and processed egg products. FSIS inspects food animals and processing plants, and establishes production standards for potential microbial and chemical contaminants. Other USDA components also have food safety responsibilities.

Positive and Negative Lists

When regulating activities or substances, agencies have the choice of issuing positive or negative lists. Negative lists allow the public to engage in any related activity or manufacture any substance that the list does not regulate. For example, the U.S. Environmental Protection Agency publishes a negative list of substances not allowed in the air above certain limits. Anyone releasing one of these substances beyond the fence line of a manufacturing site becomes subject to fines and penalties. Any substance that is not listed is allowed. In contrast, a positive list allows only activities on the list. For example, the Food and Drug Administration publishes a list of food additives. Anyone adding to food a substance that is not on this list is guilty of adulteration and becomes subject to penalties and seizure of the food.

For all foods, the Centers for Disease Control and Prevention, another part of the Department of Health and Human Services (HHS), investigates foodborne disease outbreaks and maintains a nationwide system of foodborne disease surveillance. The National Oceanic and Atmospheric Administration, part of the U.S. Department of Commerce, inspects and certifies fishing vessels and seafood processing. The U.S. Customs Service, part of the Department of Treasury, works with the other federal agencies to ensure that all foods entering and exiting the country comply with regulations. The U.S. Department of Justice prosecutes violations of all food safety laws and seizes unsafe foods not yet in the marketplace through the U.S. Marshals Service. The Federal Trade Commission protects consumers from deceptive advertising of foods.

Other federal agencies may become involved in solving specific regulatory problems related to foods. The Consumer Product Safety Commission enforces the *Poison Prevention Packaging Act*, the Federal Bureau of Investigation enforces the *Federal Anti-Tampering*

Act, the Department of Transportation enforces the *Sanitary Food Transportation Act*, and the U.S. Postal Service enforces laws against mail fraud involving foods.

Federal agencies coordinate their activities with state and local governmental organizations. Several organizations represent the aggregate interests of state and local governments, including the National Governors Association. Some organizations promote more specialized interests, including the Association of Food and Drug Officials, the National Association of City and County Health Officials, the Association of State and Territorial Public Health Laboratory Directors, the Council of State and Territorial Epidemiologists, and the National Association of State Departments of Agriculture. Food safety in the United States also involves close interactions between government and industry.

Because of the fragmentation of authority in the United States, the idea of a single agency that would assume all of these responsibilities arises from time to time. However, coordination of the existing legislation and agencies has achieved a safe food supply. Right now, the major causes of foodborne illness are the improper handling, cooking, and storage of foods by consumers. How a single agency for food regulation would solve this problem is not clear.

Ethical Systems and Values

The United States functions under a system of representative democracy. Legislators are not supposed to be disinterested philosophers. Instead, each elected official represents the interests of a voting constituency within a geographic area. We assert that this system is not the only possibility, just the best one. One major problem with this system is the lurching, erratic attention it brings to regulatory needs. Usually, a highly visible, often emotional incident focuses public attention on a certain situation, and Congress writes regulatory laws in response.

By comparison, the now defunct USSR operated under a very different legislative system. Yet many aspects of their regulatory system resembled ours. The Soviet system issued regulations and enforced them. Their government had a system to resolve disputes and to penalize wrongdoers. However, some of their quirky standards reflected political needs, not scientific data. The citizens seldom observed them. We like to think that our system is based on the needs, attitudes, objectives, and values of our citizens. These are reflected in our cultural values and our system of ethics. Thus, we rely greatly on voluntary compliance.

We try to regulate environmental pollution, simply because we think that it is the right thing to do. Most of us believe that we have some responsibility for all living things around us. (See more discussion of this principle in Chapter 11 on risk management and values. We derived the land ethic in this book from Aldo Leopold's *A Sand County Alma-*

nac. This biocentric view is based upon the idea that we are part of nature and that we value things like beauty, peace, preservation, and conservation.)

Fairness is an important public value. The costs of environmental regulation should be borne in an equitable way. That is one of the principles of our form of government. Thus, EPA bought the homes on top of contaminated land at Love Canal and Times Beach. However, the question is: how do we decide what fairness is? Two fundamental ethical principles underlie most regulatory debates: utilitarian and libertarian.

The utilitarian philosophy is based on the principle of doing the most good for the most people in a society. It deliberately sacrifices the interests of a few to better the interests of all. To implement a utilitarian system, some individual or group must decide what the interests are and what actions will secure the interests of the most people. Plato called this decision-maker the "philosopher-king," someone with power in a society who decides what is best for the society. A utilitarian form of government has some advantages. It often will perform many routine functions and operations efficiently. It will readily generate rules that the society can observe, letting everyone get on with life without having to think much about government. It lets people lead more predictable, sometimes more comfortable lives. Unfortunately, utilitarian forms of government also tend to degenerate into dictatorships. If the philosopher-king has power over people, the people may have no ready way to take that power away if it is used against them.

In contrast, libertarian philosophy is based on personal freedom and self-interest. Each person is the sole arbiter of his own behavior. Libertarian philosophy leads to minimalist forms of government that are intrinsically responsive, quickly adapting to new circumstances. Libertarian forms of government emphasize more personal freedom, but impose more personal responsibility. Government tends to become chaotic and messy. Life becomes dominated by market forces, which often are unintuitive, difficult to understand, and even more difficult to control.

Our country tends toward a libertarian republic. We prize individual liberty and individual rights. Yet, on any issue, a democracy allows a majority of the voters to impose their will on the minority. But the individual rights of the minority often are secured by law in this country. In effect, the law takes a time-out to ask people what they would like to avoid, especially if they fall into the minority on any issue. The law then attempts to secure these rights for all. It is not by accident that our form of government assigns equivalent power to the legislature and the courts. It reflects the deliberate intent of the designers of our government to seek an equilibrium between the will of the majority and the rights of the minority.

In this country, regulation lies in an uneasy equilibrium between the libertarian and utilitarian approaches. Particularly for health and safety purposes, our society does not believe that each individual can acquire the information and skills needed to make competent decisions. We expect an assurance that someone else, often a government expert, has done this for us. However, having a regulatory agency serve as philosopher-king and make these decisions means that we sometimes have to give up prized personal freedoms. The agencies have to find ways to cope with the hazards of modern life, like radiation and chemical pollution, that our citizens cannot detect on their own, but observe personal and property rights at the same time.

The Legislative Component

We elect our representatives to Congress to work for our needs. We generally ask them to help us. In turn, our representative seeks cooperation from other representatives. They are constantly torn between party discipline and representing their constituents. House members are constantly running for the next election. The pressures and stresses of perpetual campaigning, along with budgets, the press, and public attitudes consume much of their time. Since these politicians seldom are scientists, and since few of their staffers have technical skills, the details of environmental regulation often are delegated through legislation to the regulatory agencies.

All too often, laws are passed in response to some perceived crisis, such as Superfund (Love Canal), EPCRA (Bhopal), and SDWA (organic materials in the New Orleans' water supply). The technical details in these laws have increased in complexity, often bewildering the experts. The technical jargon sometimes reflects a desire of congressional staff to appear knowledgeable about matters they may not understand. In contrast, the same laws often reflect a lack of adequate monitoring data, exposure descriptions, health effects information, and remediation methods. These voids require research and technology development. A research agency, often one attached to (or a component of) a regulatory agency, could perform these tasks, but this requires patience, time, and resources, commodities our political systems lacks. Instead, we often seem to proceed backwards, by "fire, ready, aim."

In our system, congress is best placed to exercise oversight. Congress does so in a continuous way through committee and subcommittee hearings. The annual budget authorization and appropriation gives Congress control over the regulatory process. Congress has become increasingly dissatisfied with the administrative implementation of the environmental laws it has passed and has tried to institute some agency-forcing mechanisms. The concern is with how much discretion the agencies should have in implementing environmental laws. The classic devices are the authorization of citizen lawsuits and list-and-hammer provisions. By authorizing members of the public to sue, Congress delegates part of the enforcement of regulatory laws away from an agency and back to the public (usually some advocacy group working on behalf of the public). Citizens can also use the Administrative Procedures Act (APA) (Section 553(e)) to petition for the initiation of rulemaking proceedings. List-and-hammer provisions force an agency to regulate a list of substances or activities by certain target dates or else onerous regulatory provisions automatically go into effect.

Congress uses three general models to control agency discretion: the coercive model, the proscriptive model, and the ministerial model (Percival et al., 1996). In the coercive model, the one most commonly used, the discretion to regulate or not is removed, and the agency is permitted to choose the appropriate method of regulation. The proscriptive model gives the agencies the discretion about whether to regulate, but if they choose to regulate, they have to follow detailed substantive criteria. The proscriptive model is rarely used because it could deter agencies from regulating. Like list-and-hammer provisions, the ministerial model couples a deadline with a detailed substantive standard defining a specific way to regulate. This latter model has been used commonly in recent years, as most agencies fail to meet the deadlines.

The Administrative Component

The President controls the regulatory apparatus through political appointees, executive orders, and management controls, mostly through the Office of Management and Budget (OMB), which is part of the White House. One of the tools used by U.S. Presidents since Jimmy Carter has been the regulatory budget. The idea behind this budget is that regulatory agencies impose costs on the public. Their cost is not just the funds authorized by Congress to be spent on agency personnel and activities. Right now, for example, some authorities estimate that we spend approximately $180 billion per year in public funds to comply with environmental regulations.

Yet, agency resources (information, personnel, and funds) do limit the regulatory process. Thus, any agency requires some method to screen and set priorities, to focus attention on the problems of greatest importance within its domain. Particularly in the area environmental regulation, however, our overall regulatory system has no systematic method.

Data limitations are a constant problem in the evaluation of environmental problems. EPA is the only federal agency that does not have a bureau of statistics. Most of the environmental data is anecdotal in nature and much of it is not digitized. One of the environmental problems with the highest population risks involves approximately 10,000 lung cancer fatalities annually (see Chapter 8). However, this major problem was detected by accident when Stanley Watras, a worker in a nuclear plant that was not yet in operation, set off the portal monitors when he arrived at the plant. It turned out that he and his clothing were covered with the decay products (progeny) of radon from his home.

The organization of EPA works against overall coordination of environmental regulation. The offices that deal with air, water, solid waste, pesticides, and toxic substances are separate, and seldom interact in developing regulations. A few attempts have been made to develop a multimedia approach, particularly in the enforcement area. Think tanks like the National Environmental Policy Institute have focused on the need for an integrated approach to environmental problems.

EPA has considered using risk as a way to determine the priorities. The idea would be to fund the activities that would save the most lives or most reduce the prevalence of disease. A major study of the agency's priorities related them to population risk. The study found a rigorous quantitative assessment of risk, other than for cancer, to be impossible, but it did qualitatively rank four sources of cancer risk (EPA, 1987; EPA, 1990). Outside experts have enthusiastically endorsed the concept that EPA should use risk in setting priorities and developing environmental regulations. However, EPA's legal obligations closely align with public opinion, which unfortunately is largely the inverse of the experts' rankings.

Today, we promulgate most environmental regulations through informal rulemaking. Section 4 of the Administrative Procedures Act (APA) governs this process. It requires that agencies provide (1) public notice in the *Federal Register* of proposed rulemaking actions, (2) an opportunity for the public to comment, and (3) publication of the final rules in the *Federal Register*, accompanied by a statement of purpose and basis.

The first step that the public sees often is an advanced notice of proposed rulemaking (ANPR) published in the *Federal Register*. The idea here is to solicit input from the public. This can be in a public hearing or written comments. A specific time period is stated for comments and the exact location and time of the hearing(s) are published. If there is some urgency, the step of the advanced notice is skipped.

The agency reviews the comments in preparation to promulgating the proposed rule. In more complex rulemakings, the agency may contract help from an outside consulting firm to sort and analyze the comments. Draft notices of proposed rules are reviewed by a steering committee composed of representatives from the major offices in the agency. In the EPA this is called "red border" review. After reviewing the comments, the agency determines the form of the proposed rule and publishes that in the *Federal Register*. This will contain the rule and some explanatory information concerning the details, such as analysis of the comments from the ANPR. Also included are the location(s) and time(s) for the public hearings, and the time frame for public comment. A complete rulemaking docket is maintained throughout the rulemaking procedure and for the future in case of judicial review.

The agency, with possible help from a consultant, reviews the pubic comments and any new information and data concerning the proposed rule. Then the final rule is published in the *Federal Register* along with an analysis of the public comments concerning the proposed rule.

The procedures for informal rulemaking have become enormously difficult for regulatory agencies. A number of studies have shown a kind of ossification has occurred

(Percival et al., 1996). For the EPA, over 80 percent of the deadlines were missed. Agencies face the constraints of limited resources and lack of easy access to the relevant information and data. Thus informal rulemaking has become burdensome and time-consuming. Alternatives are being sought.

One of the regulatory reforms that has met with some success to overcome the ossification and burden described above is negotiated rulemaking. Here the major private groups interested in a proposed rulemaking action attempt to resolve their differences through face-to-face negotiations prior to the promulgation of the proposed rule. The agency actually appoints these representatives to a committee and holds meetings under the umbrella of the Federal Advisory Committee Act (FACA). An agency-appointed mediator may help to facilitate this group. There are no formal rules for this procedure, and the group can create rules as they go along.

FACA imposes very few requirements on an agency. Most of these are designed to make the process transparent to the public, not to allow the public to influence the proceedings. The Freedom of Information Act (FOIA) also gives the public access to an agency's documents. However, it has sharp limitations. The requestor generally must know about the existence of a document to have much hope of obtaining it. FOIA restrictions include the national defense, foreign policy, internal personnel rules and practices, trade secrets, confidential commercial or financial information, internal memoranda, unwarranted invasions of privacy, law enforcement, regulation or supervision of financial institutions, and geophysical or geological information, such as maps. In addition, some agencies simply stonewall requests. Recently the Department of Justice has demonstrated a serious will to comply with the law, but agencies can simply let requests gather mold, refusing to respond.

Adopting interim standards under the APA speeds up rulemaking. Interim standards may follow informal rulemaking but have a limited scope. All too often these acquire a life of their own and are used so long that many forget that they are only intended to be temporary. Additional procedures include alternative dispute resolution and expert peer review.

A reform was instituted in 1993 to change the regulatory system to be more workable for the American people in the form of Executive Order 12,866, Regulatory Planning and Review. The general idea was for the agencies to identify the problems they face and review existing regulations to eliminate those that are not working, are too costly, or have limited benefits.

The Judicial Component

Our courts view regulatory rulemaking as an activity that requires a congressional intent. Congress devises rules in the form of regulations that are mandated by laws, even if responsibility for rulemaking is delegated to an agency. Further, the courts discern whether an agency has interpreted a statute properly, according to the congressional intent. If the statute is silent or ambiguous about intent, the courts usually accord the agency discretion, although they sometimes will go beyond the statute to include the legislative history.

Agencies may not act in arbitrary, capricious, or abusive ways. This standard comes from the APA. The court may take an in-depth look at the record that an agency generates. An agency has the responsibility of making the record available, usually in a docket. In technical and scientific matters, the courts usually will defer to the expertise of an agency. The court will look to see whether an agency has gathered substantial evidence and whether sufficient information supports a regulation. The standard of review for regulations is fact finding by the court. This same standard applies to a risk assessment in support of a regulation, including scientific data and scientific interpretation.

Judicial reviews of regulations occur only rarely before an agency takes a final action. Reviews of risk assessments are one of the few exceptions. Several agencies have tried to separate the risk assessment process from the rulemaking process. During the risk assessment process, the agency tells stakeholders that it has not made final regulatory decisions. Thus, the risk assessment is not ready for judicial review. Later, when the agency concludes the regulatory proceedings, it informs the courts and the stakeholders that the risk assessment has already gone final, and thus is not subject to review. Fortunately the courts mostly have seen through this sham. In addition, some jurists have pointed out that a risk assessment can have the effect of a regulatory action if it promotes actions by the public. For example, a federal risk assessment can set off state and local regulations.

Judges and juries lack much scientific expertise. In addition, neither courts nor agencies cope well with decision making under uncertainty, which is a common feature of risk assessments. Thus, judicial reviews of regulations based on risk assessments are somewhat uneven. It is worth recalling that anyone can bring a lawsuit about product liability. The courts may throw the case out if the plaintiff lacks standing; for example, if the plaintiff cannot prove an injury.

Additional Information about Federal Regulation

In this chapter we have introduced some information about the regulatory panorama. James V. DeLong has noted that a human being, doing nothing else but reading the *Federal Register*, could not keep up. Even worse, only reading the regulations is not adequate. The same person would need to read the support documents, memoranda, clarifying interpretations, letters, lawsuits, policy statements, guidelines, enforcement proceedings, and other documents that amplify and explain the regulations. Obviously, understanding the regulatory matters relevant to a business is a specialized affair, and knowing what you can and cannot do has become difficult. Initially we have chosen to introduce you to the general framework and give a slightly more in-depth introduction to one area, food regulation, just to illustrate the complexity.

Help is available. You can obtain some information in compressed form from summaries of different specialized areas. For instance, since 1973 the publisher of this book, Government Institutes, has provided continuing education and practical information about federal regulatory activities. The services include reference books, such as the *Code of Federal Regulations*, in both bound and electronic formats, as well as topical books on regulatory topics in the environmental, health, and safety fields, such as the one you are reading. Thus Government Institutes is good source of government information that otherwise is difficult to obtain in a convenient and understandable format. Government Institutes also conducts continuing educational programs that combine legal, regulatory, technical, and management features of regulatory areas like toxic substances control, environmental audits, occupational safety, pollution prevention, clean air, clean water, and international standards. Leading authorities from industry, business, and government teach students how to solve practical problems in specialized areas. Government Institutes provides a schedule of the course offerings at its Internet site, and onsite training, videotapes, and self-paced instruction are available for most topics. The Internet site for Government Institutes is www.govinst.com.

Other organizations also provide extensive coverage of the many areas of regulatory practice, including newsletters and reports, such as the publications of the Bureau of National Affairs (BNA) or the courses offered by Executive Enterprises. Within any area, more specialized resources exist. Several organizations maintain list servers that allow the participants to monitor or discuss recent events. For example, the Department of Agriculture supports an interesting email list about food safety, called *FoodSafe*. To subscribe, send the message "subscribe foodsafe", followed by your email address, to majordomo@nal.usda.gov. Similarly, the University of Guelph in Canada sponsors *Fsnet*. To subscribe, send the message "subscribe fsnet-L firstname lastname" to listserv@listserv.uoguelph.ca. The following Internet sites belong to the agencies that work in the area of food safety cited in this chapter:

- The Center for Food Safety and Nutrition - www.cfsan.fda.gov/list.html
- The Center for Veterinary Medicine - www.fda.gov/cvm/default.htm
- The Centers for Disease Control - www.cdc.gov
- The Food Safety Inspection Service - www.fsis.usda.gov
- The U.S. Environmental Protection Agency - www.epa.gov
- The Bureau of Alcohol, Tobacco and Firearms - www.atf.treas.gov/core/alcohol/alcohol.htm

- The Customs Bureau - www.customs.ustreas.gov
- The Department of Justice - www.usdoj.gov
- The Federal Trade Commission - www.ftc.gov

The Food Law Institute is a leading think tank interested in federal regulation of foods and substances added to foods, which you can access at www.fdli.org. Some of the leading professional societies are the Association of Official Analytical Chemists (www.aoac.org) and the Institute of Food Technologists (www.ift.org).

More generally, several serial publications cover regulation overall, including the *National Journal*, *Regulation* (published by the Cato Institute), and the *Yale Journal of Regulation*. Many law reviews also cover regulatory topics. Several think tanks are interested in federal regulation more generally, and they sponsor the following Internet sites (some of these sites focus more on economic regulation and financial risks):

- The Annapolis Center - www.annapoliscenter.org
- The Cato Institute - www.cato.org
- The Competitive Enterprise Institute - www.CEI.org
- The Center for Regulatory Effectiveness - www.thecre.com
- Resources for the Future - www.rff.org

For background reading about the social forces that drive regulation, many good books are available. We recommend Francis Fukuyama's *The Great Disruption : Human Nature and the Reconstitution of Social Order*. It is very readable, but neither comprehensive nor authoritative about regulation.

References

DeLong, J. V. 1997. *Property Matters*. New York: The Free Press.

Fukuyama, F. 1999. *The Great Disruption: Human Nature and the Reconstitution of Social Order*. New York: The Free Press.

Leopold, A. 1968. *A Sand County Almanac*. Oxford University Press.

Percival, R. V., A. S. Miller, C. H. Schroeder, and J. P. Leape. 1996. *Environmental Regulation: Law, Science and Policy*, 2nd ed. New York: Little, Brown and Company.

U.S. EPA. Office of Policy, Planning, and Evaluation. 1987. *Unfinished Business: A Comparative Assessment of Environmental Problems*. Washington, D.C.

U.S. EPA. Science Advisory Board. 1990. *Reducing Risk: Setting Priorities and Strategies for Environmental Protection*. SAB-EC-90-021. Washington, D.C.

Exposure Assessment

Exposure

Exposure and dose are interrelated concepts, but they are different. Even experts often confuse exposure and dose with each other and sometimes carelessly use these terms in an interchangeable way. Thus, we need to consider the definitions of exposure and dose carefully. To start, we define exposure as a potential dose. Even dose is an ambiguous concept. Listed below are different kinds of doses commonly found in the literature:

- absorbed dose
- active dose
- accumulated dose
- administered dose
- applied dose
- biologically effective dose
- delivered dose
- effective dose
- intake dose
- internal dose
- potential dose
- temporally or spatially averaged dose

Thus, defining exposure as dose is a tricky procedure. To prevent confusion in this chapter, we define a dose as the amount of a substance absorbed by the body. For dermal exposure, the amount absorbed would be the quantity of the substance just inside the skin. For inhalation exposure, the amount absorbed would be quantity of the substance appearing inside the membranes of the lung. For ingestion exposure, the amount absorbed would be quantity of the substance just inside the membranes of the gastrointestinal tract. These amounts can act locally on the skin, lungs, or gastrointestinal tract at the site of absorption. The circulatory system can distribute the substance to distant sites in the body.

At the beginning of Chapter 5, we again take up the confusion between exposure and dose. To further complicate the matter, personal exposures can differ in the same environment. The concentration of a substance at a location—a physical space defined in three dimensions—and over different durations of time, is the concentration-time field. That the local concentration can vary within different regions of the field and over time is implicit in this definition. Analytical chemists can define a concentration-time field by sampling an environment at different times and locations.

You may experience the same concentration-time field as another person. Yet different amounts can get into your two bodies and distribute to the sites where damage occurs. The two of you experience different doses and different toxicities from apparently the same concentration-time field.

How is this difference possible? Our personal habits differ in our movements, breathing rates, eating habits, and the times we spend in different environments. Our eating habits differ considerably in quantity, quality, and selection. For these and other reasons, the amount of a substance that enters our bodies can differ, although we experience the same concentration-time field. The linkage of typical behavior to a common concentration-time field yields an exposure scenario.

In this book we are mostly concerned with health risks from chemical substances and, to a lesser extent, from microorganisms and ionizing or nonionizing energy. We define *exposure* as the potential dose in contact with the outer membranes of an organism. Exposure to a process or activity, like coal mining or driving a car, also generates risks, and you can estimate these risks within the same framework we apply in Chapters 1 and 2. The regulatory activities that interest us have a specific content that does not relate, for example, to the risk of bankruptcy.

Beginning with this chapter, we will concentrate on chemical substances but will include some information about microorganisms and radiated energy. Without these exposures, no dose or health effect can occur. Without adverse effects, no risks exist. To understand this subject, you will need to find out how a chemical substance, microorganism, or energy enters the environment and how an organism makes contact with it.

The physical presence of a chemical substance, microorganism, or radiation in space and time does not adequately portray exposure to us. We must contact the substance within a concentration-time field. Thus, we can change our personal exposure within a concentration-time field by our behavior. A description of behavior is an essential component of exposure.

The concept of exposure as a potential dose is new. It implies that exposure does not occur until we experience substances or energy in the four dimensions of space and duration in a concentration-time field. Thus, data about the exposure-related behaviors of subjects become part of an exposure assessment. This new concept does create problems, however. If we define *exposure* as potential dose, and *dose* as absorbed energy, microbes, or substance, exposure and dose become interdependent concepts. At times, the relationship between them seems almost circular. The important reasons for separating them will become apparent by the end of Chapter 5. So, the approach in this chapter is mostly descriptive.

The field of exposure assessment is in a state of rapid flux. Risk assessors only recently achieved consensus about an authoritative definition of exposure that is mathematically coherent (Zartarian et al., 1997). Under this definition, *exposure* is the contact between an "agent" and a "target." *Instantaneous point exposure* is the joint occurrence at a point on the target that is in contact with the concentration or intensity of the agent for some brief moment of time. Instantaneous point exposure is important; it subsumes the parallel concepts of *average exposure* and *integrated exposure*. Similarly, these concepts lead directly to the measurement of inhalation, dermal, and ingestion exposure.

In addition, the field of exposure assessment has begun to emphasize *total*, or *aggregate*, exposure (Sexton et al., 1995). In other words, even with an established drinking water exposure, the question often arises of how much additional exposure a population experiences from inhalation of a volatile substance from the drinking water during showering, or from ingestion of a substance in food cooked with the drinking water. Scientists and regulators often want to know how much inhalation of the same substance contributes to biomarkers of exposure such as urine, feces, and mother's milk. These questions have led to a recent emphasis on multimedia, multipathway exposure assessments. Legal requirements also have increased the emphasis on multimedia-multipathway exposure assessment. For example, the Food Quality Protection Act of 1996 (FQPA) mandates that EPA base pesticide tolerances on aggregate exposures.

We can address the different aspects of exposure and dose by modeling or by direct measurement. We can obtain measurements through personal monitors, (see U.S. EPA, 1987 and Wallace, 1993), tissue sampling, or other biomarkers. To understand these measurements better, we can link personal monitoring to survey information from personal diaries and questionnaires (see Figure 4.1 for more detail). Exposure assessors use these methods to learn more accurately what gets into an organism in the exposure scenario under consideration.

Table 4.1 is useful to sort out exposure assessment. The approach we present is new; the table integrates different concepts. Environmental exposure assessment now uses the approach consistently. The flow of a contaminant in Table 4.1 runs from the top (sources) to the bottom (adverse effects). Each stage represents a different concept and a

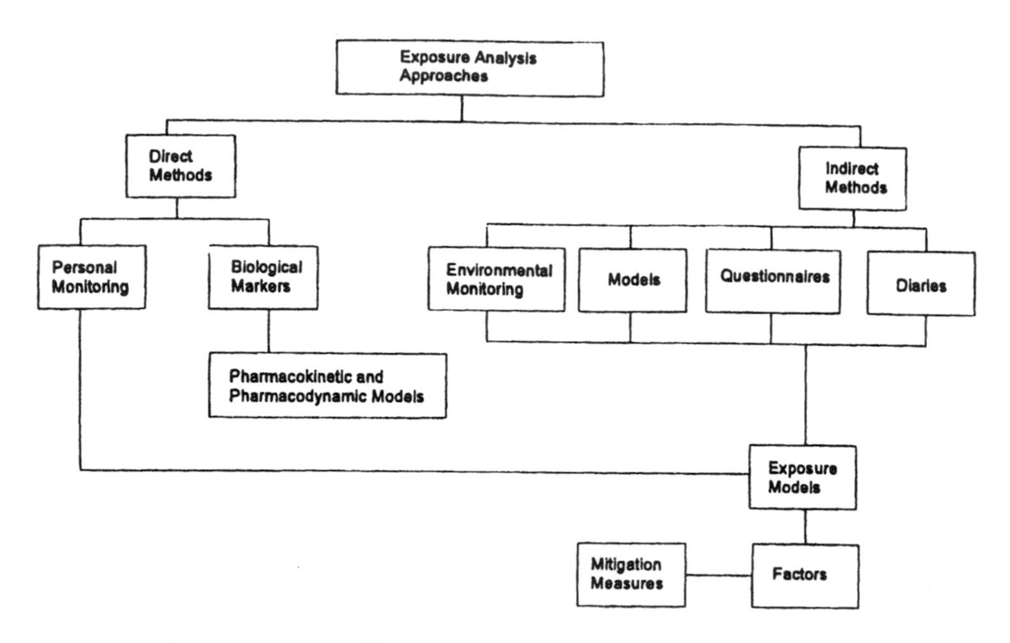

Figure 4.1 Possible approaches to measuring or estimating exposures

kind of measurement. Overall, we think of exposure as the quantities above the dotted line labeled "absorption barriers." At that point the contaminant enters the organism. Above the dotted line we have the exposure, and below the dotted line we have the dose. We will discuss dosimetry and dynamics in later chapters. *Dosimetry* refers to actions that occur between a potential dose and biologically effective dose. *Dynamics* refers to the biological reactions that follow a biologically effective dose.

Some models can convert human exposure to absorbed dose. One such model uses the media, micro-environments, and exposure pathways shown in Table 4.2.

Exposure can involve chemical, microbial, or physical processes. Thus, exposure can have units of concentration, infectivity (which is like concentration conceptually), or intensity. Some substances are composed of radionuclides—isotopes of atoms that are unstable and disintegrate, releasing energy. Thus, exposures to substances that contain radionuclides combine both chemical and radiation exposure. We treat radionuclides as a special case, and we emphasize radon as an example in the case studies, epidemiology, and population risk chapters.

We define *exposure* as:

$$\Delta E = C(t) \times \Delta t$$

where ΔE is the exposure of a person to the contaminant concentration $C(t)$ for a duration Δt. If the exposure is over a finite period of time, the general expression is a sum or integral:

$$E = \sum C_i \, \Delta t_i$$

Table 4.1 A Scheme for Describing the Pathways and Stages a Contaminant Takes from the Source to Creating an Adverse Effect in an Organism*

Emission source	Control of the source
Environmental pathways	
Exposure concentration	
Potential dose	
Pharmacokinetics	
	<—————————————| bioavailability
Applied Dose	
absorption	
absorption	<——————————————- |disposition
	<————————————|metabolism
	barriers | elimination
Internal Dose	|
	<——————————————|
Delivered Dose	|
	<——————————————|
Biologically Effective Dose	|
	<————————————|Pharmacodynamics
Biological Effect(s)	| compensation
	<——————————————-|damage
Adverse Effect(s)	| repair

**Modified from Sexton et al., 1995*

where *i* denotes the *i*th time period. For continuous exposure we can integrate over time:

$$E = \int C(t)\, dt$$

Generally we do not have the value of the concentration as a continuous function of time. Usually, at best, we have an average of sorts, sometimes as a time-weighted average (TWA). In that case:

$$E = C_{TWA}\, T$$

where T is the total time.

We think of exposure as the process of measuring or estimating the intensity, frequency, and duration of the exposures to harmful contaminants for a population. A major source of complexity in an exposure assessment is the strong influence of individual

Table 4.2 Media, Microenvironments, and Exposure Pathways in the BEADS Model*

Medium	Exposure Route	Micro-environments and Exposure Scenarios	
air	inhalation	outdoors	
		indoor	outdoor penetration
			indoor sources
			cigarette smoking
			attached garage
		active smoking	
		automobile cabin	roadway air
			cigarette smoking
		gasoline stations	refueling, other
		work/school	outdoor penetration
			indoor sources
			cigarette smoking
water	ingestion	tap water consumption	
	dermal uptake	bathing	
	inhalation	volatilization: during and after shower	

**Modified from MacIntosh, 1995*

personal habits. For example, food storage, food preparation, and dietary habits create different exposures from the same food supply. Our movements in the external environment differ from person to person. Infants and children are not just small adults. They differ quantitatively. Children move closer to the ground than adults do and often put objects from their surroundings into their mouths. A nursing newborn, an infant in diapers crawling on the ground, or a toddler who has just learned to walk will obtain qualitatively and quantitatively different exposures in the same room as an adult.

To be comprehensive, our exposure assessment must describe the spatial distribution of the contaminant in terms of its magnitude and route of exposure, its temporal distribution in terms of duration, and the schedule, size, and nature of sensitive subpopulations.

We will see that monitoring methods play a central role in exposure assessments. They include direct measurements in the air, water, and soil, as well as remote techniques, like aerial photography and multispectral overhead imagery. A most useful methodology involves direct methods like personal exposure monitors and biomarkers, such as urine, feces, hair, and mother's milk.

In the exposure assessment process, we need to characterize the physical setting by considering the following features:

- climate (temperature and precipitation)
- meteorology (wind speed and direction)
- geologic setting (location and surrounding and underlying strata)
- vegetation (unvegetated, forested, grassy)
- soil type (sandy, organic, basic)
- groundwater hydrology (depth, direction, and type of flow)
- location and description of surface water (type, flow rates, salinity)

We also have to characterize the populations subject to exposure by taking these factors into account:

- current land use (residential, commercial/industrial, recreational—as revealed by zoning maps and census data)
- activity patterns
- future land use
- subpopulations of concern (children, the elderly, the sick and disabled—by examining locations of schools, day care centers, nursing homes, hospitals)
- distant populations affected by contamination of water supplies, fish, and agricultural products

For infectious agents, you can monitor the prevalence of a disease or reactions to infection, such as serum antibodies against a microorganism. For energy, sometimes you can monitor the prevalence or amount of reactions with body macromolecules, such as altered hemoglobin molecules or the excretion of oxidation products of DNA bases. Metabolites of chemical substances in blood or urine have been particularly informative over the years. Because these biomarkers of exposure integrate exposures from all sources, they may not be as useful in a regulatory context. For example, if control of drinking water contamination is the objective, but most of the exposure comes from diet, measures of urinary concentrations of a metabolite will not prove useful as it is not directly related. Biological markers include substances found in tissues and fluids (e.g., lead in the blood or mother's milk).

You can find exposure data at the Centers for Disease Control (CDC), which conducts a National Health and Nutrition Evaluation Survey (NHANES). This survey is linked to the census and therefore is particularly useful in understanding exposures in the United States. CDC has completed several of these national surveys. The survey data from self-administered questionnaires are available at the National Center for Health Statistics website. Some of these studies also include satellite studies in representative subpopulations. These studies mostly look at the levels of chemical substances of interest from a public health perspective. The nutritional and other survey components that accompany the measurements are particularly of interest in understanding dietary exposures and behavioral effects.

The CDC has carried out several NHANES studies. NHANES II was the first survey to include analysis of biological fluids for chemical substances. The study population for NHANES II represented a random sample of the census for citizens aged 20 and over. More recently, CDC has made efforts to study children, the elderly, and pregnant women. NHANES III covered a sampling frame from 1988 to 1994, related to the 1990 census, covering a four-hour time period for each subject. NHANES IV is now underway. Because CDC chooses the subpopulations through random or stratified selection, primarily based on the census, the results fairly represent the U.S. population. Of the many surveys available to us, CDC regards biological measurements of exposure as most likely to correlate with potential health effects.

The NHANES studies describe the population distributions of adult levels of various biomarkers of exposure, but they do not measure exposures directly. If we get lucky, we may find a relationship, for example, between calcium in urine and some measure of thyroid status. However, we lack ways to link the biomarker values directly to exposure pathways. Thus the results have powerful but limited national policy implications for us.

Planning an Exposure Assessment

We conduct exposure assessments for a variety of purposes. For each assessment, however, we could use a similar set of planning questions, and by addressing these questions the assessor will be better able to decide what is needed to perform the assessment and how to obtain and use the information required. To facilitate this planning, we must first answer some important questions:

Purpose

Why is the study being conducted? What questions will the study address and how will the results be used?

Scope

Where does the study area begin and end? Will inferences be made on a national, regional, or local scale? Who or what is to be monitored? What chemicals and what media will be measured, and for which individuals, populations, or population segments will estimates of exposure and dose be developed?

Level of Detail

How accurate must the exposure or dose estimate be to achieve the purpose? How detailed must the assessment be to properly account for the biological link between exposure, dose, effect, and risk, if necessary? How is the depth of the assessment limited by resources (time and money), and what is the most effective use of those resources in terms of level of detail of the various parts of the assessment?

Approach

How will exposure or dose be measured or estimated, and are these methods appropriate given the biological links among exposure, dose, effect, and risk? How will populations be characterized? How will exposure concentrations be estimated? What is known about the environmental and biological fate of the substance? What are the important exposure pathways? What is known about expected concentrations, analytical methods, and detection limits? Are the presently available analytical methods capable of detecting the chemical of interest and can they achieve the level of quality needed in the assessment? How many samples are needed? When will the samples be collected? How frequently? How will the data be handled, analyzed, and interpreted?

By addressing each of these questions, we will develop a clear and concise definition of study objectives that will form the basis for further planning.

Purpose of the Exposure Assessment

The particular purpose for which an exposure assessment will be used will often have significant implications for the scope, level of detail, and approach of the assessment. Because of the complex nature of exposure assessments, we usually will need a multidisciplinary approach that encompasses the expertise of a variety of scientists. We should seek assistance from other scientists when we lack the expertise necessary in certain areas of the assessment.

Using Exposure Assessments in Risk Assessment

The National Research Council (NRC, 1983) described exposure assessment as one of the four major areas of risk assessment (the others are hazard identification, dose-response assessment, and risk characterization). Our primary purpose for an exposure assessment in this application is often ultimately to estimate dose, which is combined with chemical-specific dose-response data (usually from animal studies) in order to estimate risk (see Chapter 5 for more information on dose).

If our exposure assessment is part of a risk assessment to support regulations for specific chemical sources, such as point emission sources, consumer products, or pesticides, then the link between the source and the exposed or potentially exposed population is important. In this case we will find it necessary to trace chemicals from the source to the point of exposure by using source and fate models and exposure scenarios. By examining the individual components of a scenario, assessors can focus their efforts on the factors that contribute the most to exposure, and perhaps use the exposure assessment to select possible actions to reduce risk. For example, exposure assessments are often used to compare and select control or cleanup options. Most often we will use the scenario evaluation to estimate the residual risk associated with each of the alternatives under consideration. We then compare these estimates to the baseline risk to determine the relative risk reduction of each alternative. We can use these types of assessments to make screening decisions about whether to further investigate a particular chemical. We can verify these assessments through the use of personal or biological monitoring techniques.

If our exposure assessment is part of a risk assessment performed to set standards for environmental media, usually the concentration levels in the medium that pose a particular risk level are important. We will find that these assessments place less emphasis on the ultimate source of the chemical and more emphasis on linking concentration levels in the medium with exposure and dose levels of those exposed. A combination of media measurements and personal exposure monitoring could help us in assessments for this purpose, since what is being sought is the relationship between the two. Modeling may also support or supplement these assessments.

If we need an exposure assessment to determine the need to remediate a waste site or chemical spill, we emphasize calculating the risk to an individual or small group, comparing that risk to an acceptable risk level, and if necessary, we can use it to determine the appropriate cleanup actions to reach an acceptable risk. The source of chemical contamination may or may not be known. We can use modeling and scenario development as the primary techniques to access these exposures. We will usually emphasize linking sources with the exposed individuals.

If we need the exposure assessment as a screening device for setting priorities, we emphasize the comparative risk levels, perhaps with the risk estimates falling into broad categories (e.g., semi-quantitative categories such as high, medium, and low). For such quick-sorting exercises, we rarely need any techniques used other than modeling and scenario development.

Using Exposure Assessments for Status and Trends

We also can use exposure assessments to monitor status and trends. Here we emphasize what the actual exposure (or dose) is at one particular time, and how the exposure changes over time.

We define exposure status as a snapshot of exposure at a given time, usually the exposure profile of a population or population segment (perhaps a segment or statistical sample that can be studied periodically). We describe exposure trends as showing how this profile changes with time. Normally, in status and trends studies we make use of statistical sampling strategies to assure that changes can be interpreted meaningfully.

You will find that measurement is critical to these assessments. We can use personal monitoring to give the most accurate picture of exposure, but biological or media monitoring can indicate exposure levels, provided a strong link is established between the biological or media levels and the exposure levels. Usually we establish this link first by correlating biological or media levels with personal monitoring data for the same population over the same period.

Using Exposure Assessments in Epidemiologic Studies

We discuss in Chapter 6 the use of exposure assessments as important components of epidemiologic studies, where the emphasis is on using the exposure assessment to establish exposure-incidence (or exposure-effect) relationships. For this purpose, we can use personal monitoring, biological monitoring, and/or scenario development. If we find that the population under study is being currently exposed, personal monitoring or biological monitoring may be particularly helpful in establishing exposure or dose levels. If we determine that the exposure took place in the past, biological monitoring may provide useful data, provided the chemical is amenable to detection without interference or degradation and that the pharmacokinetics are known. More often, however, we use scenario development techniques to estimate exposure in the past, and often the accuracy of the estimate is limited to classifying exposure as high, medium, or low. This type of categorization is rather common, but sometimes it is very difficult to determine who belongs in a category and to interpret the results of the study.

Scope of the Assessment

By the scope of an assessment, we mean its comprehensiveness. For example, an important limitation in many exposure assessments relates to the specific chemical(s) to be evaluated. Although this seems obvious to us, the exposure to multiple chemicals or mixtures is possible. It is not always clear whether assessing "all" chemicals will result in a different risk value than assessing only certain significant chemicals and assuming that the others contribute only a minor amount to the risk. We find that this may also be true for cases where degradation products have equal or greater toxicological concerns. In these cases, we may need a preliminary investigation to determine which chemicals are likely to be in high enough concentrations to cause concern, with the possible contribution of the other chemicals discussed in the uncertainty assessment. We must also determine geographical boundaries, population exposed, environmental media to be considered, exposure pathways, and routes of concern.

Level of Detail of the Assessment

By the level of detail or depth of the assessment we mean the amount and resolution of the data used and the sophistication of the analysis employed. We determine this by the purpose of the exposure assessment and the resources available to perform the assessment. Although in theory the level of detail we need can be established by determining the accuracy of the estimate required, this is rarely the case in practice. To conserve resources, we conduct most assessments in an iterative fashion, with a screening done first; successive iterations add more detail and sophistication. After each iteration we ask: Is this level of detail or degree of confidence good enough to achieve the purpose of the assessment? If the answer is no, successive iterations continue until the answer is affirmative, new input data are generated, or, as is the case for many assessments, the available data, time, or resources are depleted. We evaluate resource-limited assessments in terms of what part of the original objectives have been accomplished and how this affects the use of the results.

Our level of detail for an exposure assessment can also be influenced by the level of sophistication or uncertainty in the assessment of health effects to be used for a risk assessment. If only very weak health information is available to us, a detailed, costly, and in-depth exposure assessment will be wasteful in most cases because the most detailed information will not add significantly to the certainty of the risk assessment.

Determining the Approach for the Exposure Assessment

The intended use of the exposure assessment will generally favor, for us, one approach to quantifying exposure over the others, or suggest that two or more approaches be combined. These approaches to exposure assessment can be viewed as different ways for us to estimate the same exposure or dose. Each has its own unique characteristics, strengths, and weaknesses, but the estimate should theoretically be the same, independent of the approach taken.

The point-of-contact approach requires us to measure chemical concentrations at the point where the individual is exposed and to record the duration of the exposure at each concentration. Some integrative techniques are inexpensive and easy to use (radiation badges), while others are costly and may present logistical challenges (personal continuous-sampling devices), and require public cooperation.

The scenario evaluation approach requires us to determine the chemical concentration and duration-of-contact data, as well as information on the exposed persons. We determine the chemical concentration by sampling and analysis or by use of fate and transport models (including simple dilution models). Models can be particularly helpful to us when some analytical data are available, but resources for additional sampling are limited. Information on human behavior and physical characteristics may be assumed or obtained by interviews or other techniques from individuals who represent the population of interest.

For the reconstruction of dose approach, we usually use measured body burden or specific biomarker data, and select or construct a biological model that uses these data to account for the chemical's behavior in the body. If we use a pharmacokinetic model, additional data on metabolic processes will be required (as well as model validation information). Information on exposure routes and relative source strengths is also helpful to us.

One of our goals in selecting the approach should include developing an estimate which has an acceptable amount of uncertainty. In general, estimates based on quality-assured measurement data, gathered to directly answer the questions of the assessment, are likely to have less uncertainty than estimates based on indirect information. The approach we select for the assessment will determine which data are needed. All three approaches also require data on intake and uptake rates if the final product of the assessment is a calculated dose.

Sometimes we will use more than one approach to estimate exposure. For example, the TEAM study combines point-of-contact measurement with the microenvironment (scenario evaluation) approach and breath measurements for the reconstruction of dose approach (U.S. EPA, 1987a). If more than one approach is used, we should consider how using each approach separately can verify or validate the others. In particular, point-of-contact measurements can be used as a check on assessments made by scenario evaluation.

Establishing the Exposure Assessment Plan

Before starting work on an exposure assessment, we should have determined the purpose, scope, level of detail, and approach for the assessment, and should be able to translate these into a set of objectives. These objectives will be the foundation for our exposure assessment plan. Our exposure assessment plan need not be a lengthy or formal document, especially for assessments that have a narrow scope and little detail. For more complex exposure assessments, however, it is helpful to have a written plan.

For exposure assessments being done as part of a risk assessment, the exposure assessment plan should reflect (in addition to the objectives) an understanding of how the results of the exposure assessment will be used in the risk assessment. For some assessments, three additional components may be needed: the sampling strategy, the modeling strategy, and the communications strategy.

Planning an Exposure Assessment as Part of a Risk Assessment

For risk assessments, we must link exposure information to the hazard identification and dose-response relationship (or exposure-response relationship). We realize that the toxic endpoints (e.g., cancer, reproductive effects, neurotoxic effects) can vary widely, and along with other aspects of the hazard identification and dose-response relationships, can have a major effect on how our exposure information must be collected and analyzed for a risk assessment. Some of these aspects include implications of limited versus repeated exposures, dose-rate considerations, reversibility of toxicological processes, and composition of the exposed population.

Limited vs. Repeated Exposures

Often the carcinogen risk models we use have lifetime time-weighted average doses in the dose-response relationships owing to their derivation from lifetime animal studies. This does not mean cancer cannot occur after single exposures (witness the A-bomb experience), merely that exposure information must be consonant with the source of the model. We note that some toxic effects, however, occur after a single or a limited number of exposures, including acute reactions such as anesthetic effects and respiratory depression, or certain developmental effects following exposure during pregnancy. For developmental effects, for example, lifetime time-weighted averages have little relevance, so different types of data must be collected, in this case usually shorter-term exposure profile data during a particular time window. Consequently, we and the scientists who conduct monitoring studies need to collaborate with those scientists who evaluate a chemical's hazard potential to assure the development of a meaningful risk assessment. If short-term peak exposures are related to the effect, then instruments used should be able to measure short-term peak concentrations. If cumulative exposure is related to the effect, long-term average sampling strategies will probably be more appropriate for us.

Dose-Rate Effects

By using average daily exposure values in a dose-response relationship, we assume that within some limits, increments of C x T (exposure concentration multiplied by time) that are equal in magnitude are equivalent in their potential to cause an effect, regardless of the pattern of exposure. This is the so-called Haber's Rule. (See Atherley, 1985.) In those cases where toxicity depends on the dose rate, we may need a more precise determination of the length of time people are exposed to various concentrations and the sequence in which these exposures occur.

Reversibility of Toxicological Processes

The averaging process for daily exposure assumes that repeated dosing continues to add to the risk potential. In some cases, after cessation of exposure, toxicological processes are reversible over time. In these cases, exposure assessments must provide enough information so that we can account for the potential influence of episodic exposures.

Composition of the Exposed Population

For some substances, the type of health effect may vary as a function of age or sex. Likewise, certain behaviors (e.g., smoking), diseases (e.g., asthma), and genetic traits (e.g., glucose-6-phosphate dehydrogenase deficiency) may affect the response of a person to a chemical substance. Special population segments, such as children, may also call for a specialized approach to data collection. (WHO, 1986)

Establishing the Sampling Strategy

If we are to meet the objectives of the assessment using measurements, it is important to establish the sampling strategy before samples are actually taken. Our sampling strategy should include setting data quality objectives, developing the sampling plan and design, using spiked and blank samples, assessing background levels, developing quality assurance project plans, validating previously generated data, and selecting and validating analytical methods.

Data Quality Objectives

All of our measurements are subject to uncertainty because of the inherent variability in the quantities being measured (e.g., spatial and temporal variability) and analytical measurement variability introduced during the measurement process through sampling and analysis. Some sources of variability can be expressed quantitatively, but others can only be described qualitatively. The larger the variability associated with individual measurements, the lower the data quality, and the greater the probability of errors in interpretation. Data quality objectives (DQOs) describe the degree of uncertainty that we and other scientists and management are willing to accept.

Realistic DQOs are essential for us. Data of insufficient quality will have little value for us in problem solving, while data of quality vastly in excess of what is needed to answer the questions provide us with few, if any, additional advantages. DQOs should consider data needs, cost-effectiveness, and the capability of the measurement process. The amount of data required depends on the level of detail necessary for the purpose of the assessment. Estimates of the number of samples to be taken and measurements to be made should account for expected sample variability. Finally, DQOs help us clarify study objectives by compelling us to establish how the data will be used before they are collected.

We establish data criteria by proposing limits (based on best judgment or perhaps a pilot study) on the acceptable level of uncertainty for each conclusion to be drawn from new data, considering the resources available for the study. DQOs should include these elements:

- a clear statement of study objectives that includes an estimation of the key study parameters, an identification of the hypotheses being tested, the specific aims of the study, and how the results will be used;
- the scope of study objectives, including the minimum size of subsamples from which separate results may be calculated and the largest unit (area, time period, or group of people) the data will represent;
- a description of the data to be obtained, the media to be sampled, and the capabilities of the analytical methodologies;
- the acceptable probabilities and uncertainties associated with false positive and false negative statements;
- a discussion of statistics used to summarize the data; any standards, reference values, or action levels used for comparison; and a description and rationale for any mathematical or statistical procedures used; and
- an estimate of the resources needed.

Sampling Plan

The sampling plan specifies how our sample is to be selected and handled. An inadequate plan will often lead to biased, unreliable, or meaningless results. Good planning, on the other hand, makes optimal use of limited resources and is more likely to produce valid results.

Our sampling design specifies the number and types of samples needed to achieve DQOs. Factors to be considered in developing the sampling design include study objectives, sources of variability (e.g., temporal and spatial heterogeneity, analytical differences) and their relative magnitudes, relative costs, and practical limitations of time, cost, and personnel.

A study may use a combination of survey and experimental techniques and involve a variety of sampling procedures. A summary of methods for measuring worker exposure is found in Lynch (1985). Smith et al. (1987) provide guidance for field sampling of pesticides. Relevant EPA reference documents include *Survey Management Handbook, Volumes I and II* (U.S. EPA, 1984); *Soil Sampling Quality Assurance User's Guide* (U.S. EPA, 1990); and *A Rationale for the Assessment of Errors in the Sampling of Soils* (U.S. EPA, 1989). A detailed description of methods for enumerating and characterizing populations exposed to chemical substances is contained in *Methods for Assessing Exposure to Chemical Substances, Volume 4* (U.S. EPA, 1985a).

Factors that we need to consider in selecting sampling locations include population density, historical sampling results, patterns of environmental contamination and environmental characteristics such as stream flow or prevailing wind direction, access to the sample site, types of samples, and health and safety requirements.

The frequency and duration of sample collection will depend on whether we are concerned with acute or chronic exposures, how rapidly contamination patterns are changing, ways in which chemicals are released into the environment, and to what degree physical conditions are expected to vary in the future.

There are many sources of information for us on methods for selecting sampling locations. Schweitzer and Black (1985) and Schweitzer and Santolucito (1984) give statistical methods for selecting sampling locations for groundwater, soil, and hazardous wastes. A practical guide for ground-water sampling (U.S. EPA, 1985b) and a handbook for stream sampling (U.S. EPA, 1986d) are also available.

The type of sample to be taken and the physical and chemical properties of the chemical of concern usually dictate the sampling frequency. For example, determining the concentration of a volatile chemical in surface water requires a higher sampling frequency than necessary for groundwater because the chemical concentration of the surface water changes more rapidly. Sampling frequency might also depend on whether the health effects of concern result from acute or chronic exposures. More frequent sampling may be needed to determine peak exposures versus average exposure.

We often use a preliminary survey to estimate the optimum number, spacing, and sampling frequency. Factors to be considered include technical objectives, resources, program schedule, types of analyses, and the constituents to be evaluated. Shaw et al. (1984), Sanders and Adrian (1978), and Nelson and Ward (1981) discuss statistical techniques for determining the optimal number of samples.

The sampling duration we choose depends on the analytical method chosen, the limits of detection, the physical and chemical properties of the analyte, chemical concentration, and knowledge of transport and transformation mechanisms. Sampling duration may be extended to ensure adequate collection of a chemical at low concentration or curtailed to prevent the breakthrough of one at high concentration. Sampling duration is directly related to selection of statistical procedures, such as trend or cross-sectional analyses.

We normally run storage stability studies with periodic sample analysis concurrently with the storage of treated samples. However, in certain situations where chemicals are prone to break down or have high volatility, it is advisable to run a storage stability study in advance so that proper storage and maximum time of storage can be determined prior to sample collection and storage. Unless storage stability has been previously documented, we should analyze samples as soon as possible after collection to avoid storage stability problems. Individual programs may have specific time limits on storage, depending on the types of samples being analyzed.

Quality Assurance Samples

We should plan sampling to ensure that the samples are not biased by the introduction of field or laboratory contaminants. If sample validity is in question, all associated analytical data will be suspect. Field- and laboratory-spiked samples and blank samples should be analyzed concurrently to validate results. The plan should provide instructions clear enough so that each worker can collect, prepare, preserve, and analyze samples according to established protocols.

Background Level

Background presence may be due to natural or anthropogenic sources. At some sites it is significant and must be accounted for. We should try to determine local background concentrations by gathering data from nearby locations clearly unaffected by the site under investigation. When differences between a background (control area) and a target site are to be determined experimentally, the control area must be sampled with the same detail and care as the target.

Quality Assurance and Quality Control

Quality assurance (QA) assures that a product meets defined standards of quality with a stated level of confidence. QA includes quality control. Quality assurance begins with our establishment of DQOs and continues throughout the measurement process. Each laboratory should have a QA program and, for each study, a detailed quality assurance project plan, with language clear enough to preclude confusion and misunderstanding. The plan should list the DQOs and fully describe the analytes, all materials, methods, and procedures used, and the responsibilities of project participants. The EPA has prepared a guidance document (U.S. EPA, 1980) that describes all these elements and provides complete guidance for plan preparation. Quality control (QC) ensures that a product or service is satisfactory, dependable, and economical. A QC program should include development and strict adherence to principles of good laboratory practice, consistent use of standard operational procedures, and carefully-designed protocols for each measurement effort. The program should ensure that errors have been statistically characterized and reduced to acceptable levels.

Quality Assurance and Quality Control for Previously Generated Data

We may use previously generated data to fulfill current needs. Any data developed through previous studies should be validated with respect to both quality and extrapolation to current use. We should consider how long ago the data were collected and whether they are still representative. The criteria for method selection and validation should also be followed when analyzing existing data. Other points considered in data evaluation include the collection protocol, analytical methods, detection limits, laboratory performance, and sample handling.

Selection and Validation of Analytical Methods

There are several major steps we need to take in the method selection and validation process. First, we establish methods requirements. Next, we revise existing methods for suitability to the current application. If a new method must be developed, it is subjected to field and laboratory testing to determine its performance; these tests are then repeated by other laboratories using a round robin test. Finally, the method is revised as indicated by laboratory testing. *Guidance for Data Usability in Risk Assessment* (U.S. EPA, 1990b) provides an extensive discussion of this topic.

Finally, you need to review the examples of types of measurements that characterize exposure-related media and parameters in your development of an exposure plan (see Table 4.3).

Sources of Exposure

You will note that sources of environmental risk may involve release into the air, soil, surface water, or groundwater. These releases can come from a point source such as a smokestack, a line source such as an automobile moving along a highway, or an area source such as farms (e.g., pesticides and nutrients). They may be due to normal operations or accidents. These risk agents may be physical, biological, radiological, or in the form of energy.

Table 4.4 lists some important categories of risk agents for us and Table 4.5 lists some chemical risk agents. Most of the units for these contaminants are well known. However, those for radioactivity may involve unfamiliar units like those for activity (Ci, Bq, working level), energy (gray, rad), and damage (Sv, rem) (see Cothern and Smith, 1987).

Fate and Transport Models

For many exposure studies, you will need fate and transport models of contaminants in the environment for the many pathways from the source to the organism you are interested in. Generally we consider the air, water, and soil pathways. The contaminant can be transformed physically by volatilization, chemically by oxidation, or biologically by methylation. An illustration of the exposure pathways for a waste site is shown in Figure 4.2

There are several important physical and chemical environmental fate parameters. Some of these are listed in Table 4.6. A partition coefficient such as K_{oc} and K_d is the ratio of chemical concentration in the different phases under equilibrium conditions. K_{oc} is the ratio of the concentration of the contaminant in soil to that in water. For more details about fate and transport, see Hemond and Fechner, 1994

Table 4.3a Examples of Types of Measurements to Characterize Exposure-Related Media and Parameters (U.S. EPA, 1992)[a]

Type of Measurement (sample)	Usually Attempts to Characterize (whole)	Example	Typical Information Needed to Characterize Exposure
A. FOR USE IN EXPOSURE SCENARIO EVALUATION			
1. Fixed-Location Monitoring	Environmental medium; samples used to establish long-term indications of media quality and trends.	National Stream Quality Accounting Network (NASQAN)[b], water quality networks, air quality networks.	Population location and activities relative to monitoring locations; fate of pollutants over distance between monitoring and point of exposure; time variation of pollutant concentration at point of exposure.
2. Short-Term Media Monitoring	Environmental or ambient medium; samples used to establish a snapshot of quality of medium over relatively short time.	Special studies of environmental media, indoor air.	Population location and activities (this is critical since it must be closely matched to variations in concentrations due to short period of study); fate of pollutants between measurement point and point of exposure; time variation of pollutant concentration at point of exposure.
3. Source Monitoring of Facilities	Release rates to the environment from sources (facilities). Often given in terms of relationships between release amounts and various operating parameters of the facilities.	Stack sampling, effluent sampling, leachate sampling from landfills, incinerator ash sampling, fugitive emissions sampling, pollution control device sampling.	Fate of pollutants from point of entry into the environment to point of exposure; population location and activities; time variation of release.

Table 4.3a ***(continued)***

Type of Measurement (sample)	Usually Attempts to Characterize (whole)	Example	Typical Information Needed to Characterize Exposure
A. FOR USE IN EXPOSURE SCENARIO EVALUATION			
4. Food Samples (also see #9 below)	Concentrations of contaminants in food supply.	FDA Total Diet Study Program, market basket studies, shelf studies, cooked-food diet sampling.	Dietary habits of various age, sex, or cultural groups. Relationship between food items sampled and groups (geographic, ethnic, demographic) studied. Relationships between concentrations in uncooked versus prepared food.
5. Drinking Water Samples	Concentrations of pollutants in drinking water supply.	Groundwater Supply Survey[d], Community Water Supply Survey[e], tap water.	Fate and distribution of pollutants from point of sample to point of consumption. Population served by specific facilities and consumption rates. For exposure due to other uses (*e.g.*, cooking and showering), need to know activity patterns and volatilization rates.
6. Consumer Products	Concentration levels of various products.	Shelf surveys, e.g., solvent concentration in household cleaners.[f]	Establish use patterns and/or market share of particular products, individual exposure at various usage levels, extent of passive exposure.

[a] Intake or uptake information also is needed to characterize dose.
[b] U.S. EPA (1985c)
[c] U.S. EPA (1986a)
[d] U.S. EPA (1985c)
[e] U.S. EPA (1985d)
[f] U.S. EPA (1985a)

Table 4.3a *(continued)*

Type of Measurement (sample)	Usually Attempts to Characterize (whole)	Example	Typical Information Needed to Characterize Exposure
A. FOR USE IN EXPOSURE SCENARIO EVALUATION			
7. Breathing Zone Measurements	Exposure to airborne chemicals.	Industrial hygiene studies, occupational surveys, indoor air studies.	Location, activities, and time spent relative to monitoring locations. Protective measures/avoidance.
8. Microenvironmental Studies	Ambient medium in a defined area, e.g., kitchen, automobile interior, office setting, parking lot.	Special studies of indoor air, house dust, contaminated surfaces, radon measurements, office building studies.	Activities of study populations relative to monitoring locations and time exposed.
9. Surface Soil Sample	Degree of contamination of soil available for contact.	Soil samples at contaminated sites.	Fate of pollution on/in soil; activities of potentially exposed populations.
10. Soil Core	Soil including pollution available for ground-water contamination; can be an indication of quality and trends over time.	Soil sampling at hazardous waste sites.	Fate of substance in soil; speciation and bioavailability, contact and ingestion rates as a function of activity patterns and age.
11. Fish Tissue	Extent of contamination of edible fish tissue.	National Shellfish Survey[9].	Relationship of samples to food supply for individuals or population of interest; consumption and preparation habits .

[8] U.S. EPA (1986a)

Table 4.3b Examples of Types of Measurements to Characterize Exposure-Related Media and Parameters (U.S. EPA, 1992)[a]

Type of Measurement (sample)	Usually Attempts to Characterize (whole)	Example	Typical Information Needed to Characterize Exposure
B. FOR USE IN POINT-OF-CONTACT MEASUREMENT			
1. Air Pump/Particulates and Vapors	Exposure of an individual or population via the air medium.	TEAM study,[h] carbon monoxide study. Breathing zone sampling in industrial settings.	Direct measurement of individual exposure during time sampled. In order to characterize exposure to population, relationships between individuals and the population must be established as well as relationships between times sampled and other times for the same individuals, and relationships between sampled individuals and other populations. In order to make these links, activities of the sampled individuals compared to populations characterized are needed in some detail.
2. Passive Vapor Sampling	Same as above.	Same as above.	Same as above.
3. Split Sample Food/ Split Sample Drinking Water	Exposures of an individual or population via ingestion.	TEAM study.[j]	Same as above.
4. Skin Patch Samples	Dermal exposure of an individual or population.	Pesticide Applicator Survey.[k]	1) Same as above. 2) Skin penetration.

[h] U.S. EPA (1987a)
[i] U.S. EPA (1987a)
[j] U.S. EPA (1987a)
[k] U.S. EPA (1987b)

Table 4.3c Examples of Types of Measurements to Characterize Exposure-Related Media and Parameters (U.S. EPA, 1992)[a]

Type of Measurement (sample)	Usually Attempts to Characterize (whole)	Example	Typical Information Needed to Characterize Exposure
C. FOR USE IN EXPOSURE ESTIMATION FROM RECONSTRUCTED DOSE			
1. Breath	Total internal dose for individuals or population (usually indicative of relatively recent exposures).	Measurement of volatile organic chemicals (VOCs), alcohol. (Usually limited to volatile compounds).	1) Relationship between individuals and population; exposure history (*i.e.*, steady-state or not) pharmacokinetics (chemical half-life), possible storage reservoirs within the body. 2) Relationship between breath content and body burden.
2. Blood	Total internal dose for individuals or population (may be indicative of either relatively recent exposures to fat-soluble organics or long term body burden for metals).	Lead studies, pesticides, heavy metals (usually best for soluble compounds, although blood lipid analysis may reveal lipophilic compounds).	1) Same as above. 2) Relationship between blood content and body burden.
3. Adipose	Total internal dose for individuals or population (usually indicative of long-term averages for fat- soluble organics).	NHATS,[1] dioxin studies, PCBs (usually limited to lipophilic compounds).	1) Same as above. 2) Relationship between adipose content and body burden.

[1] U.S. EPA (1986b)

Table 4.3c ***(continued)***

Type of Measurement (sample)	Usually Attempts to Characterize (whole)	Example	Typical Information Needed to Characterize Exposure
C. FOR USE IN EXPOSURE ESTIMATION FROM RECONSTRUCTED DOSE			
4. Nails, Hair	Total internal dose for individuals or population (usually indicative of past exposure in weeks to months range; can sometimes be used to evaluate exposure patterns).	Heavy metal studies (usually limited to metals).	1) Same as above. 2) Relationship between nails, hair content and body burden.
5. Urine	Total internal dose for individuals or population (usually indicative of elimination rates); time from exposure to appearance in urine may vary, depending on chemical.	Studies of tetrachloroethylene[m] and trichloroethylene.[n]	1) Same as above. 2) Relationship between urine content and body burden.

[m] U.S. EPA (1986c)
[n] U.S. EPA (1987c)

Table 4.4 Some Important Categories of Risk Agents and Sources (Covello and Merkhofer, 1993)

Criteria Air Pollutants (Clean Air Act)

EXAMPLE RISK AGENTS	SOURCE/DESCRIPTION	EXAMPLE CONCERNS
carbon monoxide	automobiles, other fossil fuel combustion	headaches, dizziness, drowsiness, nausea/ vomiting, coma, death
lead	leaded gasoline and automobile exhaust	neurologic and behavioral disorders, especially in children
ozone	formed from atmospheric reactions involving volatile organic compounds (VOCs) and nitrogen dioxide	respiratory effects, materials damage (esp. textiles, paints, and elastomers), reduced forest growth
nitrogen dioxide	fossil fuel combustion	respiratory effects, photochemical smog, acid deposition with effects on forests and other ecosystems
particulate matter (various sizes and chemically composed particles)	electric utilities construction activities, mining, windblown dust	aggravation of chronic respiratory and cardiac disease, increased acute respiratory illness, decreased visibility
sulfur oxides	fossil fuel combustion	effects on lung function, acid deposition with effects on aquatic and other ecosystems, materials damage

Other Toxic Air Pollutants

volatile organic compounds	underground petroleum storage tanks, solvents	cancer
metals such as arsenic, cadmium, and chromium	industrial boilers and furnaces burning recycled oil, municipal waste incinerators	Lung damage, cancer
organic compounds such as PCBs and	industrial boilers and furnaces burning recycled oil, municipal waste incinerators, electrical capacitors and transformers	cancer, effects on fish and wildlife, bioaccumulation
polycyclic aromatic hydrocarbons (PAHs)	coal fired electric power plants, forest fires, tobacco smoke	cancer

Table 4.4 *(continued)*

Greenhouse Gases

EXAMPLE RISK AGENTS	SOURCE/DESCRIPTION	EXAMPLE CONCERNS
chlorofluorocarbons (CFCs)	compounds used as aerosol propellants, foam blowing agents, refrigerants, solvent	skin cancer, ecological effects, global climate change
carbon dioxide	combustion of fossil fuels, deforestation	global climate rise in sea level, changed crop yields
Radioactivity		
radon	radioactive gas emitted by soil, rock, and building materials	lung cancer, property values
other naturally occurring radioisotopes	mining, industrial processing of raw materials	cancer
radioactive production products	nuclear reactors, weapons production, particle accelerators, medical instruments	cancer
Pesticides		
approximately 50,000 pesticide products derived from about 600 basic chemicals	agriculture, industry and household use	various effects on human health, wildlife
Point-Source Discharges to Surface Waters		
chlorination products, ammonia, thermal pollution	industrial operations, municipal waste management facilities, power plants	damage to aquatic systems, loss of water recreational facilities
Nonpoint Discharges to Surface Waters		
sediment, excess nutrients, pesticides, acids from mining operations	runoff from precipitation, agricultural runoff, urban runoff	erosion, silting of reservoirs, clogging of shipping channels, deposition of toxic pollutants attached to sediments
Hazardous Waste		
toxics, radioactive waste, particulates, excess nutrients, microbes	municipal waste management facilities, abandoned toxic waste sites, landfills, surface impoundments, incinerators, solvent recovery facilities	groundwater, surface water, and air contamination

Table 4.4 (continued)

EXAMPLE RISK AGENTS	SOURCE/DESCRIPTION	EXAMPLE CONCERNS
Structural Accidents		
collapse of engineering structures	failure of dams, mines, in-progress concrete and masonry structures, energy installations, underground tunnels, covered excavations	lost lives, property damage, destruction of fish or wildlife habitat, soil erosion
Chemical Spills		
PCBs, ammonia, chlorine	chemical releases from truck, barge, rail and ship transportation, industrial production and storage facilities, electric utilities	worker safety, public health
petroleum products	spills from pipelines, storage tanks, production facilities	effects on public health, wildlife
Viruses		
HIV (AIDS), Herpes simplex, Rhinovirus (common cold)	submicroscopic infective agents capable of attacking and destroying living cells	infectious diseases, including AIDS, flu, and the common cold
Explosives and Combustibles		
propane, gasoline, dynamite, rice flour	petroleum refineries and distributors chemical processing plants, marine terminals, distilleries, grain-handling facilities	fire, injuries, lost lives, property damage, crop loss
Genetically Engineered Organisms		
ice minus	bacterium with excised gene having ability to promote formation of ice crystals below 0°C	feared side effects, such as impacts on rainfall, toxicity to local crops
strain of *Pseudomonas*	oil-degrading organisms	gene transfer, toxic metabolites, impacts on ocean organisms

Table 4.5 Some Chemical Risk Agents (Covello and Merkhofer, 1993)

Agent	Source/Description	Primary Concerns
acetone	fingernail polish remover, glues, solvents	accidental poisoning, irritation of eyes and mucous membranes
acrylonitrile	plastic resin in pipes, textiles	cancer, releases hydrogen cyanide when burned
aflatoxin	mold toxin on peanuts and corn	cancer of the liver
Alar (daminozide)	sprayed on apples, grapes, cherries, and peanuts to make crop ripen, improve appearance, and increase storage life	cancer
alcohol	alcoholic drinks, medications	birth defects, neurological impairment
aldicarb	carbamate pesticide	poisoning through skin, spontaneous abortions
Aldrin/ dieldrin	organochlorine insecticide	neurological effects, cancer environmental effects (fish, birds)
aluminum	cooking pots, additive in food, nonprescription drugs	possible role in Alzheimer's disease
ammonia	household cleaning solutions, industry and farming accidents	irritation of skin, eye, and respiratory passages, pulmonary, edema, pneumonia
antibiotics in animal feed	improving growth rate of food-producing animals	increase in occurrence of drug-resistant pathogens
arsenic	smelting, used oil, pesticides wood preservative	skin, liver and lung cancer, poisoning
asbestos	fireproof insulation, ceiling tiles, brake linings	lung cancer, mesothelioma, asbestosis
benzene	petrochemical and refining industries, constituent of gasoline, tobacco smoke	bone marrow damage, leukemia, blood disorders, upper respiratory tract irritation, cancer
benzidene	manufacture of dye, paper, textiles, and leather	cancer of the bladder
benzo[a]pyrene (B[a]p)	forest fires, cigarettes, wood burning stoves, cooked foods	cancer
beryllium	oil and coal combustion, mining, production of beryllium metal, cement plants, alloys, ceramics, rocket propellants	dermatitis, ulcers, inflamm-tion of mucous membranes, chronic berylliosis, cancer

Table 4.5 Some Chemical Risk Agents *(continued)*

Agent	Source/Description	Primary Concerns
cadmium	fossil fuel combustion, fertilizers, shellfish, electroplating, batteries	lung cancer, emphysema, heart disease, kidney and liver disease
carbaryl (bees)	carbamate pesticide	environmental effects
carbon black	photocopy toner, paint pigment, manufacturing	provides transport vehicle for polycyclic aromatic hydrocarbons (PAHs) and other mutagenic contaminants
carbon tetrachloride	dry cleaning fluid, pesticide fumagant	cancer, liver and kidney disease
chlorine	drinking water disinfectant, tanker truck spills	pulmonary edema, carcinogenic by-products of chlorination
chloroform	sewage treatment plants, showering with chlorinated water	cancer, liver and kidney disease
chloromethane	manufacture of polymers and resins, antiknock compounds for gasoline	possible cardinogen, mutagen
chromium	chrome plating, paint pigments, leather tanning, wood preservatives	lung cancer, liver and kidney damage
cobalt	foam stabilizer in beer	heart damage
cyclamate	artificial sweetener	cancer
2,4-D	herbicide	forms nitrosamines in intestinal tract, environmental effects (fish)
DDT	organochlorine insecticide	cancer, reproductive effects, environmental damage (eggshell thinning)
Diazinon	organophosphate insecticide	environmental effects (birds, fish)
Dibromochloropropane (DBCP)	soil fumigant	sterility
Diethylstilbesterol (DES)	drug to prevent miscarriage	birth defects, cancer of the vagina
dioxin	contaminant in herbicide, cardboard containers	chloracne, cancer, birth defects
ethylenebisdithiocarbamates (EBDCs)	class of carbamate pesticides	cancer, birth defects

Table 4.5 Some Chemical Risk Agents *(continued)*

Agent	Source/Description	Primary Concerns
ethylene dibromide (EDB)	fumigant for grain, dye production, scavenger in leaded gasoline	bromide poisoning, cancer
formaldehyde	foam insulation, pressed wood materials	irritation of eyes and respiratory system, headaches, cancer
heptachlor/ chlordane	organochlorine insecticide	cancer, environmental effects (plants, bees, fish)
hexachlorohene	disinfectant	neurological effects in infants
hydrogen cyanide	plastic manufacturing, petroleum	acute cyanide poisoning
Kepone	organochlorine insecticide	neurological effects, sterility, environmental effects
Lindane	organochlorine pesticide fumagant	neurological effects, possible carcinogen, environmental effects (fish, birds)
Malathion	Organiophosphate pesticide	environmental effects (fish)
mercury	thermometers, dental fillings, trace contaminants of ores and fuels	memory losses, tremors, neurological damage
methyl ethyl ketone (MEK)	lacquers, varnishes, industrial solvent	irritation of eyes, mucous membranes, skin
methylene chloride (DCM)	paint strippers and aerosol paints, used in decaffeination of coffee	cancer
methyl-mercury	organic compound of mercury, industrial and commercial releases accumulated in fish and shellfish	neurologic damage, reproductive effects, bioaccumulation
naphthalene	mothballs	accidental poisoning of infants
nickel	coins, metal products, metal refineries, food	skin reaction, cancer
nitrobenzene	production of aniline	destruction of blood hemoglobin
nitrosamines	formed from precursors, amines nitrogen oxides, and nitrates in air, water, and food	cancer
parathion	organic phosphate pesticide	respiratory effects, environmental effects (bees, fish)

Table 4.5 Some Chemical Risk Agents *(continued)*

Agent	Source/Description	Primary Concerns
phenol	manufacturing, disinfectant in petroleum and other industries	skin and eye irritation
polybrominated biphenyls (PBBs)	fire retardant, plastic parts	mislabeled animal feed
saccharin	artificial sweetener	cancer
tetrachloro-ethylene (PCE)	dry cleaning fluid, chlorin-ation of acetylene, degreasing metals	smog, eye irritation, possible carcinogen, liver damage
thalidomide	sedative	birth defects
toluene	petrochemical and refining industries, constituent of gasoline, cigarette smoke, model glue	irritation to skin and eyes, kidney and liver damage
trichloroethane (TCA)	drain cleaners, shoe polish, spot removers	possible mutagen
trichloro-ethylene (TCE)	metal degreasing	cancer
2,4,5-Trichloro-phenoxy acetic acid (2,4,5-T)	herbicide	chloracne, birth defects, cancer, environmental effects (fish)
TRIS	flame retardant in children's sleepwear	cancer
Vinyl Chloride	production of polyvinyl chloride, PVC pipes, plastic packaging	angiosarcoma (a rare type of tumor), lung cancer, birth defects

Several important considerations necessary for our understanding of fate and transport:

- What are the principal mechanisms for change or removal in each of the environmental media (air, water, soil)?
- How does the chemical behave in air, water, soil, and biological media? Does it bioaccumulate or biodegrade? Is it absorbed or taken up by plants?
- Does the agent react with other compounds in the environment?
- Is there intermedia transfer? What are the mechanisms for intermedia transfer? What are the rates of the intermedia transfer or reacting mechanism(s)?
- How long might the chemical remain in each environmental medium? How does its concentration change with time in each media?

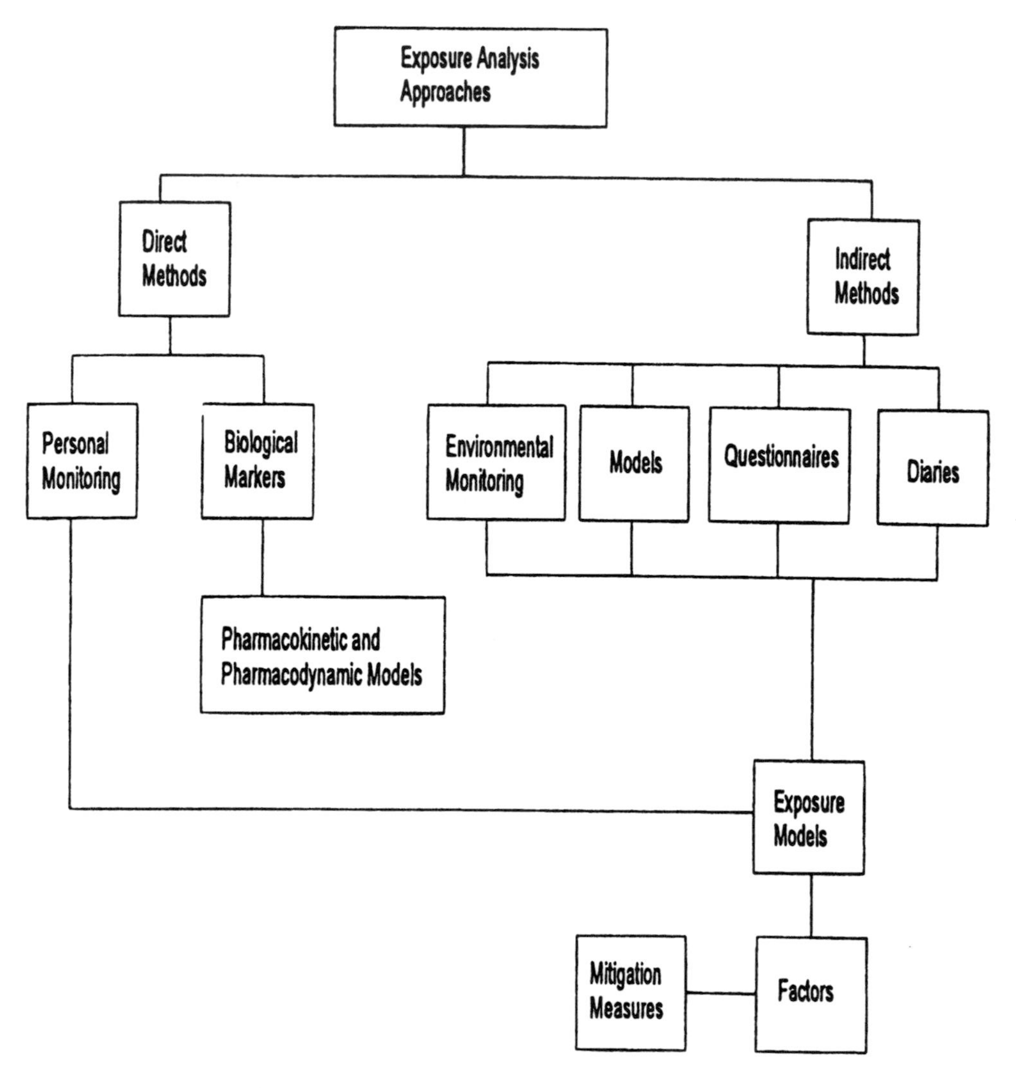

Figure 4.2 Illustration of exposure pathways for a waste site

- What are the products into which the agent might degrade or change in the environment? Are these products potentially of concern?
- Is a steady-state concentration distribution in the environment or in specific segments of the environment achieved?

Figures 4.3, 4.4, and 4.5 show flow charts for our fate and transport assessments for the atmosphere, surface waters and sediments, and soils and groundwater respectively. (Willard Chappell, University of Colorado at Denver, private communication)

Table 4.6 Important Physical and Chemical Environmental Fate Parameters

K_{oc} Provides a measure of the extent of chemical partitioning between organic carbon and water at equilibrium. The higher the K_{oc}, the more likely a chemical is to bind to soil or sediment than to remain in water.

K_d Provides a soil or sediment specific measure of the extent of chemical partitioning between soil or sediment and water, unadjusted for dependence upon organic carbon. To adjust for the fraction of organic carbon present in soil or sediment (f_{oc}), use $K_d = K_{oc} \times f_{oc}$. The higher the K_d, the more likely a chemical is to bind to soil or sediment than to remain in water.

Solubility is an upper limit on a chemical's dissolved concentration in water at a specified temperature. Aqueous concentrations in excess of solubility may indicate sorption onto sediments, the presence of solubilizing chemicals such as solvents or the presence of non-aqueous phase liquid.

Henry's Law Constant provides a measure of the extent of chemical partitioning between air and water at equilibrium. The higher Henry's Law constant, the more likely a chemical is to volatilize then to remain in the water.

Vapor Pressure is the pressure exerted by a chemical vapor in equilibrium with its solid or liquid form at any given temperature. It is used to calculate the rate of volatilization of a pure substance from a surface or in estimating a Henry's Law constant for chemicals with low water solubility. The higher the vapor pressure, the more likely a chemical is to exist in a gaseous state.

Diffusivity describes the movement of a molecule in a liquid or gas medium as a result of differences in concentration. It is used to calculate the dispersive component of chemical transport. The higher the diffusivity, the more likely a chemical is to move in response to concentration gradients.

Bioconcentration Factor (BCF) provides a measure of the extent of chemical partitioning at equilibrium between a biological medium such as fish or plant tissue and an external medium such as water. The higher the BCF, the greater the accumulation in living tissue is likely to be.

Media-Specific Half-Life provides a relative measure of the persistence of a chemical in a given medium, although actual values can vary greatly depending on site-specific conditions. The greater the half-life, the more persistent a chemical is likely to be.

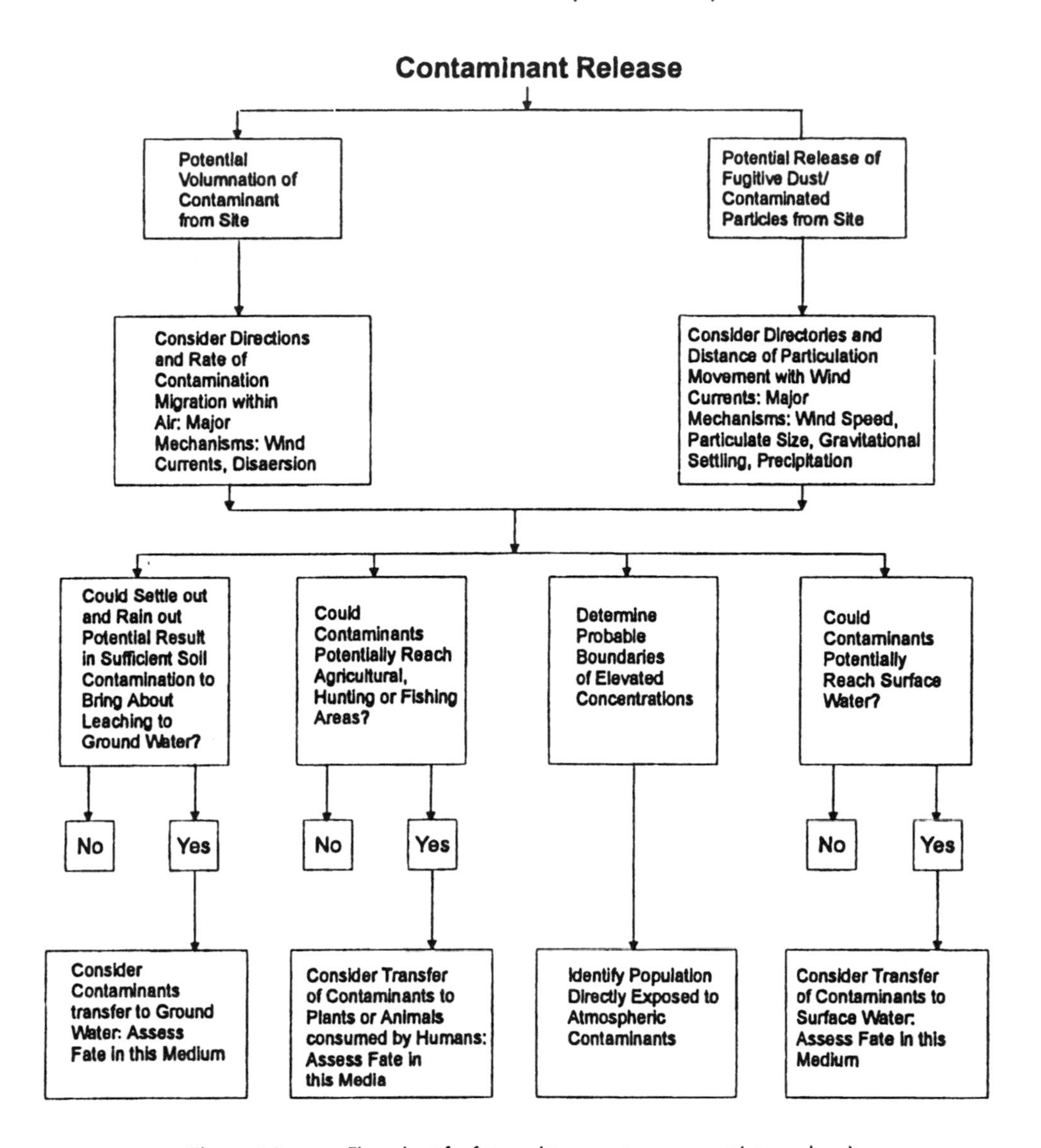

Figure 4.3 Flow chart for fate and transport assessment (atmosphere)

Exposure Models

Our models for exposure assessment represent simplified versions of the essential elements of the exposure process. They range from back-of-the envelope calculations to main frame computer models (see Hemond and Fechner, 1994; Paustenbach, 1985). There are seven factors that differentiate various pollutant transport and fate models:

- transport media: air, surface water, soil, groundwater, or biota;
- geographic scale: global, national, regional, or local;

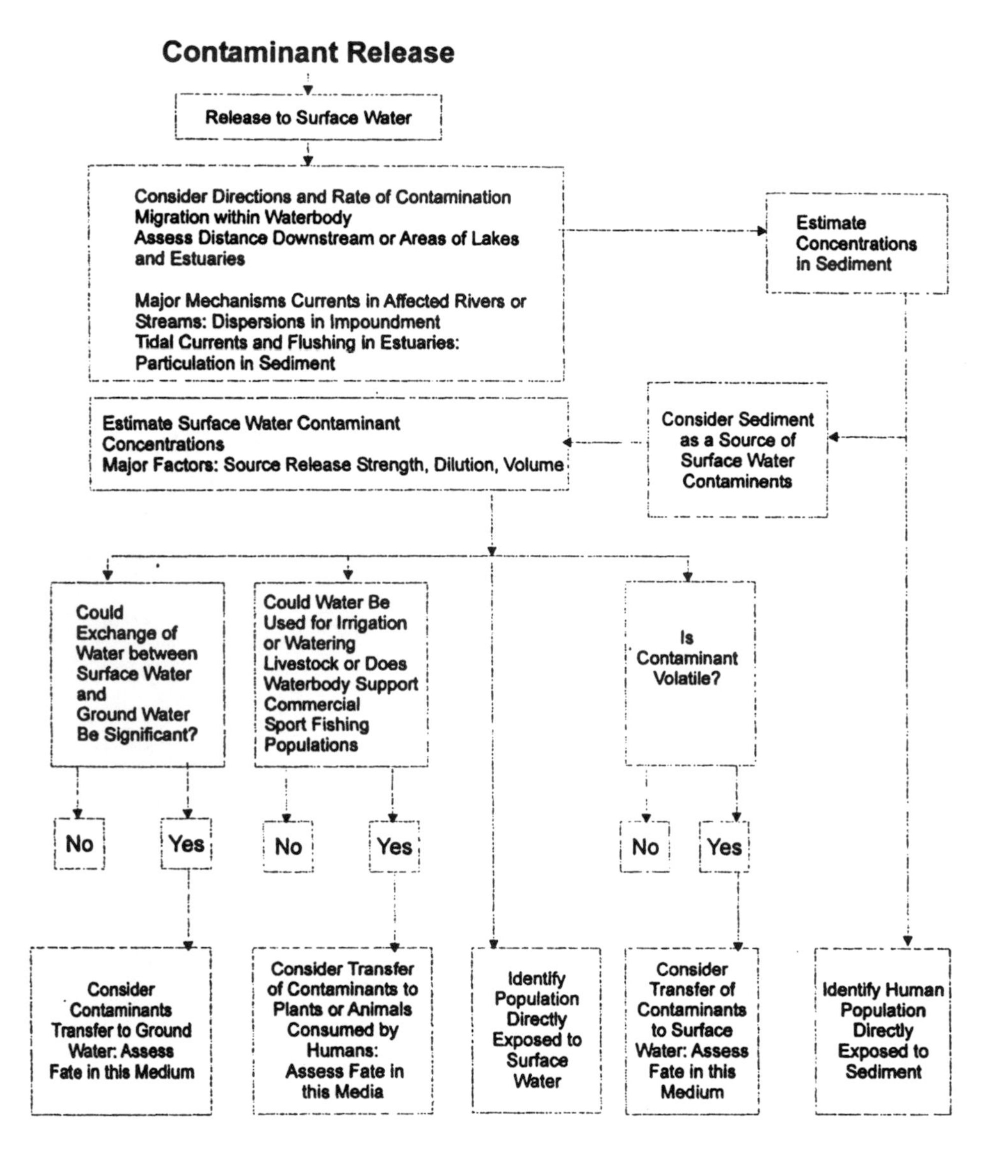

Figure 4.4 Flow chart for fate and transport assessment (surface water and sediments)

- pollutant source characteristics: continuous or instantaneous, industrial, residential, or commercial;
- risk agents involved;
- receptor populations: humans, animals, plants, microorganisms, industrial, residential, or commercial;
- exposure routes: ingestion, inhalation, dermal; and
- time frame.

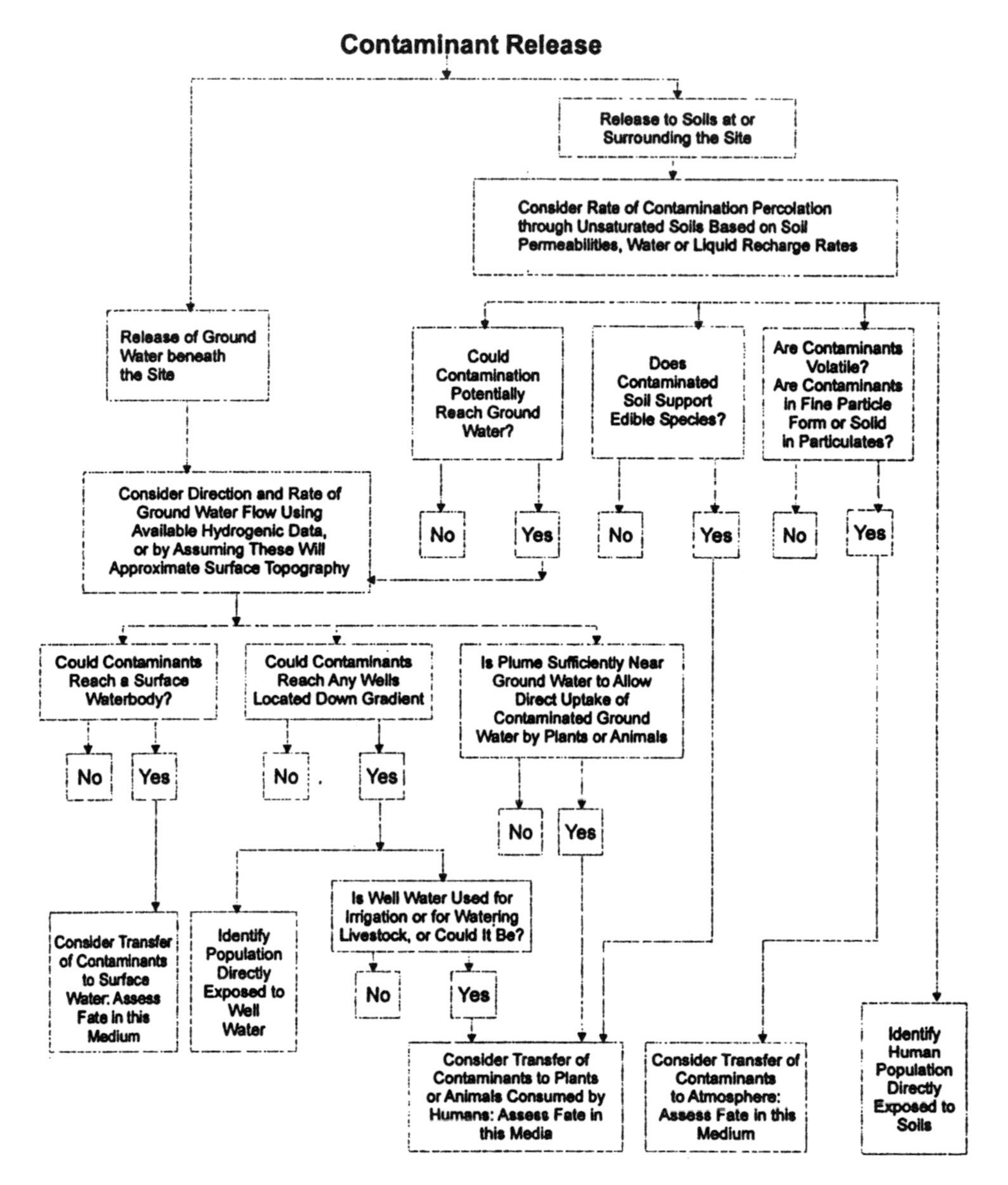

Figure 4.5 Flow chart for fate and transport assessment (soils and groundwater)

Models of exposure need to:

- characterize the physical setting;
- characterize the potentially exposed populations;
- identify pathways;
- include the source and mechanism of release;
- describe the retention or transport mechanism;

- estimate the concentration at the point of contact with humans or other life in the environment;
- describe the exposure route (ingestion, inhalation, dermal);
- explain fate and transport; and
- describe transformations involved.

Three things can happen to a chemical that is in the environment: it can remain, be carried elsewhere by a transport process, or be eliminated by being chemically transformed into another chemical. For more detail concerning fate and transport models, see Hemond and Fechner, 1994. The origin of a contaminant is called the *source* and its destination is called a *sink*.

Our study of the physical transport of chemicals starts with the two mechanisms of advection and transport. *Advection* is the bulk movement of air and water, as seen in blowing wind or flowing streams. *Transport* can involve molecular diffusion or turbulent diffusion. *Molecular diffusion* describes the situation where a chemical moves from where its concentration is relatively high to another location where its concentration is lower. *Turbulent diffusion* is the situation involving random motion of air or water that carries the chemical and is described as *Fickian*.

Ficks first law in one dimension is: $J = - D(dC/dx)$ (1)

where J is the flux density (M/L^2T), D is the Fickian mass transport coefficient (L^2/T), C is the chemical concentration (M/L^3), and x is the distance over which a concentration change is being considered (L). When we write this law in three dimensions it becomes:

$J = - \nabla C$, where ∇ is the gradient operator and D is equal in all directions.

The rate of change of a chemical at any point in space, dC/dt, equals the rates of chemical input and output by physical means plus the rate of net internal production (sources minus sinks). The inputs and outputs that occur by physical means (advection and Fickian transport) are expressed in terms of fluid velocity (V), the diffusion/dispersion coefficient (D), and the chemical concentration gradient in the fluid (air or water) (dC/dx). The input or output associated with internal sources or sinks of the chemical is represented by r. In one dimension this is:

$$dC/dt = -V\,(dC/dx) + d/dx[\,D\,(dC/dx)] + r \quad (2)$$

Changes in the concentration can also occur if a source or sink process, such as a chemical or biological reaction, is introducing or removing the compound of interest (r). In three dimensions this is written as:

$$dC/dt = -\,V\,\nabla C + \nabla\,D(\nabla C) + r$$

Physical Transport in Surface Waters

Contamination of surface waters can be from point or non-point sources. Pollutants may dissolve, float on the surface, or volatilize. Sometimes, especially for metals, we find that the particulates sink to the bottom. Thus we find the currents and local disturbances of the sediments important. For shallow lakes, the effect of the wind can be important.

The two equations that we commonly use in analyzing transport in surface waters are:

$V = C\,[RS]^{1/2}$, the Chezy equation

where V is the velocity (L/T), C is the Chezy coefficient ($L^{1/2}/T$), R is the hydraulic radius (L), and S is the slope of the river (L), and

$V = [1.486R^{2/3}S^{1/2}]/n$, the Manning equation

where n is the Manning coefficient describing river channel roughness ($T/L^{1/3}$).

The coefficients C and n are largely determined experimentally for different types of stream channel dimensions. If we have Fickian mixing; then the solution to Equation 2 is:

$$C(x,t) = [M/\sqrt{4\pi D\,t}\,]e^{-((x\,-\,Vt)^2/4D\,t} \qquad (3)$$

where C is the concentration of the chemical (M/L^3), M is the mass of the chemical injected per cross-sectional area of the river (M/L^2), x is the distance downstream of the injection location (L), V is the river velocity (L/T), t is the time elapsed since injection (T), and D is the Fickian mixing coefficient (including effects of turbulent diffusion and dispersion) (L^2/T).

Similar equations can be derived for lakes, estuaries, and wetlands. Another important concept is the degradation of chemicals biologically. This is favored in reducing conditions that occur where oxygen has been depleted. Bioaccumulation describes the broad processes by which a pollutant chemical is stored in organisms, and bioconcentration describes situations where the chemical in an organism accumulates to a concentration higher than that found in the external environment.

Finally we need to realize that chemicals in surface waters can also be degraded by light or through hydrolysis, elimination reactions, and nucleophilic substitutions

Groundwater

Groundwater contamination is determined largely by underground waterflow patterns. Our analysis will involve boundary conditions describing the flow and transport along the perimeter of the modeled area. Starting from the earth's surface, we generally encounter an unsaturated zone, the groundwater table, the saturated zone, and the volume below that.

In general, if groundwater is contaminated, it is nearly impossible to pump and treat because of the cost. The best approach is to prevent contamination of groundwater aquifers. Another aspect of groundwater is the use of injection wells. Although this is a local source, the contamination can be widespread because aquifers can be large.

Atmosphere

For the atmospheric situation, we find that the solution to Fick's equation (1) is:

$$\chi = \frac{Q}{\pi u \sigma_y \sigma_z} e^{-\frac{1}{2}(\frac{H}{\sigma_z})^2} \qquad (4)$$

where χ is the concentration (M/L^3), H is the stack height (L), Q is the emission rate (M/T), μ is the wind speed (L/T), and σ_x, σ_y, and σ_z are the standard deviations of the concentrations in the three dimensions (L), with z being the vertical direction, x being the direction in which the plume is moving, and y being the breadth of the plume (see Figure 4.6).

Intake Calculations

Once the concentrations are known, we can calculate the intakes by knowing the exposure factors. These factors are listed in detail in the EPA *Exposure Factors Handbook* (U.S. EPA, 1998). The purpose of the *Exposure Factors Handbook* is to (1) summarize data on human behaviors and characteristics which affect exposure to environmental contaminants, and (2) recommend values for use for these factors. In this section we provide formulas for some standard exposure processes.

Inhalation

The air inhalation intake rate is a product of the contaminant concentration, inhalation rate, absorption factor, body weight, and exposure factor, as follows:

air inhalation intake = (C x IR x AF x EF)/BW

where:

air inhalation intake is in units of mg/kg/day

C = contaminant concentration (mg/M^3)

IR = inhalation rate (M^3/day)

AF = absorption factor (unitless)

EF = exposure factor (unitless)

BW = body weight (kg)

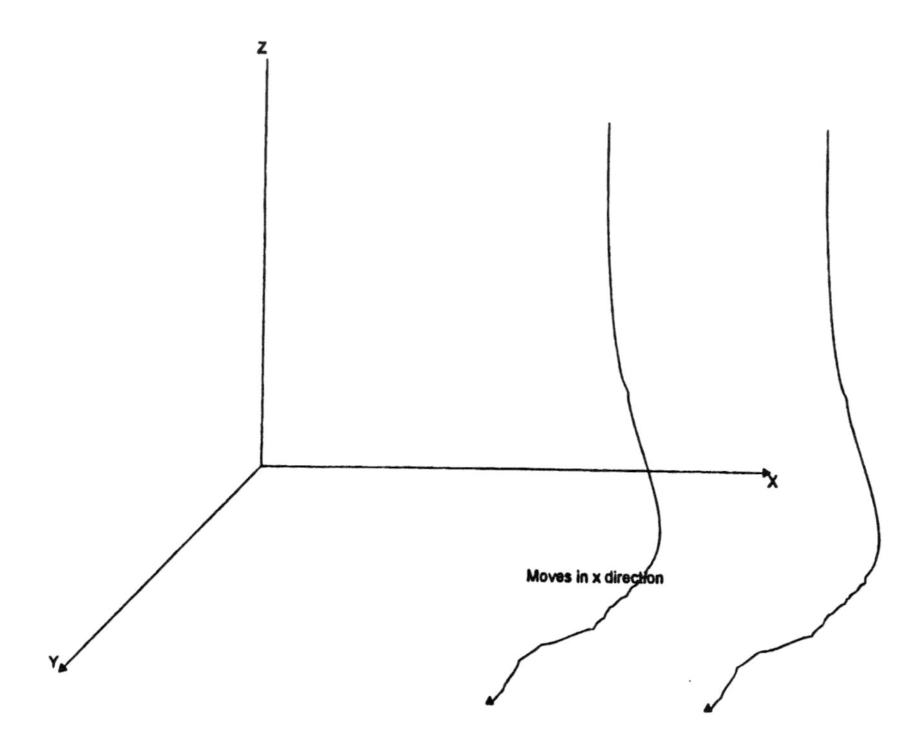

Figure 4.6 Example of a plume moving in the x direction and expanding

The *exposure factor* is a product of exposure frequency (events/year) and exposure duration (years/lifetime). If exposure is constant over a lifetime, EF = 1. An absorption factor of 1 is assumed unless specific data are available indicating less. The above is for chronic exposures; if we need daily inhalation rates we can use concentration for that day. If we are considering particulates, the formula is modified by a factor in the denominator, which is the fraction of airborne particulates depositing in the deep lung. We know that the cilia in our respiratory tract removes particulates larger than 10 microns. If they are smaller than 2 microns, Brownian motion removes them, and if they are between 2 and 10 microns, they remain in the lung.

Water Ingestion

For water ingestion, the intake is determined as follows:

water ingestion intake = (C x IR x AF x EF)/BW

where water ingestion rate is in units of mg/kg/day, and

C = contaminant concentration (mg/L)

IR = ingestion rate (L/day)

AF = absorption factor (unitless)

EF = exposure factor (unitless)

BW = body weight (kg)

Absorption factors are often unknown, in which case they are assumed to be equal to 1 (100%). The exposure factor is the product of exposure frequency (days/year) and exposure duration (years/total days exposure is averaged over). EPA assumes an average daily intake rate of 2 L/day for adults and 1 L/day for children. This factor often is a matter of controversy, e.g., when looking at exposure in an epidemiology study of exercising field workers who probably drink an average of 6 L/day.

For example, consider a situation where the primary water supply is contaminated with 367 mg/L of methyl chloride. Compute lifetime intake for an adult (70 kg) and convert to dose, assuming 100% absorption:

$$ID_W = (C \times IR \times AF \times EF)/BW = 367 \text{ mg/L} \times 2 \text{ L/day} \times 1 \times 1/70 \text{ kg}$$

$$= 10 \text{ mg/kg/day (round off 10.49 to 10)}$$

Soil Ingestion

For soil intake, we have the following:

$$\text{soil ingestion intake} = (C \times IR \times AF \times EF \times 10^{-6})/BW$$

where soil ingestion intake is in units of dose mg/kg/day, and

C = contaminant concentration (mg/kg)

IR = soil ingestion rate (mg/day)

AF = absorption factor (unitless)

EF = exposure factor (unitless)

BW = body weight (kg)

The conversion factor of 10^{-6} converts the result to mg/kg/day. The soil exposure factor reflects the individual's access to possible exposure and is a product of exposure frequency (days/year) and exposure duration (years/total days exposure duration).

ATSDR recommends these soil ingestion values:

Age (years)	Weight (kg)	Ingestion rate (mg/day)	
		Maximum	Average
0–1	10	250	50
1–6	16	500	100
6–11	30	250	50
12–17	50	100	50
> 18	70	100	50

If adults do not exhibit frequent hand-to-mouth activity (most nonsmokers and nonnail-biters), it is probably reasonable for us to use an average value of 25 mg. Also note that pica children represent a group that have much larger intakes. Because children are in contact with indoor dust more frequently than outdoor soils, the relative contributions of soil and dust are important. Dusts often have higher contaminant concentrations.

As an example of soil ingestion, assume there is soil with contaminant concentration of 100 mg/kg. Assume that an individual is on site 5 days per week, 50 weeks per year, for 30 years.

EF = frequency x duration

[5 days/week x (1 week/7days) x 50 weeks x (1 yr/52 wk)] x [30 years/lifetime x (lifetime/70 years)] = 0.29

Assuming AF = 1.0, the soil ingestion can be determined as follows:

soil ingestion intake = (C x IR x AF x EF x 10^{-6})/BW = (100 mg/kg x 25 mg/day x 1.0 x 0.29 x 10^{-6})/70 kg = 1 x 10^{-5} mg/kg/day

Dermal Absorption

Dermal absorption can come from soil or water. The factors involved are the area of exposed skin, duration of contact, chemical concentration, chemical-specific permeability, and skin condition. Also, we note that this absorption can occur during bathing, showering, or swimming. The permeability of our skin to a chemical is influenced by the physical and chemical properties of the substance, including its molecular weight, electrostatic charge, hydrophobicity, and solubility in aqueous and lipid media. In general, chemicals with high skin permeability have lower molecular weights, are non-ionized, and are lipid soluble. Dermal absorption can be significantly increased by skin abrasions and pathological conditions, such as psoriasis or eczema.

Dermal Absorption from Water

For dermal absorption from water, the intake is determined as follows:

water dermal intake = (C x P x SA x ET x EF x ED) x (1 liter/1,000 cm^3)/BW

where water dermal intake is in units of mg/kg/day, and

C = contaminant concentration in water (mg/L)

P = permeability constant (cm/hr)

SA = exposed body surface area (cm^2)

ET = exposure time (hours/day)

EF = exposure frequency (days/year)

ED = exposure duration (years/total days exposure)

BW = body weight (kg)

Shown below are standard values:

50th percentile total body surface area (cm²)

Age (years)	Male	Female
3–6	7,280	7,110
6–9	9,310	9,190
9–12	11,600	11,600
12–15	14,900	14,800
15–18	17,500	16,000
18–70	19,400	16,900

50th percentile body part specific areas for males (cm²)

Age (years)	Arms	Hands	Legs
3–4	960	400	1,800
6–7	1,100	410	2,400
9–10	1,300	570	3,100
18–72	2,300	820	5,500

We should also note, for example, that it has been found with chloroform that, in showers, dermal exposure is roughly equal to the inhalation exposure and each is about ⅓ of the ingestion exposure. After a shower we are breathing out a contaminant such as chloroform in the water at a concentration of

$$C_{bs} = 3.1 + 0.43 \times C_W$$

where

C_{bs} = breath concentration after shower (mg/m³), and

C_W = tap water (shower) concentration (mg/L)

Dermal Exposures from Soil

Dermal absorption of contaminants from soil or dust depends on the area of contact, the duration of contact, the bond between the contaminant and the soil, and the ability of the contaminant to penetrate the skin. Factors such as lipophilicity, polarity, volatility, molecular weight, and solubility influence the amount absorbed. An important factor is adherence, the quantity (mg/cm) of soil on the skin; ATSDR recommends a value of two mg/cm² (other values range from 0.5 to three mg/cm²).

Some Standard Values for Soil Dermal Exposure

Age (Years)	Body Weight (kg)	Surface Area (cm^2)	Area Exposed (%)	Area Exposed (cm^2)	Adhered Soil (mg)
0–01	10	3,500	30	1,050	2,100
01–11	30	8,750	30	2,625	5,250
12–17	50	15,235	28	4,300	8,600
18–70	70	19,400	24	4,700	9,400

The dermal intake from soils is determined as follows:

soil dermal intake = CC x A x BF x EF x 10^{-6})/BW

where soil dermal intake is in units of mg/kg/day

CC = contaminant concentration in soil (mg/kg)

A = total soil adhered (mg)

BF = bioavailability factor (unitless)

EF = exposure factor (unitless)

BW = body weight (kg)

The factor of 10^{-6} kg/mg is for conversion. The total soil adhering to the dermal surface is estimated as the product of the exposed dermal area and the soil adherence concentration. The exposure factor is the product of the exposure frequency (days/year) and exposure duration (years/total days exposure).

Example Calculation

Estimate the intake of an 11-year-old child exposed to a soil contaminant at a rate of 100 mg/kg every day from birth through 11 years of age. Assume that the average exposed skin surface area during this time is 30%.

soil dermal intake = exposure age 0–1 + exposure age 1–11

soil dermal intake = (C x A x BF x EF x 10^{-6})/BW

= [(100 x 2100 x 1 x (1/1 1) x 10^{-6} /10] + [(100 x 5250 x 1 x (10/1 1) x10^{-6}/30]

= 0.02 mg/kg/day intake through food ingestion

Food Absorption

Contaminants in air can be deposited on soil and plant surfaces and incorporated with plants either through root uptake or foliar absorption. Contaminants can also be incorporated directly into the soil by amendments such as sewage sludge, fertilizer, and pesticides. Contaminants in water used for irrigation are also an important source of food contamination. Finally, there are areas where natural levels of chemicals in the soil may lead to a toxicity or nutritional deficiency. Livestock as well as fish can incorporate contaminants in what they eat.

Absorption of contaminants from food can be determined as follows:

$$\text{food intake} = \Sigma(CL_i \times CR_i \times EF)/BW$$

where food intake is in units of mg/kg/day, and

CL_i = concentration of contaminants in the ith food group (mg/kg/day)

CR_i = consumption rate of the ith food group (g/day)

EF = exposure factor (unitless)

BW = body weight (kg)

Example of Food Absorption

Listed below are the results of our calculation of the food ingestion dose for cadmium entering the human pathway through a garden crop. (PH is the percentage of food that is homegrown.)

Food	CL	CR	PH	EF	BW	(mg/kg/day)
potatoes	0.02	88	9.30	1	70	0.0023
dark green vegetables	0.01	15	21.20	1	70	0.0005
deep yellow vegetables	0.51	15	21.20	1	70	0.0232
tomatoes	0.24	38	21.20	1	70	0.0276
other vegetables	0.01	136	21.20	1	70	0.0041
Total						0.0577

In looking at the transport of contaminants through food chains, it is important to be able to estimate the concentration of contaminant in a plant that is growing in a soil with a known or estimated concentration of the contaminant. One other issue is to estimate the concentration in a fish based on the concentration of the contaminant in the water.

Multimedia-Multipathway Exposures and Models

The fourteen steps in conducting a multimedia-multipathway study are described below (problem formulation, background review, forecast, detailed protocols, institutional review, survey conduct, data tabulation, quality assurance, data analysis, data interpretation, report preparation, peer review, and publication). The dividing lines between these steps are not sharp. Others might categorize the tasks differently, but cover approximately the same content (Lioy and Pellizzarri, 1995). Overall, these steps cover the essential elements of a good quality exposure study.

In principle, we find that multimedia-multipathway studies do not differ much from single pathway or single medium exposure assessments because every multimedia-multipathway study is a composite of several single pathway, single medium exposure assessments. Multimedia-multipathway studies do impose additional requirements on investigators, and as a consequence they require stronger management. Multimedia-multipathway studies have greater complexity. They generate more data and require more sophisticated data storage and statistical analysis.

It often is not clear how you will aggregate data between media or pathways. Study elements may interact; for example inhalation sampling may interfere with dermal protocols. The same interactions can also create opportunities for greater efficiency; for example, a multimedia-multipathway study can use the same study subject for multiple measurements. So, investigators have to manage more complex sampling and interviews, but these interviews produce more data for the same resource expenditures than several single medium-single pathways studies. Similarly, investigators can gain analytical efficiency with high through-put methods, which become cost-effective with the greater numbers of analyses in multimedia-multipathway studies.

Study objectives often call for the estimation of several chemically related substances, so analytical chemists can develop methods to measure the analytes simultaneously in one chromatography procedure. Different time frames for exposure to the same substance through different pathways will complicate your analysis. The time frame problem becomes especially prominent when investigators look at more than one substance because of different rates of transport and degradation, and if biomarkers of exposure are used, for different elimination half-lives.

The steps in planning and conducting a multimedia-multipathway study are summarized in more detail below. They are not autonomous steps to carry out sequentially. Instead, the steps relate to an overall study in an integrated way, sometimes leading to an iterate of this overall approach in a more detailed way, after a preliminary effort. The rationale for this approach relates in part to the logical requirements for high-quality study design, but it also relates to experience with later steps. For example, the interpretation of data for a risk assessment or preparation of a report for a special audience strongly influences the planning of earlier steps. As a practical matter, this approach yields studies that are more relevant and thorough. While the initial planning costs may be somewhat higher, overall this approach eventually yields a much more cost-effective final product.

1. *Formulate the problem (design intent).*

   At the outset, understanding the question(s) being asked is crucial. Different approaches are appropriate for hypothesis testing studies than for descriptive studies, which may produce data that will generate new hypotheses. An exposure assessment may call for a cluster study, supplementation of an ongoing survey, a look at population-based variance in exposure, or even a literature review, to obtain the needed data from existing information. Matching of the study design to the exposure assessment question(s) is critical. The exposure assessment also may contribute to an effort to link exposures to health effects, as part of an epidemiology study, which will impose additional design requirements.

2. *Review the background.*

   Adequate literature review is essential to a good exposure assessment study. The literature often helps to establish whether a new study is even necessary, whether the necessary data can be obtained as part of a larger planned or ongoing survey, and what design problems can be anticipated.

3. *Forecast the necessary study population and survey content (including questionnaire and analytes, if any).*

   The investigator needs to select a study population that is capable of answering the study's needs. Establishing appropriate criteria, such as the number of subjects, whether to include both sexes, and so forth, is part of the design process. Prior statistical review is essential, and simulation studies sometimes are possible, which can greatly enhance the efficiency of design and anticipation of problems. A Quality Control plan is part of a good exposure assessment study, and it should be developed in this step.

4. *Field test procedures and/or conduct a preliminary study.*

   Pre-testing of survey instruments, testing of sampling procedures, and validation of analytical protocols often help to avoid problems that would otherwise crop up too late during data analysis and interpretation. If all of these elements of an exposure assessment study can be assembled into a preliminary study, so much the better, because the preliminary study will evoke problems of integrating the study components. Given the complexity of multimedia-multipathway studies, this can be particularly instructive.

5. *Prepare detailed study protocols and plans.*

   A written proposal with appended protocols for component procedures is necessary to obtain reviews. The investigators will have technical need for explicit directions. Most experienced investigators use a checklist for completeness, like that in EPA's Survey Management Handbook.

6. *Obtain institutional reviews.*

   Depending on the setting, various organizations will have a right-of-review and/or approval before a study is initiated. For surveys, the Office of Management and Budget has obligatory oversight. Often they are helpful in bringing the study objectives in line with the conduct and data to be obtained. Other agencies may necessarily be involved, such as the Department of Agriculture or the Bureau of Wages and Standards in the Department of Labor. Institutional Review Board review of plans for human studies is obligatory. In addition, prior review of a draft plan by other groups likely to be involved, such as state or local public health departments, can enhance the conduct of a study. Prior review by the Quality Assurance team will decrease potential problems of data tabulation and validation. Similarly, a prior consultation about the draft plan with potential peer reviewers is a good idea.

7. *Conduct the survey.*

8. *Tabulate the data.*

   Multimedia-multipathway studies generate large amounts of data. Tabulation is best accomplished in terms of the primary unit of observation. In human studies, this is each individual study subject. The secondary unit of observation usually is a time and location for interview or sampling of the individual.

9. *Secure quality assurance.*

10. *Analyze the data.*

    The appropriate methods of analysis will depend on the kind of study. The statistical methods appropriate for hypothesis testing studies differ from the methods for descriptive studies.

11. *Interpret the data.*

    If the study aims to test one or more hypotheses, interpretation is fairly simple. The data either substantiate or disprove the hypotheses. The interpretation of descriptive studies is more difficult. One approach is to search the tabulated data for all possible associations, rank the plausible interpretations in terms of statistical power, and regard the most powerful associations as hypotheses to test in future studies. This approach regards descriptive studies as hypothesis formulating studies.

12. *Prepare reports.*

13. *Obtain peer review.*

14. *Publish and/or present the results.*

Most Important Design Elements in Multimedia-Multipathway Studies

A single design element is of overwhelming importance in designing a multimedia-multipathway exposure assessment study: Will the conduct of the study answer the questions posed?

The questions you ask can vary, and thus the design elements will vary. For example, if linkage to health effects is most important, then measurement of a biomarker of exposure will become most important. Biomarkers of exposure, such as blood or urine concentrations of an analyte, are most amenable of association with health effects observed in the same population. These measures automatically enforce integration across multiple media and multiple pathways onto a study. Such studies do not, however, provide any insight into the sources or pathways of exposure that led to the biomarker concentrations.

Obtaining a population distribution of exposure by multiple media and multiple pathways is particularly difficult. The relative contribution of one pathway, for example, dermal exposure, to total exposure may vary from the least exposed to the most exposed members of the population. Therefore, simply obtaining the mean and variance of exposure within the study population is difficult, and the mean and variance of exposure must meet the study objectives. If the aim of the study is to elicit some cutoff point of exposure in the population, such as the most exposed 10%, the population size requirements become the dominant design variable.

National Human Exposure Assessment Survey

EPA's recent National Human Exposure Assessment Survey (NHEXAS) studies are the best examples of important design elements in multimedia-multipathway exposure assessments. NHEXAS also has important implications for the field of exposure assessment.

NHEXAS is a freestanding survey, not directly linked to other federal population surveys, such as the Center for Disease Control and Prevention's National Health and Nutrition Evaluation Survey (NHANES). To represent the U.S. population accurately, the survey needs to choose study subjects randomly with respect to many characteristics, including age, sex, ethnicity, geographic location, income, height, and weight. This goal represents a very difficult study design problem.

Different consortia, mostly of university-based investigators, but including Westat, have carried out three preliminary studies. The current field studies involve three areas, each corresponding to one of the different consortia: Baltimore, Maryland, the state of Arizona, and EPA's Region 5 (Illinois, Indiana, Michigan, Minnesota, Ohio, and Wisconsin).

NHEXAS investigators presently are conducting preliminary studies to insure that their methods will work before carrying out a more extensive population study of the entire United States. Thus, EPA has not yet attempted a national-scale exposure survey. Instead, NHEXAS is in a methods research mode. The NHEXAS consortia are testing survey design strategies, working out analytical methods, developing a knowledge of baseline exposure levels, and advancing the state of the art.

NHEXAS has strong policy implications for EPA's implementation of the Food Quality Protection Act (FQPA). FQPA requires aggregate exposure assessments for pesticide tolerances. While the congressional intent may seem straightforward, aggregate exposure still is a fuzzy concept. Right now, the exposure assessment community cannot agree on how to add together exposures from different pathways. Clearly, adding together exposures with different units, adding together total masses from each pathway for a person, or summing the relative ranks of exposure for each pathway for an individual in a population are not sensible approaches.

Even integration through relative contribution to some biomarker of exposure, such as the concentration of an analyte in blood or urine, is arguable. For example, if the toxicity of a substance relates to concentration in the liver, the contribution of dietary exposure to liver concentration relative to blood concentration will differ from the contribution of inhalation exposure. After dietary exposure, the substance will pass through the liver first before entering the circulation, but after inhalation the substance will enter the circulation first, and the liver will extract only a portion of total blood flow. In contrast, if the toxicity of a second substance relates to lung concentration, inhalation exposure will contribute more toxicity, relative to blood concentration, than dietary exposure.

The data from NHEXAS form a large, multidimensional matrix of study subjects, survey data, substances, and analyte data. In some instances, subjects have been resampled, and time of sampling adds another dimension. In addition, in some instances households, instead of individuals, have been the primary focus of the studies. The survey data summarizes behaviors, characteristics, some health status parameters, and an accounting of their exposures to designated substances during a short interval of time, usually twenty-four hours. In the preliminary studies, the consortia and EPA designated the substances for relevance, representativeness, and practicality.

EPA contracted with the Cadmus Group for literature reviews of health effects of candidate substances and used the reviews to prioritize substances for the survey based on public health concern. The Agency filtered this initial list both for the ability of investigators to analyze the substances chemically and for the pervasiveness of exposures. Even so, NHEXAS investigators generally represent that they chose the analytes for public health reasons (Sexton et al., 1995b). Some of the NHEXAS data represent only a few peaks from chromatographic runs. In these instances, a substantial number of peaks remain unevaluated. Anyone reviewing the NHEXAS data about organophosphates may want to look at the primary charts from the runs, which represent currently unevaluated exposures. Thus, substantial additional data are available for data analysis.

Other federal organizations participate with EPA in NHEXAS; most notable are various agencies with the Department of Health and Human Services (DHHS), including the Agency for Toxic Substances and Disease Registry, CDC, FDA, National Center for Health Statistics, National Cancer Institute, National Institute of Environmental Health Sciences, and National Institute of Occupational Safety and Health. Non-DHHS agencies participating include the National Institute of Standards and Technology and the Department of Energy.

EPA undertook the preliminary NHEXAS studies to test their survey design and analytical approach in the field. According to the official view, work started in 1991, but EPA only had a serious plan to take to the Office of Management and Budget (OMB) in 1994. OMB initially vetoed the plan because EPA could not have met any of the stated objectives with the presented survey design. Subsequently, OMB worked with EPA to change the survey's objectives to accord more with the plan. Only in 1995, after OMB intervention, did EPA change the study goals to accord with what the Agency wanted to do, and obtain funding.

Three government committees oversee the effort: a coordinating committee across the federal agencies, a coordinating committee within EPA that represents various programs (air, water, research) but from conceptual, not programmatic perspectives, and an executive committee with representatives from the few EPA offices directly involved in the study.

The Baltimore project is providing time series data, in part to probe the relationship between short-term and long-term time frames. The Arizona project is testing field study methods and strategies. In parts of their work, the Arizona consortium assessed the exposures of all members of the same household where a study subject lived. At present they have not analyzed the data for these non-subject household members. Thus, more extensive survey data are available for data analysis.

The Region 5 project also is testing field study methods and strategies (Pellazzari et al., 1995). EPA's Region 5 covers Illinois, Indiana, Michigan, Minnesota, Ohio, and Wisconsin.

Many of the exposure distributions obtained by NHEXAS consortia are not log-normally distributed. One explanation for these highly skewed populations is that the exposure distributions covered several distinct subpopulations, each with a log-normal profile. To understand these findings, some NHEXAS studies have or will have components focused on special populations, such as children's exposure.

The primary problem NHEXAS has at present is that the national survey needs to use established methods to measure exposures by all relevant routes (inhalation, ingestion, and dermal) from all media (e.g., household dust, drinking water). Such methods were generally not available at the time EPA began to plan NHEXAS; thus, the effort would have exceeded the state of the art. The consortia are establishing these methods as they proceed with the preliminary studies. The NHEXAS studies really are basic research into exposure assessment, more than scoping studies for a national survey.

Based on the overall results from the three consortia, EPA eventually intends to measure exposures of the U.S. population to hazardous substances in a systematic way. Ultimately, EPA will conduct special studies to test hypotheses suggested from the national survey, after it is completed, such as oversampling of populations too sparse for statistical accuracy in a random trial with limited numbers of persons. An extensive survey would be conducted over a few years, given potential survey designs, lists of substances, analytical methods, and data processing plans.

EPA intends to release a summary of the current NHEXAS data, which they describe as a database, by the end of 1999. The Agency has thought about placing the database on the Internet. How this will impact the regulation of pesticides remains to be seen. Some kind of meta-analysis might prove possible that combined the data from all three consortia. At present, no method is available to format the data in a unitary way.

In essence, EPA's overall plan for preliminary field testing has gone quite well. Their time line and budget came up short only in two areas: insufficient resources to analyze the data from the initial studies and insufficient time to interact with OMB to obtain approval. The Agency's plan has worked far better than comparable survey efforts. The enthusiasm generated from the overall success, and the scientific advances, suggest that NERL will tap into strong support for a national survey within the regulatory programs in EPA and other federal agencies.

The data are undergoing scientific peer review, and NHEXAS will have the advantage of IHEC input well before initiating a national survey. IHEC is a long-standing group within SAB. It started by reviewing indoor air quality research at the request of the Administrator.

Potential Problems in Multimedia-Multipathway Studies

Exposure assessment is a part of risk assessment. Risk assessors need multimedia-multipathway exposure assessments when they need to understand the contributions of different sources of exposure to the probabilities of health effects. Risk assessors need multimedia-multipathway exposure assessments when they need to relate biomarkers of exposure to source contributions. Risk assessors need multimedia-multipathway exposure assessments when they develop models of risk, and exposure assessors need multimedia-multipathway studies when they develop or validate exposure models.

Because the data available for multimedia-multipathway models are scarce, every multimedia-multipathway study can contribute to modeling progress by considering programming and validation needs in designing field studies. Some of the more prominent problems encountered in the current studies are as follows:

1. *Does the study use validated methods?*
   Some studies either cannot or do not use established questionnaires or analytical methods to measure exposures by all relevant routes (inhalation, ingestion, and dermal) from all relevant media (e.g., household dust, drinking water). If existing methods are available for the substances of interest, their use will reduce the pressure on the design process. It will not have to meet research as well as information needs. In some surveys, it is not clear that a need existed to extend the state of the art. If not, resources went to waste. If so, hopefully the basic research will advance the state of the art in exposure assessment.

2. *Does the statistical analysis appropriately correct for the extensive numbers of observations made on each study subject?*
   If not, many "statistically significant" associations attributed to the analytes and questionnaire responses will be spurious.

3. *If the study tests a specific hypothesis, is there a clear statement of the hypothesis prior to study design and are statistical criteria of acceptance and rejection a part of the study design?*
   If not, the exposure assessment community may regard the interpretation as data dredging. If the study is descriptive in nature, the plan should include appropriate resources for data analysis and include the analytical approach as one of the design criteria.

4. *Does the study have adequate numbers of subjects to meet its objectives?*
   Does each question in the survey instrument and analyte measurement in the sampling contribute to an objective? If not, the study will waste resources and may fail to yield an interpretable result.

5. *If the study substantially involves a community's resources, does the study plan include a strategy to obtain cooperation and input from the community? Does the study include elements that will inform the community, obtain participation, and inform appropriate regulatory authorities of the results?*
   If not, the study may meet with resistance, and even have difficulty obtaining adequate response rates and subject interaction.

6. *Does the study plan adequately budget for the analysis of the data?*
   Multimedia-multipathway studies particularly generate large amounts of data, and problems of scale exist in tabulating and analyzing this data. Leaving lots of unanalyzed data will call into question any results obtained from the analyzed portion.

7. *Does the study plan include adequate amounts of replicate data to estimate the variance in measurement?*
   For example, the season of observation may change general exposure values. Similarly, time of observation will affect biomarker levels. For example, measurements of some urinary metabolites vary between NHANES and NHEXAS in apparently similar populations. Because NHANES uses first morning void urine, whereas NHEXAS uses twenty-four hour urine collection, the time of collection may explain some of the differences. High variations in metabolite levels have been found both within individuals and between individuals in some current multimedia-multipathway studies. A study plan that does not allow for estimation of this variation will generate results that will have difficulty obtaining acceptance.

8. *Will the study generate data to estimate the effects of sampling time frames?*
   Acute exposure measurements in a population do not necessarily reflect the population's chronic exposure (Buck et al., 1995). The inaccuracy of acute measurements will vary from substance to substance, depending on the stability of sources, and for biomarkers of exposure, depending on pharmacokinetics of a substance.

9. *If the study aims to measure central tendencies and associate variation in a population's exposure to a substance, will the data also be used to estimate some cutoff in the distribution?*
   For example, many regulatory discussions focus on the extremes of distributions, such as the most exposed 5% of the population. A study with an adequate sampling frame for central tendency and variation may fail to fill the regulatory need. Many of the exposure distributions obtained by NHEXAS consortia are highly skewed but not log-normally distributed.

10. *Is the study's objective a highly exposed or special subpopulation?*
    If so, resources directed to the general population may provide no information. One explanation for some of the highly skewed profiles in NHEXAS is that the population's exposure distributions covered several distinct subpopulations, each with its own log-normal profile of exposure. After completion of initial survey, it may prove possible to focus the design criteria such that only subjects at risk for high exposures are studied. If a special population is the object of study, such as children, immunosuppressed persons (organ transplants, AIDS), elderly, pregnant women, persons with special personal habits, occupational group, or resi-

dence location (urban or rural), the plan should specify the prevalence of (and pathways to obtain a sufficient number of) study subjects with the required characteristics. If an oversampling strategy is employed, the plan should specify how the data interpretation will relate the study population to the total population in the geographical area of the sampling frame. However, a multimedia, multipathway model is more likely to produce estimates that will correlate with biomarker studies. Both these models and the biomarkers inherently integrate exposures from different sources and routes.

Sources of Additional Information

We adapted this section from the *Guidelines for Exposure Assessment* developed by the U.S. EPA. These guidelines can be obtained on the Internet at www.epa.gov/ncea.

The two papers about exposure assessment that we recommend most are "A Decade of Studies of Human Exposure: What Have We Learned?" (Wallace, 1993) and "Quantitative Definition of Exposure and Related Concepts" (Zartarian, et al., 1997).

References

Atherly, G. 1985. A critical review of time-weighted average as an index of exposure and dose, and of its key elements. *Am. Ind. Hyg. Assoc. J.* 46:481–487.

Buck, R. J., K. A. Hammerstrom, and P. B. Ryan. 1995. Estimating long-term exposure from short-term measurements. *J. Exposure Analysis and Environmental Epidemiology* 5: 297–325.

Cothern, C. R. and J. E. Smith, Jr. 1987. *Environmental Radon*. New York: Plenum Press.

Covello, V. T. and M. W. Merkhofer. 1993. *Risk Assessment Methods: Approaches for Assessing Health and Environmental Risks.* New York: Plenum Press.

Hemond, H. F. and E. J. Fechner. 1994. *Chemical Fate and Transport in the Environment.* New York: Academic Press.

Lebowitz, M. D., M. K. O'Rourke, S. Gordon, D. J. Moschandreas, T. Buckley, and M. Nishioka. 1995. Population-based exposure measurements in Arizona: A Phase I field study in support of the National Human Exposure Assessment Survey. *J. Exposure Analysis and Environmental Epidemiology* 5: 297–325.

Lioy, P. J. and E. Pellazzari. 1995. Conceptual framework for designing a national survey of human exposure. *J. Exposure Analysis and Environmental Epidemiology* 5: 425–444.

Lynch, J. R. 1985. Measurement of worker exposure. In: Cralley, L. J.; Cralley, L. V. , eds. *Patty's Industrial Hygiene and Toxicology. Volume 3a: The Work Envionment.* 2nd ed. New York: Wiley-Interscience.

National Research Council. Committee on the Institutional Means for Assessment of Risks to Public Health. Commission on Life Sciences. NRC.1983. *Risk Assessment in the Federal Government: Managing the Process.* Washington: National Academy Press.

MacIntosh, D. L., J. Xue, H. Ozkaynak, J. D. Spengler, and P. B. Ryan. 1995. A population-based exposure model for benzene. *J. of Exposure Analysis and Environmental Epidemiology* 5:375–403.

Nelson, J. D. and R. C. Ward. 1981. Statistical considerations and sampling techniques for ground-water quality monitoring. *Ground Water* 19:617–625.

Paustenbach, D. J. 1985. Occupational exposure limits, pharmacokinetics, and usual work schedules. In *Patty's Industrial Hygiene and Toxicology. Vol. 3a: The Work Environment.* 2nd ed., eds. L. J. Cralley and L. V. Cralley. New York: John Wiley & Sons.

Pellazzari, E., P. Lioy, J. Quackenboss, R. Whitmore, A. Clayton, N. Freeman, J. Waldman, K. Thomas, C. Rodes, and T. Wilcosky. 1995. Population-based exposure measurements in EPA Region 5: A Phase I field study in support of the National Human Exposure Assessment. *J. Exposure Analysis and Environmental Epidemiology* 5: 327–358.

Sanders, T. G. and D. D. Adrian. 1978. Sampling frequency for river quality monitoring. *Water Resources Research* 14:569–576.

Schweitzer, G. E. and S. C. Black. 1985. Monitoring statistics. *Environ. Sci. Technol.* 19:1026–1030.

Schweitzer, G. E. and J. A. Santolucito. 1984. Environmental sampling for hazardous wastes. ACS Symposium Series Number 267. Washington: American Chemical Society.

Sexton, K., D. E. Kleffman, and M. A. Callahan. 1995a. An introduction to the National Human Exposure Assessment Survey (NHEXAS) and related Phase I field studies. *J. Exposure Analysis and Environmental Epidemiology* 5:229–232.

Sexton, K., M. A. Callahan, E. F. Bryan, C. G. Saint, and W. P. Wood. 1995b. Informed decisions about protection and promoting public health: Rationale for a national human exposure assessment survey. *J. Exposure Analysis and Environmental Epidemiology* 5:233–256.

Sexton, K., K. Callahan, and E. F. Bryan. 1995. Estimating exposure and dose to characterize health risks: The role of human tissue monitoring in exposure assessment. *Environmental Health and Perspectives,* Suppl 3:13–30.

Shaw, R. W., M. V. Smith, and R. J. J. Pour. 1984. The effect of sample frequency on aerosol mean-values. *J. Air Pollut. Control Assoc.* 34:839–841.

Smith, C. N., R. S. Parrish, and R. F. Carsel. 1987. Estimating sample requirements for field evaluations of pesticide leaching. *Environ. Toxicol. Chem.* 6:345–357.

U.S. EPA. Office of Monitoring Systems and Quality Assurance. Office of Research and Development. 1980. *Interim Guidelines and Specifications for Preparing Quality Assurance Project Plans.* QAMS-005/80. Washington, D.C.

U.S. EPA. Office of Policy. Planning and Evaluation. 1984. *Survey Management Handbook. Volumes I and II.* EPA-230/12-84/002. Washington, D.C.

U.S. EPA. Office of Toxic Substances. 1985a. *Methods for Assessing Exposure to Chemical Substances; Volume 4: Methods for Enumerating and Characterizing Populations Exposed to Chemical Substances.* EPA-560/5-85/004, NTIS PB86-107042. Washington, D.C.

U.S. EPA. Robert S. Kerr Environmental Research Lab. Office of Research and Development. 1985b. *Practical Guide for Ground-Water Sampling.* EPA-600/2-85/104, NTIS PB86-137304. Ada, OK.

U.S. EPA. Office of Toxic Substances. 1985c. *Methods for Assessing Exposure to Chemical Substances. Volume 2: Methods for Assessing Exposure to Chemicals in the Ambient Environment.* EPA-560/ 5-85/002, NTIS PB86-107067. Washington, D.C.

U.S. EPA. Office of Toxic Substances. 1985d. *Methods for Assessing Exposure to Chemical Substances. Volume 5: Methods for Assessing Exposure to Chemical Substances in Drinking Water.* EPA-560/5-85/006, NTIS PB86-1232156. Washington, D.C.

U.S. EPA. Office of Toxic Substances. 1986a. *Methods for Assessing Exposure to Chemical Substances. Volume 8: Methods for Assessing Environmental Pathways of Food Contamination.* EPA-560/5-85/008. Washington, D.C.

U.S. EPA. Office of Toxic Substances. 1986b. *Analysis for Polychlorinated Dibenzeo-p-dioxins (PCDD) and Dibenzofurans (PCDF) in Human Adipose Tissue: Method Evaluation Study.* EPA-560/5- 86/020. Washington, D.C.

U.S. EPA. Office of Health and Environmental Assessment. Office of Research and Development. 1986c. *Addendum to the Health Assessment Document for Tetrachloroethylene (Perchloroethylene): Updated Carcinogenicity Assessment for Tetrachloroethylene (Perchloroethylene, PERC, PCE).* Review Draft. EPA-600/8-82/005FA, NTIS PB86-174489/AS . Washington, D.C.

U.S. EPA. Office of Acid Deposition, Environmental Monitoring and Quality Assurance. Office of Research and Development. 1987a. *The Total Exposure Assessment Methodology (TEAM) Study. Volume I: Summary and Analysis.* EPA-600/6-87/002a. Washington, D.C.

U.S. EPA. Office of Pesticide Programs. Office of Pesticides and Toxic Substances. 1987b. *Pesticide Assessment Guidelines for Applicator Exposure Monitoring - Subdivision U.* EPA-540/9-87/127. Washington, D.C.

U.S. EPA. Office of Health and Environmental Assessment. Office of Research and Development. 1987c. *Addendum to the Health Assessment Document for Trichloroethylene: Updated Carcinogenicity Assessment for Trichloroethylene.* Review Draft. EPA-600/8-82/006FA, NTIS PB87-228045/AS. Washington, D.C.

U.S. EPA. Office of Research and Development. 1989. *A Rationale for the Assessment of Errors in the Sampling of Soils.* EPA-600/4-90/013. Washington, D.C.

U.S. EPA. Environmental Monitoring Systems Laboratory. Office of Research and Development. 1990. *Soil Sampling Quality Assurance User's Guide.* EPA-600/8-89/046. Washington, D.C.

U.S. EPA. 1990b. Guidance for useability in risk assessment. Interim Final. Office of Emergency and Remedial Response. EPA-540/G-90/008. Washington, D.C.

U.S. EPA. Office of Health and Environmental Assessment (RD-689). 1992. *Guidelines For Exposure Assessment [FRL-4129-5]. Final Guidelines for Exposure Assessment.* Washington, D.C.

U.S. EPA. Risk Assessment Forum. 1997. *Guiding Principles for Monte Carlo Analysis.* EPA/630/R-97/001. Washington, D.C.

U.S. EPA. National Center for Environmental Assessment. 1998. *Exposure Factors Handbook: Volume I - General Factors, Volume II - Food Ingestion Factors, and Volume III - Activity Factors.* Update to *Exposure Factors Handbook.* EPA/600/8-89/043 (May 1989). Washington, D.C.

Wallace, L. A. 1993. A decade of studies of human exposure: What have we learned? *Risk. Anal.* 13:125–143.

World Health Organization (WHO). 1986. Principles for evaluating health risks from chemicals during infancy and early childhood: The need for a special approach. Environmental Health Criteria 59:26–33. Geneva: World Health Organization.

Zartarian, V. G., W. R. Ott, and N. A. Duan. 1997. Quantitative definition of exposure and related concepts. *J. of Exposure Anallysis and Environmental Epidemiology* 7:411–437.

Chapter Five ....

Risk Assessment: Dosimetry

Dose (and some terminology)

As we learned in the chapter on exposure, dose begins where exposure leaves off. Once an organism absorbs a substance (and the substance enters a membrane), dosimetry begins. Often, for risk assessment purposes, if you get the exposure right, analytically the dose just happens. As part of our discussion of dosimetry, we will learn about *pharmacokinetics* and physiological pharmacokinetics, a part of biologically based modeling. Pharmacokinetics is the timing of dosimetry—how fast it happens.

As we have seen in the introductory chapter, a conceptual model of a process is a way to gain understanding. You can always convert a conceptual model into a mathematical model. In this book (and, we think, in reality), exposure is the amount of a substance outside of an organism available for a dose, and dose is the amount of the substance that enters the organism. However, once absorbed, dose usually is best described as concentration, e.g. mg/kg (body weight). Many of the conceptual and mathematical problems you will encounter in risk assessment relate to the transitions between amount (e.g., mg) and concentration, and back again.

The best measure of the effect of dose relates most closely to the biological effect of interest (premature death, weight loss, tumor induction, increased blood pressure, nausea, or increased respiration rate). Thus, we are primarily concerned with the bioactive substance, whether the absorbed chemical or its metabolite, when it gets to its target. Note that the biological effects of some substances change their own dosimetry (e.g., nausea after ingestion, or increased respiration rate after inhalation). For this reason, no single concept of dose works for all risks. Understanding the *dose metric* becomes an important part of biological risk assessment. If we understand enough about the biological effect to know, for example, that the peak concentration of a metabolite over time most closely relates to toxicity, we have solved a difficult part of our overall risk assessment problem.

Note the confusing terminology: exposure is not dose. If you think that an exposure is a dose, you have no reason to try to understand dose. The relationship between dose and exposure is not constant. Dose does not increase or decrease exactly as exposure increases or decreases, because the fraction absorbed may change. The relationship between dose and exposure is not always constant between different routes of exposure, different species, or two different substances that produce the same biological effect. The relationship between dose and exposure may not be constant between two different biological effects produced by the same substance. Also, the relationship between dose and exposure may not be constant over time, particularly when data about acute exposures are used to estimate the risks of chronic exposures.

Below we provide definitions for some of the confusing terminology used in the risk assessment literature, using the overall concepts we employ in this book:

Applied dose = an exposure

Absorbed dose = a dose

Administered dose = an exposure

Bioactive dose = a dose (the dose of the portion of a substance or its metabolites that produces a biological effect of interest)

Bioavailable dose = an exposure

Delivered dose = a dose (the dose of a substance or its metabolites available for interaction of interest with the cell type or organ that produces a biological effect of interest)

Dermal dose = usually an exposure, unless corrected for bioavailability

Equivalent dose = a dose (the dose producing the same extent of a biological effect in an organism, or prevalence in a group of organisms or another species, as a dose of some related substance, or a dose of the same substance after a different duration)

Inhaled dose = usually an exposure, unless corrected for bioavailability

Injected dose = a dose

Internal dose = a dose

Intramuscular dose = a dose

Intraperitoneal dose = a dose

Intravenous dose = a dose

Metabolized dose = a dose (the portion of the dose metabolically transformed, not necessarily to a biologically active substance)

Oral dose = usually an exposure, unless corrected for bioavailability

Pharmacological dose = an exposure, unless by the intravenous or intraperitoneal routes

Subcutaneous dose = a dose

Surrogate dose is the dose in the general circulation or in the interstitial water of an organ that best correlates with the dose of biologically active substance in the vicinity of receptor macromolecules (Reitz et al., 1990).

In general, risk assessors use the term *dosimetry* to mean more descriptive information, for example, descriptions of metabolic pathways, dose units, the dose metric, experimental measurements, or the amounts of a substance entering and leaving the body after an exposure. Sometimes dosimetry is used in a way that implies that time is not involved. More commonly, dosimetry incorporates the parameter of time through kinetics. For example, dosimetry includes not only that portion of an exposure an animal absorbs, but how rapidly the animal absorbs this dose; not just the metabolic transformations of a substance, but how fast they occur. Overall, dosimetry is a description of the effect of an organism on a substance.

Extrapolation and *interpolation* are similar processes. Extrapolation implies that you have modeled a process beyond the range of your observations. Interpolation implies that you have modeled a process within your range of observations, but you have used a modeled relationship, not an experimental measurement.

If the tissue sensitivity is the same between two species, then you can extrapolate between species by attributing the same biological effect to the same dose delivered to the same organ in both species. If a particular metabolite produces a biological effect at one dose, you can interpolate to lower exposures by understanding the relationship between exposure and the production of this particular metabolite.

If we know the relationship between, for example, the mouse and human dose, how can we relate the endpoint or health effect for the mouse with that for the human? Consider the parallelogram in Figure 5.1. Is the relationship a linear one? Can we take direct ratios? Often this relationship is not linear. That is why we must consider kinetic relationships and will in this chapter develop the ideas underlying pharmacokinetics and physiological pharmacokinetics. For more details, see Rhomberg, 1995 and Lindstedt, 1987.

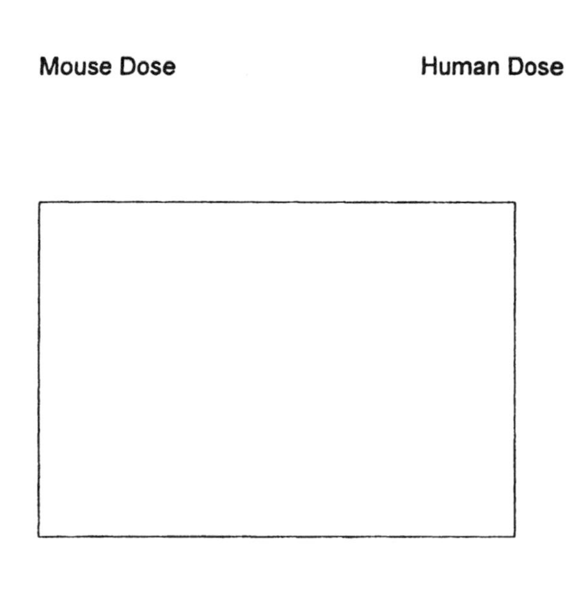

Figure 5.1 Parallelogram relating mouse and human dose, mouse and human effects

We can get some idea of the problem risk assessors confront by considering how to scale the relationships of the mouse to the human. Some consider the human as just a big mouse. This, in general, is not true. For example, mice live approximately two years, whereas humans live approximately 70 years. If a chronic dose in a mouse somehow produced the same magnitude of effect in a human over a human lifetime, we might take a direct ratio of the body weights of the mouse and the human to extrapolate dose. The U.S. Food and Drug Administration (FDA) looks at interspecies conversion of doses in this way. For example, FDA would view a 20 μg dose given to a 20 gram mouse as: 20 μg/20 gm, or 1 mg/kg. If this is a daily dose, FDA would express it as 1 mg per kilogram-day or 1 mg/(kg-day). This body-weight specific dose then scales directly to body weight in any other species. In a 70 kg human, the equivalent dose would be 70 mg [Human dose = mouse dose (human weight/mouse weight) = 20 (3500) = 70,000].

Alternatively, we could scale between the mouse and human by using the two thirds power of body weight. The two thirds power of body weight is not really the surface area of the body; sometimes it is an indicator of metabolic rate. If we were to wrap the body of an animal in Saran-wrap and then unfold it, the Saran-wrap would not be a measure of surface area because we would miss many folds and crevices, particularly on a microscopic scale. However, many persons refer to the two thirds power of body weight as body surface area, and they believe that metabolic rate is proportional to body surface area. Using this conceptual model, the body weights of the mouse and human would scale by the two thirds power of the body weight; (body weight)$^{2/3}$. Using this relationship, the human body weight equivalent dose to 1 mg/kg.d in a mouse would be the same as 0.067 mg/kg.d in a human [Human dose = mouse dose (human weight/mouse weight)$^{2/3}$ = 20 (231) = 4,620 μg]. The two thirds power of body weight works better in scaling between different developmental stages of the same species than it performs in scaling between different species. Some federal agencies have used body weight to the two thirds power to extrapolate doses of carcinogens between species.

Finally, we could scale the weights using the concept of physiological time. This approach involves noting that chronological time and biological time differ; as animals get larger, lifespan increases. Using this approach, the body weights scale as the three quarters power or (body weight)$^{3/4}$. Using this relationship, the body weight equivalent dose to 1 mg/kg.d in a mouse would be approximately 0.13 mg/kg.d in a human [Human dose = mouse dose (human weight/mouse weight)$^{3/4}$ = 20 (455) = 9,100 μg]. The three quarters power of body weight works better in scaling between different species than it performs in scaling between different developmental stages of the same species.

It is important for you to understand that these relationships are indicators of average dose scaling between species. For any one substance, the dose in two different species is not likely to scale exactly as any of these three power relationships suggests. If you know the doses in both species that produce effects, use those doses. If you have to predict the probability of a chronic effect at some dose, relying on any of these three relationships will be wrong. For this reason, pediatricians refuse to estimate doses for children based on adult doses. However, many toxicologists believe that physiological time is the best basis to extrapolate chronic doses between species, on the average; it is least likely to be wrong.

Extrapolating dose is an attempt to make more accurate predictions about biological effects. Because risk assessors often have to estimate risk for one species based on data in a different species, extrapolating dose is an essential part of risk assessment. In some situations, however, the species at risk closely resembles (or is even is the same species as) the species whose exposures and potencies have been studied. If so, the risk assessor does not need to engage in extrapolation, and dose estimation may be unnecessary.

Knowledge about the components of dose has evolved from an accumulation of descriptive information. Scientists usually divide dosimetry (or pharmacokinetics) into four processes: absorption, distribution, metabolism, and elimination. You may see these represented in shorthand as ADME. For example, "ADME studies" implies that some information is available about the absorption, distribution, metabolism, and elimination of a substance. We often say that a substance is poorly absorbed, widely distributed in the body, extensively metabolized, or eliminated slowly. Later, scientists gather more precise, quantitative information about the substance. For example, we will say that 3% of a substance is absorbed after ingestion, a substance has a volume of distribution equal to total body water, only 2% is eliminated unmetabolized, or elimination takes four days. Eventually we study the timing of these quantitative events in the form of kinetics.

Dose can be measured in many different ways, depending on the quantity being measured. For chemicals, the units are mass—moles, mg, mg/(body weight), or mg/kg.day. For ionizing radiation, we use energy, or more specifically the energy per weight of the cells undergoing irradiation. Units for ionizing radiation can be in counts per time, using some device that measures energy in counts. For infectious agents, we are most interested in whether absorption occurs or not, because the agent will multiply in the infected organism. Risk assessors call this an *all or none* measure. Often, the dose of an infectious agent will be stated as *plaque forming units* (PFUs) or *colony forming units* (CFUs). These terms arise from cell culture measurements of the agent; for example, a virus may destroy some of the cells in a layer, leading to a plaque, or bacteria may grow into a colony. However, the primary measure of interest is infectivity, and sometimes *infectious units* are known. For example, if approximately 70% of mice contract a disease when given the same amount of virus preparation, the preparation had approximately one infectious unit per dose. (This relationship results from the Poisson relationship. If each mouse received an average of one infectious unit, some mice received no units, some one, some two, and so forth.)

Dose rate also is an important quantity. To get the same dose over a minute or over a lifetime can result in different outcomes. Swallowing a large bottle of aspirin at one time can result in aspirin poisoning, but one aspirin a day for one year will not. Often the outcome is more serious if the same dose is absorbed in a shorter time. The duration of doses in toxicology involves some conventional terminology: acute, subchronic, and chronic. Usually, an acute dose is a single or brief exposure, one injection or inhalation for a few hours for a human. A chronic dose implies exposure over a lifetime, or at least for several years. The duration of subchronic doses falls between chronic and acute durations.

Since the experiments that scientists can perform with human subjects are limited, regulatory agencies often have to predict human doses and adverse human outcomes with suboptimal data, often data obtained from animal or cell-culture experiments. Models support the use of these measurements by fitting the limited data into a formal mathematical structure. Throughout modern medicine, toxicology, and risk assessment, the common belief is that dose facilitates more accurate modeling than exposure; that is, dose is a more reliable indicator of outcome between species. For example, risk assessors can estimate human metabolic rates from experiments with cultured mouse liver cells more accurately if the assessor bases calculations on doses because that is the actual amount that gets to the cells. Similarly, most pharmacologists believe that blood levels (concentrations) of a substance more accurately predict outcomes than any other measure of dose.

Risk assessors increasingly use models to interpret pharmacokinetic data and to understand dosimetry (Andersen et al., 1995). Even simple one-compartment models, which represent the body as a bag of fluid, illuminate whole-body distribution and elimination rates, and ultimately help to predict effects (or their absence). Knowledge of pharmacokinetics is helpful in situations where, for example, a substance does not bioaccumulate, metabolite has a structure predicted to form adducts with DNA, or when a substance mimics a hormone.

Even simple models can help in interpreting the meaning of data from cultured cell or biochemical experiments for intact animals. *Physiological pharmacokinetic (PBPK) models* attempt to mathematically describe body processes and structures in as realistic and detailed a manner as necessary to fit data from whole-animal experiments. PBPK models incorporate the anatomical and physiological parameters necessary to integrate data and predict effects (Lutz and Dedrick, 1985). The parameters are obtained independently of the data about a chemical substance, often from published, widely accepted sources (O'Flaherty, 1989). While PBPK models are complex and resource intensive to construct, they have the advantage of directly incorporating metabolic rate constants and partition coefficients for the movement of substances between body compartments. If the predicted tissue dose from a biologically based model correlates with an adverse outcome, the predicted risk will be more accurate and scientifically justified. Thus, PBPK models become most important when we combine them with experimental tests of predicted dose. The experimental measurements are the same as the biomarkers of exposure we described in Chapter 4.

Exposure and dose are intimately connected to each other. For example, if we want to understand the impact of simultaneous ingestion and dermal exposures, we have no intuitively obvious way to integrate the experimental measurements without referring to dose. If we choose, for example, to integrate exposures with respect to blood levels of a particular metabolite, then we have a meaningful metric. The substance may absorb well in the gastrointestinal tract but not from skin. We can compare the ingestion exposure to the dermal exposure in terms of their contribution to blood level. You should note, however, that if you change the dose of reference, for example from blood levels to liver concentration of a metabolite of the substance, you would not integrate the different exposures in the same way, because the dermal exposure and the ingestion exposure to the substance will contribute differently to blood level and liver concentrations of the metabolite.

Experiments that test biological models by measuring biological markers of exposure when the exposure is known are particularly powerful. The feedback between a model used to design an experiment that is most sensitive to model parameters and the experimental data used to modify the model's parameters leads to more rapid construction of accurate predictive models.

Dosimetric Measurements

Different biological insults are measured in different ways—for example, infectious diseases as PFUs, ionizing radiation as counts per minute, or chemical substances as milligrams. To convert a conceptual model into a mathematical model, you must have quantitative measures of the dose. Measuring doses, such as substances and their metabolites in biological materials (feces, blood, urine, etc.), can be ugly, and the substances usually are present at very low levels, making chemical measurement and analysis difficult. Yet a scenario involving risk does not begin to make sense until you try to attach numerical values to your ideas. You try to manipulate these values, looking for coherence and consistency in your estimates.

When we consider infectious particle measurements, we do not have conservation of infectivity (infectivity can be created or destroyed). Similarly, absorption of radiation often converts the energy into heat. However, for chemical substances, conservation of matter (the chemical elements in the particle) allows some cross-checking of calculations. Radioisotopic labeling of a chemical substance and ADME measurements of radioactivity after exposure are very revealing of the substance's dosimetry. Radioisotopic labels allow greater sensitivity in measurements because of the greater sensitivity of radiation measurements. For example, using ^{32}P-labeled DNA or RNA in the genome of a virus, we can streak biological fluid on a photographic plate and look for "stars"—points of multiple emissions—to count the number of virus particles. Each star on the plate represents a virus particle, but not necessarily an infectious one.

The presentation of risk estimates as population distributions creates particularly important data needs in dosimetric experiments. We have established the distributions of some factors in the human population, such as body weight, food consumption, and water consumption. When animal data provide the primary basis for estimates, risk assessors often find a distressing lack of similar individual animal estimates. For example, we can report an average body weight or an average water consumption, but not individual body weight or water consumption, much less organ weights of individual animals. While average data permit the calculation, for example, of an elimination rate, risk assessors cannot readily calculate an accurate mass balance without fluid and tissue volumes. Sometimes typical data about laboratory rodents of more popular strains are known (Roth et al., 1993). For example, if water consumption increases in one animal but remains constant in another, and conversion to a metabolite appears unchanged, tissue concentrations and elimination rates must change in the animal that drank more.

Radioisotopic labels allow greater sensitivity in measurements because of the ease of measuring activity in counts per minute.

Dosimetric Routes

We have used the point of intake at the outer membrane of the organisms as the dividing line between exposure and dose. In Chapter 4 we did some calculations at these points of intake, considering different routes or pathways. We need to realize that these routes are also the places where we estimate doses. The pathways of relevance to risk assessment are dermal, oral, and inhalation; in addition, we need to cover intravenous (iv.) adminstration, because this route of dosing yields important information, particularly in comparison to the other three routes.

When the skin of an organism contacts gases, clothing, liquids, solids, or any particles in its environment, it may absorb chemical substances. Unabsorbed gases and liquids at sufficiently high concentrations, clothing, or abrasion from particles may alter the skin and change absorption. The route of exposure is not synonymous with the physical form of a substance. For example, persons in contact with benzene gas in the environment absorb much of their benzene dose through the skin, not inhalation. Experimentally, scientists can separate these routes. For example, they can isolate inhalational exposure through nose-only administration and isolate skin absorption from the air by supplying pure air through a nose-only device to animals otherwise immersed in air containing the gas of interest. Similarly, most people come in contact with rhinoviruses (the cause of the common cold) through direct skin to skin contact with other persons, or skin contact with contaminated air, which deposits viruses on the skin. However, skin is not the route of absorption. Instead, virus particles are absorbed through mucous membranes after hand to mouth and hand to eye movements. Thus, people who wash their hands frequently during cold seasons will experience far fewer colds.

In general, abrasion increases dermal absorption. Substances that directly affect skin may change absorption. For example, if a substance kills some skin cells, the skin will become more permeable and the skin cells at the site will undergo cell division and repair. If absorbed through skin, some substances or their metabolites also may circulate through the blood back to the skin, causing secondary injuries through their biological effects and also altering absorption.

When an organism inhales air, the gases, particles, and substances on the particles can come in contact with lung membranes and be absorbed directly into the lung tissues and into blood passing through the lungs. Other substances pass out of the lung through expired air without absorption. At sufficiently high concentrations, both absorbed and unabsorbed substances can directly alter lung tissues, changing absorption. If absorbed, some substances or their metabolites can also cause biological effects that alter absorption, for example by injuring lung cells or by provoking defense mechanisms. Lung defenses against particles also will remove some portion of the inhaled particles, and the organism will ingest them, leading to secondary absorption by the oral route.

Much of the progress with the Clean Air Act has come through consideration of the inhalational route of dosing. When exposed to air with gas or particulate contaminants, absorption often is both efficient and rapid. However, dosimetry differs from substance to substance, and air regulations need to take these differences into account.

When an organism ingests food, water, or particulates, chemical substances in them are absorbed directly by the tissues of the gastrointestinal tract into blood or lymph passing through the tract. Microorganisms in the tract and biochemical processes in gastrointestinal tissues may digest or metabolize ingested materials into new, absorbable substances. Other substances pass through the gastrointestinal tract without being absorbed, appearing in the feces. At sufficiently high concentrations, both absorbed and unabsorbed substances can alter the gastrointestinal tract, changing absorption. If absorbed, some substances cause physiological effects that alter absorption, for example, by changing blood flow through the gastrointestinal tract.

The ingestion pathway often combines with the inhalational route in dosimetry. Substances in the airways get trapped in a layer of fluid, called the mucociliary tree, that moves up and out of the lung and is swallowed. Experimentally, these pathways are difficult to separate. After an inhalational exposure, if ingestion contributes significantly to the absorption of a substance, the dosimetric effect is somewhat greater absorption, and the kinetic effect is slower, more prolonged absorption.

Intravenous and intraperitoneal routes of dosing are important in human medicine as therapeutic ways to get drugs into people who otherwise could not absorb the material. To a large extent, these routes serve experimentally as surrogates for all routes of dosing that involve absorption through mucous membranes. However, in practice, other routes have also become important, including intraoccular, which is important in the absorption of rhinoviruses and eye medications, and intranasal, which is important in the absorption of particular substances, such as cocaine and snuff.

Absorption, Distribution, Metabolism, and Elimination

Scientists can separate the ADME processes and study each in relative isolation, as part of the description of dosimetry. [For a typical list of ADME studies used in regulation, see the FDA "Redbook" (FDA, 1993).] We have already covered some simple concepts of absorption in the previous section, where we addressed routes of exposure and dosing, because we have emphasized the conceptual linkage between exposure and dosimetry, and we have tried to separate exposure pathways from dosimetric routes. Many of the important factors in absorption are route specific. However, the more lipid soluble a substance is, the generally greater its absorption. Lipid solubility becomes prominent in structure-activity relationships. When studying a series of related substances that have the same uses and routes, lipid solubility often gives useful insight into the efficiency of absorption. Classically, the substance is dissolved in water or olive oil, then mixed vigorously with equal amounts of both liquids. The two liquid phases are separated, and the amount of the substance is measured in both phases. Tabular lists of oil-water partition coefficients are available. Sometimes different liquids are substituted for oil or water. For example, ethyl acetate is insoluble in water and sometimes serves as a less variable substitute for olive oil.

After absorption by any route, a substance will distribute throughout the body. Since some substances do not dissolve well and do not gain access to blood after intradermal injection beneath the skin or intramuscular injection into muscle tissue, injection site toxicity sometimes becomes important. For example, intradermal iron deposits may cause sarcomas at the site of injection. Overall distribution demands an understanding of where the metabolites of a substance, not just the parent substance, locate. For some carcinogens, understanding dosimetry is crucial, as with the sodium salts of substances like saccharine or the monomeric precursors of certain plastics, like melamine. Particles accumulate in the bladder from metabolites that are insoluble in urine, leading to bladder irritation, which as a chronic condition results in local tumors.

The simplest and most direct method of understanding distribution is whole body radioautography, which still is underutilized both in experimental science and in risk assessment (Busch, 1977). One reason for the slow acceptance of these methods is the necessity of creating an isotopic label in a substance that will create tracks in photographic emulsions. Another reason is bad statistics. Naive persons often view the results of whole body radioautography as "using too few animals." Instead, the need to use few animals is one of the important advantages of these methods. In brief, at different time intervals after administration, the entire animal is frozen and sliced into thin layers. Each layer is deposited onto film and left to generate a photographic exposure. The resulting photographs illustrate the movement and concentration of the radioisotope throughout the body—an integrated picture of the parent substance and all of its metabolites. The photographic density can be estimated quantitatively.

More often in the past, however, scientists did not engage in whole body radioautography. Instead they isolated organs at different time intervals and extracted the substance and its metabolites from these organs. In theory, they can account for all of the absorbed material in this way. While radioisotopic labels also are useful, these experiments have the advantage of facilitating the separation of parent compound and various metabolites. Routes of metabolism and distribution may be very complex. For example, the body converts trichloroethylene, an industrial degreasing chemical, into a wide array of metabolites after absorption, primarily in the liver. Some are reabsorbed after elimination into the gut and further converted into new metabolites by a different organ, the kidney. One of these metabolites probably is crucial to the carcinogenic effects of trichloroethylene.

Liver is the principle but not the only site of metabolism for many substances. For substances absorbed in the gastrointestinal tract, liver metabolism is particularly important in dosimetry, because the blood flow leading from the tract passes entirely through the liver in a process called *first pass metabolism*. If the liver metabolizes a substance efficiently, in effect little of the parent substance enters the circulation. Traditionally, scientists have divided mammalian metabolism into two kinds of reactions, called phase 1 and phase 2 reactions. Phase 1 reactions involve the chemical oxidation, reduction, and cleavage of substances. These reactions generally increase water solubility at the same time as they convert the parent substance into the new metabolite. The intact organism can eliminate the more water soluble metabolites through urine and can extract the more water soluble metabolites from body compartments, such as fat. These reactions often increase the chemical reactivity of a substance and may result in a more toxic substance. Thus, phase 1 metabolic reactions both toxify and detoxify.

In contrast, phase 2 reactions conjugate a substance and/or its metabolites by adding large new chemical substituents to its structure. A variety of conjugation reactions occur, for example, creating glucuronide and sulfate derivatives. These metabolites are generally much more water soluble and much less toxic. So, phase 2 reactions tend to eliminate a substance more rapidly. A detailed description of these reactions is beyond the scope of this text. Introductory pharmacology and toxicology textbooks cover them in detail.

Elimination refers to both the removal of a substance from the body and loss of biological activity resulting from the substance. In theory, even if a substance underwent no metabolism, excretion through urine, sweat, and expired air would remove it from the body. In addition, even if a removal of a substance from the body occurred slowly in the absence of metabolism, chemical conversion to less toxic metabolites would eliminate the substance. For this reason, some scientists define metabolism as all of the ADME reactions together, but this usage is infrequent.

In human medicine, each drug has two fundamental characteristics: clearance and volume of distribution. Clearance is the apparent volume of blood removed per unit of time per unit of body weight that characterizes the elimination of the substance, usually given in units of ml/min-kg. Clearance is essential to estimate long-term doses of a drug. To achieve a steady state blood concentration of a substance, a physician needs to replace the drug at this rate. Thus the dose rate is given by the clearance multiplied by the desired blood concentration. The volume of distribution relates the amount of a substance in the intact animal to the blood concentration, and volume of distribution usually is described in units of liters per kilogram of body weight. You would need to dissolve a substance in this apparent volume of fluid to achieve the observed blood concentration. For a typical 70 kilogram man, the volume of blood is approximately 6 liters, extracellular fluid comprises another 12 liters, and total body water is approximately 42 liters. Depending on whether a substance distributes into blood, total extracellular fluid, or total body water, very different volumes of distribution would be obtained. The volume of distribution of a substance is proportional to the half-life multiplied by the clearance. However, both clearance and volume of distribution are *apparent* quantities; that is, they arrive from mathematical relationships to elimination data. It is not unusual to have an apparent volume of distribution much larger than physically possible, for example, if a substance is bound to proteins in the body.

Pharmacokinetics

One principle of pharmacokinetics is that the blood concentration of a substance or its metabolite usually predicts the magnitude of its biological effect better than the dose. Thus we often work with blood concentrations and relate them to doses. Many ways exist to measure blood concentration, and blood is not a homogeneous medium. It consists of cellular elements (red and white blood cells) and plasma. Plasma consists of serum and coagulable proteins. The risk assessor needs to know whether pharmacokinetic data refer to blood, plasma, or serum concentrations. If a substance concentrates in blood cells, these concentrations can differ dramatically. Similarly, some substances bind to proteins in the blood, particularly serum albumin. Serum albumin remains in solution in serum, so that the concentration of a bound substance will not differ much between blood, plasma, and serum.

The simplest models of pharmacokinetics assume that the exposed organism is a single compartment, as if an organism is a container of liquid. The entire dose enters this container, and processes of degradation within the container or loss from the container leads to elimination. Disappearance from blood allows estimation of a whole-body elimination rate. Because intravenous dosing usually takes place nearly instantaneously (bolus dosing) without a time lag for absorption, the elimination based on blood levels after intravenous dosing leads to more accurate estimates.

The most important measures of blood concentration are peak concentration; average concentration, which usually requires a statement about the period of observation; and the *area under the curve* (AUC). If we plot the concentration of a substance or metabolite in blood as a function of time, the resulting curve allows us to determine the integrated blood concentration. The AUC has units of concentration X time, for example, mg/l-min. The AUC is particularly useful in understanding the bioavailability of a substance. We do this by comparing the blood concentrations after iv adminstration to the blood concentrations after some other route of exposure. The AUC gives a measure of the total amount in the blood. Usually, the smaller the AUC after administration of a standard mass of a substance, the less a substance is available at target sites. The rest of the substance either did not get absorbed or was eliminated rapidly by partition into some other body compartment, such as bone, by metabolism or excretion. The ratio of the area under the curve for ingestion to that for iv injection, particularly the ratio at equivalent blood concentrations instead of equivalent doses, represents the bioavailabilty of the substance (AUC_{oral}/AUC_{iv}), because partition and elimination of the substance should be approximately equivalent by the two routes.

Pharmacokinetic studies enable the interpretation of exposure-dose relationships, doses, blood levels, and of delivered doses. Correlations between concentrations and effects have greater relevance to the estimation of risk when accompanied by pharmacokinetic information (Hawkins and Chasseaud, 1985).

Consider Fick's first law. It states that the rate of diffusion of a solute down a concentration gradient is proportional to the magnitude of the gradient, or:

$$\frac{dM}{dt} = -DA\frac{dC}{dx}$$

where M is the mass, C is the concentration, D is the diffusion constant with dimensions L^2/T, A is the cross-sectional area of the diffusion volume, and dx is the distance over which the infinitesimally small concentration dC is measured. Now let us restate this for diffusion across a membrane barrier of thickness dx, where the concentration gradient is dC and is approximated by the concentration difference across the membrane, $C_1 - C_2$, and DA/dx is the first order transfer constant k_t for diffusion across the membrane, then:

$$\frac{dM}{dt} = -\frac{DA}{dx}(C_1 - C_2)$$

$$= -k_t (C_1 - C_2)$$

where k_t has dimensions of L^3/T. A large body of evidence supports the idea that this transfer has first order kinetics. In first order kinetics, the loss of a substance occurs as a constant fraction of the concentration per unit time. For example, 20% is lost per hour. If we divide both sides of the above equation by V, the equation expresses the rate of change of concentration instead of mass, and we have a new equation:

$$\frac{dC}{dt} = -k_e C$$

where k_e is the elimination rate constant with dimensions of T^{-1}. Integrating this equation (see Chapter 2) yields:

$$C = C_o e^{-k_e t}$$

The half-life of a substance is related to k_e by:

$$T_{1/2} = 0.693/k_e$$

where 0.693 is the natural log of two.

In essence, we represent an organism as one big pool of fluid in this approach. We assume that a substance mixes thoroughly and rapidly in the body immediately after absorption. Then, elimination draws a small volume of fluid out of the sphere, and we replace the missing fluid and remix the remaining concentration in the same volume of fluid. So, concentration declines. One remarkable study is that of Dedrick and coworkers, which examined the behavior of an antitumor agent, methotrexate, in five species: mouse, rat, monkey, dog, and human (Willard Chappell, private communication). They considered single iv injections over a wide dose range of 0.1 to 450 mg/kg. Figure 5.2 shows the results. If the loss of methotrexate followed exponential decay, the lines would be straight on the semi-log plot.

When we replot this data on a different set of axes, it becomes more clear. We argue here that the volumes tend to scale according to $M^{0.25}$ and thus use $(\text{body weight})^{1/4}$. We can also rescale the ordinate data according to $C(t)/(D/M)$ which is essentially $C(t)/C_o$ as shown in Figure 5.3. Now the data fall on a single curve. The plot in Figure 5.3 relates

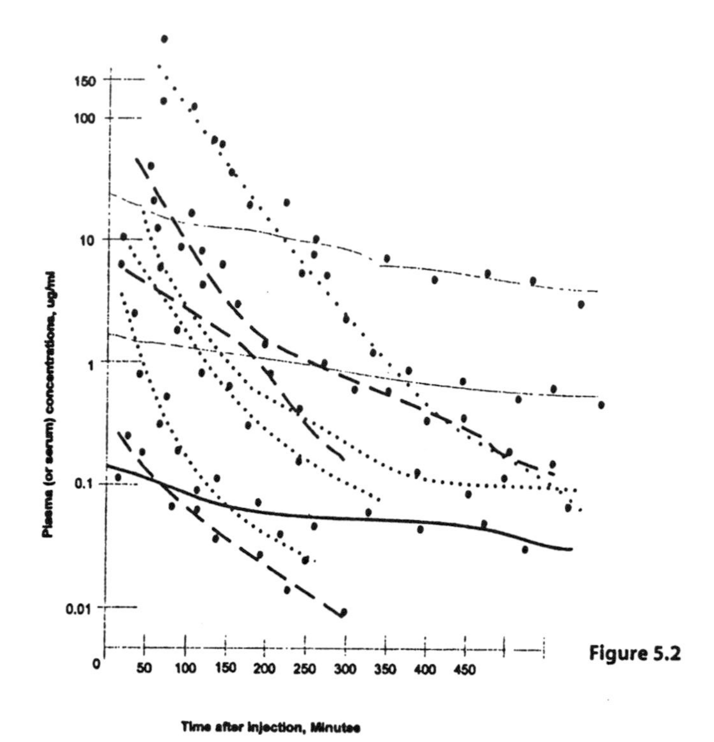

Figure 5.2 Plasma concentrations of methotrexate in several species after ip and iv injections. The solid lines are for humans.

equivalent time or pharmacokinetic time. This approach compares chronological time to biological time, as illustrated in Figure 5.4

An important quantity in this type of analysis is the area under the concentration v time curve, or AUC. This is the integral of:

$$C = C_o \, e^{-k_e t}$$

or

$$AUC = C_o/k_e = D/k_e V_1$$

For most systems, a one compartment model is not adequate. We can construct a two compartment model using a simple extension of the reasoning used for the one compartment model (see Figure 5.5):

$$\frac{dC_1}{dt} = k_{21}C_2 - (k_{12} + k_e)\, C_1$$

and

$$\frac{dC_2}{dt} = k_{12}C_1 - k_{21}C_2$$

The solution to these equations for C yields:

$$C_1 = A_oE^{-\alpha t} + B_oE^{-\beta t}$$

$$A_o = \frac{D(\alpha - k_{21})}{V_1(\alpha - \beta)}$$

$$\beta = 1/2[(k_{12} + k_{21} + k_e) - [(k_{12} + k_{21} + k_e)^2 - 4k_{21}k_e]1/2]$$

$$B_o = \frac{D(k_{21} - \beta)}{V_1(\alpha - \beta)}$$

$$\alpha = 1/2[(k_{12} + k_{21} + k_e) + [(k_{12} + k_{21} + k_e)^2 - 4k_{21}k_e]^{1/2}]$$

Scientists often try to determine the animal species that resembles humans the most in the metabolism of a substance. With no other information as guidance, most assessors will consider this species as the best biological model and, therefore, the best test species for experiments about both effects and dosimetry. Scientists may combine pharmacokinetic data with tissue-specific data about human effects—for example, effects from cultured human cells—to model dose-response relationships that better predict effects in humans than rote animal experiments.

In any kinetic experiment, scientists try to ascertain the mass balance of a substance in all tissues, although the frequent failure to observe this necessity is not surprising. An appropriate pharmacokinetic study will give a risk assessor some assurance that an experimental aberration did not make the administered substance seem to disappear by magic. For unknown reasons, many ADME studies focus on a few organs and fail to report the total amount of a substance and its metabolites remaining in the carcass, a tendency somewhat like failing to sign your income tax return. If the investigators cannot recover the mass of the dose, either as the original substance or its metabolites, the study indicates that dosimetry of the substance is not understood, and the risk assessor cannot rely on any of the kinetic measurements.

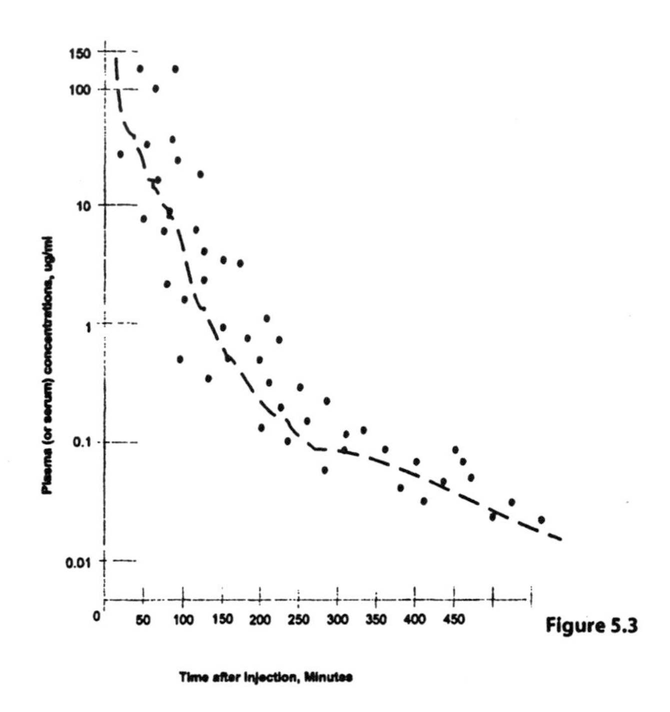

Figure 5.3 Data from Figure 5.2 after normalizing the dose and converting to equivalent time

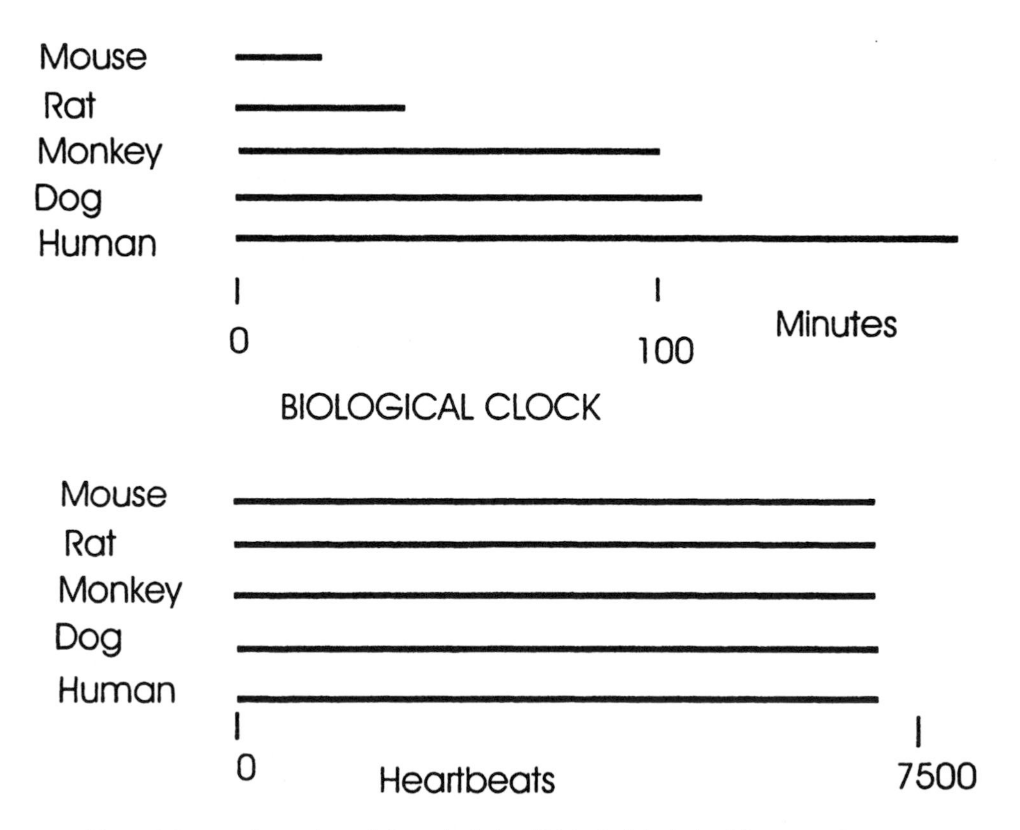

Figure 5.4 Comparison of chronological and biological clocks for different species

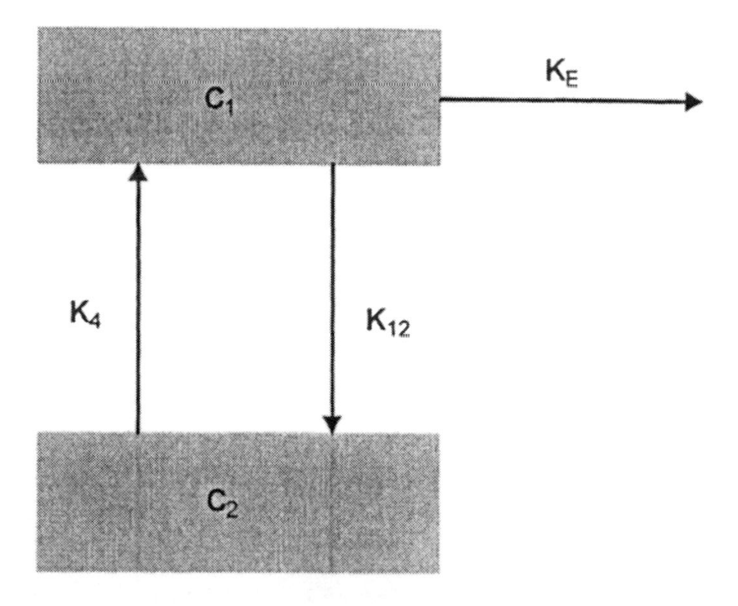

Figure 5.5 A two-compartment pharmacokinetic model

Most risk assessors assume that elimination of a substance is a first-order process, which often is true. First-order elimination also is the simplest model possible. Scientists can detect nonlinear elimination processes in animal experiments, usually zero-order elimination, where some process becomes saturated, but doing so requires many animals (Andersen, 1981). In a zero-order process, the same amount of a substance disappears from the pool during each unit of time. The risk assessor should approach all preliminary descriptions of dosimetry with skepticism. Most zero-order processes become first order processes at lower doses, particularly doses relevant to environmental exposures. Experimental studies of intravenous dosing often contribute greatly to risk assessments, although it is not an environmental route of exposure. Such data may be available with animal studies or when the chemical substance of interest also is used in human medicine.

Physiological Pharmacokinetics

The parameters in the simple models (half-life, volume of distribution, and clearance), which we described in the previous sections, are arbitrary, in the sense that we derived them only from curve-fitting of experimental data. It is not unusual to obtain a volume of distribution in such an experiment that is several times larger than the test animal. Such a model, while both useful and descriptive in a practical sense, clearly is biologically and physically impossible. A major advance in understanding pharmacokinetics processes during the past three decades directly addresses this difficulty. It is the idea of replacing these arbitrary parameters with independently obtained and biologically accurate parameters that describe the dosimetry. These models move from a small number of arbitrary compartments to a complex series of compartments that attempt to describe the entire pharmacokinetic process accurately. Because they incorporate anatomical reality, these models derive much of their information from biological measurements that are independent of the experimental information about the elimination of a specific substance.

A specific form of biological simulation, physiologically-based pharmacokinetics (PBPK) modeling, has become central to regulatory risk assessment. A PBPK model differs from an empirical, or statistically-based, pharmacokinetic model in that (1) the structure portrays actual anatomical compartments, instead of abstractions developed for the sole purpose of obtaining better fits to the data, and (2) the parameters depend on independently obtained physiological and biochemical data (Clewell and Andersen, 1985; Andersen, 1988; Andersen and Krishnan, 1994). Each model has two independent data sets—one for the species involved and one for the substance involved. Scientists can then, for example, replace the parameters in a PBPK model of one species, such as mouse, with those from another, such as human, to understand the quantitative implications of mouse dosimetry for humans. Because body processes in different species do not act identically on each substance, a PBPK model will better reflect specificity of action and thus yield a better toxicological interpretation. For example, if a particular metabolic reaction in liver is crucial to the disposition of a substance, PBPK models permits the assessor to use data about the metabolic rate of conversion by isolated mouse liver cells and human liver cells and thus obtain much better extrapolation of dose between the two species.

At present, most PBPK models aim to assess dose to a target tissue. Tissue dose improves on external dose for purposes of predicting toxicity, especially for estimating quantitative indicators of chronic effects, such as carcinogenic potency. The fundamental assumption, implicit in this operation, is that a tissue from a different species will respond equivalently to the same dose of toxic material (whether administered substance or metabolite) given all differences in cell number, life span, and so forth. In essence, PBPK models generate improved dosimetry measures and, for this reason, yield better insight into the quantitative implications of animal data for human risk.

For the purposes of this book, you should think about PBPK models as an extension of multicompartment models from the previous section, with many more compartments and with constraints on each compartment, so that their sizes and other characteristics, such as blood flow and extraction of the substance from blood, rely on independent biological measurements. Thus, at a minimum, a PBPK model must have compartments for (1) the site of absorption, such as skin or lung, including blood flow through the site of absorption, (2) the circulatory system, (3) the major organ(s) of metabolism, (4) the major organ(s) of elimination, (5) the target organ where adverse effects occur, and (6) the rest of the body. Many models divide the rest of the body into separate compartments, for example, into areas representing poorly perfused tissue and highly perfused tissues. The modeler will add or subtract compartments as necessary to model the desired objectives (e.g., the dose of a biologically active metabolite delivered to a target organ). While the modeling community has conflicting views about how much biological detail to incorporate into PBPK models, generally more is better, in the sense that modelers cannot anticipate how others will use a model. However, detailed models become mathematically complex, and some modelers have moved in the other direction, inquiring about methods to reduce the structure of PBPK models to a minimum.

Consider that the gastrointestinal tract can (1) absorb an ingested substance directly into blood or lymph, (2) digest it in the intestinal lumen, sometimes through intestinal microorganisms, to form absorbable chemicals, or (3) pass it through without absorption. In addition, the substance may alter the tract, thereby changing its own absorption (or lack of absorption). All of these processes are exposure-dependent. An experiment involving exposure to a large amount of a substance may reveal one of these processes, which may not even occur after exposure to a small amount. The only reasonable approach to modeling such a complex, dynamic process is to construct a model that includes the dose-dependent processes in its parameters. Trying to construct such a model without independent experimental data about these biological processes would be difficult, if not impossible. Trying to integrate these data into a model with arbitrary parameters would truly be impossible.

Metabolism will affect the dose of a substance absorbed from the gastrointestinal tract. Models that attempt to incorporate more details than the data allow, or to extrapolate far beyond the data, are not generally useful. In part, this problem relates to degrees of freedom. If ADME data only describe a few compartments, a model based on these data should contain only one more compartment that represents the residual mass of the substance. Sometimes, however, the assessor can insert data into an existing model of the substance or of a chemically related substance.

Assessors can use cell culture or biochemical data to augment whole-animal data. Partition coefficients, binding to components of organs, and kinetic parameters for enzymatic degradation obtained from these studies fit directly into biologically based models. Most assessors prefer to use these independent sources of data to construct a model, which they then compare to whole-animal data. This process leads to scientifically more robust models. The current tendency is to fit models to data from single exposures to predict the pharmacokinetics of chronic exposures. Because partition, binding, and enzymatic processes can vary with the age of an organism, and chemical substances can contribute to changes in these rates, comparing the predictions to data from chronically exposed organisms is important.

Even without direct human data, metabolic experiments with isolated human tissues from autopsy materials or cultured cells can supplement a biologically based model of disposition in an animal converted to human parameters. These models basically assume that a human is a gigantic rodent or dog, with only quantitative differences between the species. Most importantly, however, assessors can validate such models with only sparse human data.

Whenever possible, biologically based models should lead to population distributions of pharmacokinetics processes. If data about individual experimental animals are available, assessors can insert the appropriate parameters for an animal of the same weight and age, such as blood and tissue volumes. Some models facilitate scaling of parameters based on independent allometric data. Useful summaries of allometric data are available (ILSI, 1994; O'Flaherty, 1991).

Biologically based models generally use blood, bile, nutrient, urine, and fecal flow rates from the literature. Assessors obtain absorption rates, partition coefficients, extraction coefficients, binding coefficients, and elimination rates for the substance of experimental interest. Models that rely on fitting to whole-animal data for the latter group of parameters are generally not acceptable (Reitz et al., 1996). Most assessors also regard the acceptability of biologically based models of disposition in relation to the availability of similar models for several different substances.

Sources of Additional Information about Dosimetry

The subdisciplines of chemical biology, chemotherapy, pharmacology, and toxicology are deeply concerned with dosimetry, but in an integrated way and to such an extent that they do not separate these considerations from other information about the outcomes of chemical exposures (see Chapter 7). Many scientists characterize the two domains as kinetics and dynamics, but in practice the two kinds of information are not separated in the study of a substance. In risk assessment, with a heavy reliance on models, separation of these two areas is more typical. The assessor needs dosimetric information in different places in a model from outcome information. Several professional groups are more concerned with pharmacokinetics exclusively, including some of the divisions within the American Association of Pharmaceutical Sciences (**www.cpb.uokhsc.edu/aaps/**). Extensive free and commercial software is available to engage in pharmacokinetic curve fitting (for example, see Beal and Sheiner, 1980). Similarly, several simulation languages are available to build computational PBPK models, such as ACSL and SCoP. This practice is highly specialized, in part because much of the anatomical and physiological data about the model parameters for a species have been worked through in other models, developed historically for other substances, and in part because of the need to couple experimental work with model construction.

Extensive information about dosimetry is available from textbooks and scholarly literature. For example, for classical pharmacokinetics, see M. Gibaldi and D. Perrier, *Pharmacokinetics,* 2nd ed., New York: Marcel Dekker (1982) or W. L. Roth and J. F. Young, "Use of Pharmacokinetic Data under the FDA's Redbook II Guidelines for Direct Food Additives," *International Journal of Toxicology 17:* 355–381 (1998). A good starting point for biologically-based modeling, including PBPK modeling, is M. E. Andersen, H. J. Clewell, and C. B. Frederick, "Applying Simulation Modeling to Problems in Toxicology and Risk Assessment—A short perspective," *Toxicol. Appl. Pharmacol. 133:* 181–187 (1995).

References

Andersen, M. E. 1981. Saturable Metabolism and Its Relationship to Toxicity. *CRC Crit. Rev. Toxicol. 109:* 579–585.

Andersen, M. E., H. J. Clewell, and C. B. Frederick. 1995. Applying Simulation Modeling to Problems in Toxicology and Risk Assessment - A Short Perspective. *Toxicol. Appl. Pharmacol.* 133: 181–187.

Beal, S. L. and L. B. Sheiner. 1980. *NONMEM Users Guide.*Technical Report of the Division of Clinical Pharmacology, University of California, San Francisco.

Buelke-Sam, J., J. F. Holson, and C. J. Nelson. 1982. Blood Flow during Pregnancy in the Rat: II. Dynamics of Litter Variability in Uterine Flow. *Teratology* 26:279–288.

Busch, U. 1977. Whole-Body Autoradiography: Use for Pilot Studies of Pharmacokinetics in Rats. *Acta Pharmacol. Toxicol.* 41: 28–29.

Chen, H. G. and J. F. Gross. 1979. Estimation of Tissue-to-Plasma Partition Coefficients Used in Physiological Pharmacokinetic Models. *J. Pharmacokinet. Biopharm.* 7(l):117–125.

Clausen, J. and M. H. BICKEL. 1993. Prediction of Drug Distribution in Distribution Dialysis and in Vivo from Binding to Tissues and Blood. *J. Pharm. Sci.* 82:345–349.

FDA. 1993. Draft Toxicological Principles for the Safety Assessment of Direct Food Additives and Color Additives Used in Food - Redbook II. Center for Food Safety and Nutrition.

Gallo, J. M., F. C. Lam, and D. G. Perrier. 1987. Area Method for the Estimation of Partition Coefficients for Physiological Pharmacokinetic Models. *J. Pharmacokinet. Biopharm.* 15(3):271–280.

Gibaldi, M. and D. Perrier. 1982. *Pharmacokinetics* (2nd). New York: Marcel Dekker.

Gillette, J. R, On the Role of Pharmacokineties in Integrating Results from in Vivo and in Vitro Studies. *Food Chem. Toxicol.* 24: 711–720 (1986).

Grass, G .M. 1997. Simulation Models to Predict Oral Drug Absorption from in Vitro dDta. *Adv. Drug Delivery Rev.* 23: 199–219.

Hawkins, D. R. and L. F. Chasseaud. 1985. Reasons for Monitoring Kinetics in Safety Evaluation Studies. IN *Receptors and Other Targets for Toxic Substances. Arch. Toxicol. (Suppl.)* 8: 160–167.

International Life Sciences Institute. 1994. *Physiological Parameter Values for Physiologically-Based Pharmacokinetic Models.* Washington, D.C.: ILSI.

Irons, R. D. and E. A. Gross. 1981. Standardization and Calibration of Whole-Body Autoradiography for Routine Semiquantitative Analysis of the Distribution of 14 C-Labeled Compounds in Animal Tissues. *Toxicol. Appl. Pharmacol.* 59: 250–256.

Iwatsubo, T., N. Hirota, T. Ooie, H. Suzuki, N. Shimada, K. Chiba, T. Ishizaki, C. E. Green, C. A. Tyson, and Y. Sugiyama. 1997. Prediction of in *Vivo* Drug Metabolism in the Human Liver from in Vitro Metabolism Data. *Pharmacol. Ther.* 73(2): 147–171.

Lindstedt, S. L. 1987. Allometry: Body Size Constraints in Animal Design. IN *Pharmokinetics in Risk Assessment, Drinking Water and Health, Volume 8.* Washington, D.C.: National Academy Press.

Reitz, R. H., A. L. Mendrala, and R. A. Corley. 1990. Stimulating the Risk of Liver Cancer Associated with Human Exposures to Chloroform Using Physiologically Based Pharmacokinetic Modeling. *Toxicol. Appl. Pharmacol. 105:* 443–459.

Rhomberg, L. *What constitutes "dose"? (Definitions).* 1995. Washington, D.C.: ILSI Press.

Roth, W. L. and J. F. Young. 1998. Use of Pharmacokinetic Data under the FDA's Redbook II Guidelines for Direct Food Additives. *Int. J. of Toxicol.,* 17: 355–381.

Chapter Six

Epidemiology

Introduction to Epidemiology

Suppose that you want to estimate the probability of a future loss of human life associated with a particular disease under specific conditions. What better basis could you find than direct data under conditions similar to those that interest you; in other words, previous outcomes observed under the real world conditions? You would have the best chance of making an accurate prediction because your information would embody an environment similar to the one of your prediction. You would have information about the right species. The background risks would be similar. You would have a firsthand basis to understand a human health risk.

Much of the information that you might want to obtain to estimate the potency of a substance in a human population may already exist within the field of *epidemiology*. Literally, epidemiology means the study of populations. Epidemiology and toxicology, the subject of the next chapters, are closely intertwined subjects. From the perspective of this book, the two disciplines provide complementary ways to establish a potency estimate. Our objective of calculating risk using estimates of potency and exposure remains unchanged.

Traditionally, epidemiologists surveyed outcomes, mostly diseases or deaths, in human populations and attempted to discover the causes of these outcomes, mostly infectious diseases. This history helps to explain an apparent lack of interest in risk assessment among epidemiologists. Infectious disease experts are more concerned with whether or not microorganisms infect individuals; how much infection matters less. Thus, dose-response relationships interest them less. The identity of the infectious agent and disease etiology are paramount. In contrast, most assessments of risks from chemical substances require the amount of exposure for each individual in a population.

The history of epidemiology also explains its primarily contemplative nature. Human experimental studies with infectious agents are usually unethical. Most infectious agents are highly species-specific. So, experimental studies in a laboratory species usually contribute little useful knowledge. Epidemiology more resembles astronomy than physics. Most epidemiological studies take an observational approach. Epidemiologists search for associations between outcomes, exposures, and the factors that account for these associations. The reasonable assumption in this process is that a greater presence of association explains disease. It is a multidisciplinary science, traditionally requiring skills in both statistics, medicine, and study design. Like most hybrids, epidemiology has great vigor. It also has many practical applications.

We described exposure assessment in detail in Chapter 4, so this chapter will focus more on outcomes and design. Mathematically, epidemiologists find it useful to describe the data about associations as variables. Thus, at least one variable will be an outcome, either for each of several populations or for each individual in a population. Exposure is the second variable. A *correlation* describes the quantitative extent of the relationship between two variables, without considering whether one variable necessarily follows from the other.

If you initially classify the members of a population by exposure and later observe the outcomes, you have conducted a *cohort study*. The simplest cohort studies have only two subpopulations: exposed and not exposed. Binary classification of exposure, while important in epidemiology, is less useful in risk assessment. However, cohort studies are particularly useful in understanding what outcomes follow from exposure. You can study many outcomes with ease.

On the other hand, if you initially classify the study subjects according to an outcome, for example, subjects who have a disease or do not, and later look at the exposures of the groups, you have done a *case-control study*. Case-control studies are particularly useful in understanding which, of many, exposures might lead to a disease.

Epidemiologists elicit the patterns of a disease with correlational studies. The patterns help to identify causes and explain etiologies, much like detective work. Cohort and case-control studies form the primary content of epidemiology. For example, epidemiologists will try to understand whether disease prevalence changes between regions. If the prevalence of stomach cancer is much higher in Japan than the U.S., a subsequent study of recent immigrants might show whether some factor in the environment or a genetic factor might explain the incidence of stomach cancer. Because the prevalence of stomach cancer drops among immigrants from Japan to the U.S., we suspect an environmental factor. A genetic factor would not change with location. Because the drop occurs slowly, over several generations, we suspect some factor associated with Japanese culture. It does not matter whether we obtain this information by means of case-control studies (stomach cancer or not) or by means of cohort studies (living in the U.S. or Japan). If a cohort study confirms a case-control study, or vice-versa, we consider the information more robust.

An epidemiological study can have either a population or an individual as its unit of observation. Epidemiologists call comparisons of populations *ecological studies*. For example, if you examined the incidence of stomach cancer among communities along the Mississippi River, you would conduct an ecological cohort study. If you collect data about people living along the Mississippi River into one study and compare distance with stomach cancer, you would conduct a cohort study of individuals. Ecological studies are prone to misinterpretations, and we will not discuss them much in this book. However, ecological studies can provide useful clues early in an inquiry, because you can often obtain the necessary data from published statistics.

An epidemiological study with variables that describe events in the past is a *retrospective study*. Epidemiologists review old hospital records or ask subjects about their experiences. If you classify members of the population first and wait for events in the future, then you will have conducted a *prospective study*. Retrospective and prospective studies often get confused with hypothesis-generating and hypothesis-testing studies. Usually, a descriptive or hypothesis-generating study looks at many variables without any prior ideas about the potential relationships between variables. Instead, you try to discover correlations between variables as a matter of *data analysis*. It does not matter whether you look backward or forward in time, but the statistical techniques and assumptions appropriate for data analysis differ from those for hypothesis-testing.

A correlation does not refer to the nature of the relationship between two variables. It does not matter whether one variable necessarily follows from the other variable. In many situations, however, you can classify variables as either dependent or independent. Dependent events necessarily follow from independent events. If you believe that two variables relate to each other, you can try to judge which one follows the other by conducting a *hypothesis-testing study*. If a hypothesis-generating study or another source of information suggests, for example, that penicillin treatment leads to stomach cancer, you might compare a penicillin-treated population with a similar population not treated with penicillin. If your cohort study revealed a difference in stomach cancer prevalence between the two populations, you could analyze the difference to judge whether the difference arose merely by random chance, by conducting a *statistical analysis*. A risk analyst also will find the study more interesting if increasing penicillin exposure correlates with increasing stomach cancer prevalence. (You should know that some epidemiologists do not agree with the distinction that we draw between data analysis and statistical analysis.)

If your study looks at two variables that depend on a third, unknown variable, you can easily reach a false conclusion. For example, over several decades, decreases in stork nests each year in Germany correlated with decreases in the number of children born. You would not conclude from such data that storks bring babies! You know better. Instead, industrialization of Germany brought many changes. Home heating shifted from coal to oil, and as a result, newer homes lacked chimneys. Fewer chimneys meant fewer nesting sites for storks. Simultaneously, the shift to an industrial economy led to smaller family sizes. An association between two variables is not evidence of a direct relationship between them. Yet, almost every month brings a press report that confuses correlation with causation.

The difference between hypothesis-generating and hypothesis-testing studies also leads to confusion in interpreting epidemiological studies. Because epidemiological data are subject to chance, a fit to a hypothesis or a dramatic difference in prevalence could arise randomly. The usual criterion of *statistical significance* in a study is a probability of less than one in twenty that chance alone generated the results ($p = 0.05$). If, however, a descriptive study has forty variables, application of this statistical criterion suggests that at least two variables should appear significantly different, even if all of the study results are random [$1/20 \times 40 = 2$]. You could get a better clue about the effect of chance by correcting for the number of variables, but such corrections are more difficult and controversial. If you treat the descriptive study as hypothesis-generating and conduct another, independent study under rigorous conditions that isolates the two variables, you will have tested the hypothesis.

The application of standard tests of statistical significance to descriptive data usually violates the assumptions used to develop the tests. A full explanation of this problem lies beyond the scope of this book, but most textbooks of statistics deal with these assumptions in detail. Epidemiologists sometimes explain the difficulty of analyzing a study with many variables as the Texas sharpshooter problem. (The Texas sharpshooter points a gun at a barn and fires several rounds. Afterward, the shooter looks for bullet holes close together and draws a bullseye around them.) As a practical matter, your best hope of understanding whether the two statistically different variables tell you something about the outcome or exposure of interest would lie in repeating the study. Thus, the replication of studies is a crucial part of epidemiology.

Hypothesis-testing studies have their own intrinsic difficulties. Usually, an epidemiologist plays an active part in gathering the data in a hypothesis-testing study. Participation may unintentionally, perhaps in unobserved ways, change the observations. For example, an enthusiastic epidemiologist might inadvertently communicate a belief in a hypothesis to study subjects. In turn, the subjects might report more of the "desired" symptoms or suppress undesired outcomes, just to please the epidemiologist.

To avoid this problem, many studies *blind* the study subjects. Interviewers do not indicate who should experience the outcome and who should not. For example, a nurse might give all of the study subjects an injection and not tell the subjects who received a drug and who did not. Similarly, investigators usually give random designations to samples sent for chemical analysis. So, the chemist cannot tell which samples should have values consistent with the hypothesis. Sometimes, even blinding of a study is not sufficient to protect its integrity. For example, if the nurse knows who gets a drug and who does not, the nurse could reveal an expectation in subtle ways. Consequently, some studies resort to *double blind* processes. In double blind studies, even the field investigators are not informed of the study hypothesis. They receive randomly coded study materials. To double blind the drug study, someone would assign random codes to the injections, such that the nurse would not know who received a drug and who did not.

Resisting the temptation to reach a conclusion about a disease process is difficult when the data in front of you suggest an important relationship, perhaps one you suspected previously. It becomes easier to resist, however, if you create a specific hypothesis and plan to gather new data to test the hypothesis. You can still conduct a test of the hypothesis without conducting a prospective study. Further, not all prospective studies test hypotheses. However, if the study is prospective, you will control the sequence of events. The time relationship between independent and dependent variables is crucial. You also will control the study conditions, and so a double blind study becomes possible.

A 2 x 2 table presents epidemiological data in a useful way. You can easily calculate the *odds ratio* from the table. For example, suppose our hypothesis leads us to expect certain results in a penicillin-treated cohort if stomach cancer prevalence does not change after treatment. We could develop our expected results by applying the background stomach cancer prevalence from a large population to a population of the same size as our study population. We addressed background rates earlier, when we discussed life tables in Chapter 1.

The odds ratio is the chance of a disease occurring in an exposed population divided by the chance of the disease occurring in an unexposed population. The patterns that you would observe, in looking at many potential exposure factors, would arise from comparisons of odds ratios.

	Disease	*No disease*	*Prevalence*
Exposed	A	B	A/A+B
Unexposed	C	D	C/C+D
Ratio	A/C	B/D	

You can calculate the odds ratio most easily as (A x D) / (B x C). For example,

	Stomach cancer	*No stomach cancer*	*Prevalence*
Exposed	2	9,998	2×10^{-4}
Unexposed	10	99,990	1×10^{-4}
Ratio	0.2	0.1	

Here, the odds ratio for stomach cancer is 2 x 99,990 / 10 x 9998 = 2. (This difference is not statistically significant because of the small number of observed cases.)

If you like to gamble, you will immediately realize that a relative risk of 0.05 (1 in 20) is the same as odds of one to 19. Any risk [x/y] is equivalent to an odds ratio, adjusted by subtracting the numerator from the denominator [x/(y-x)]. In many situations, you will deal with large populations and low risks. If so, the odds ratios will be the same as the relative risks. Usually, either statistic is a good measure of the magnitude or strength of an association.

The patterns that you find in examining the odds ratios of potential exposure factors could arise from other factors associated with the factor you investigated. Particularly with exposures to chemical substances, a contaminant or a break down product of the suspect substance may cause the outcome. For example, dioxin is a highly potent substance that forms when someone heats organic materials containing chlorine to high temperatures. Initially, epidemiologists suspected that chlorinated phenols led to specific outcomes, such as a skin condition called chloracne. Eventually, they found that dioxin contaminants caused chloracne, not the pesticides.

Confounding factors may complicate investigations of causation, although confounding factors matter less for risk assessment purposes. For example, if dioxins contaminate chlorinated phenols, the risk associated with dioxin will correlate with exposure to the chlorinated phenols. To explain an odds ratio through a confounding factor associated with the exposure you study, the confounding factor must have a higher odds ratio than the study variable. This means that, if you study an association with a high odds ratio, a confounding factor is unlikely. With more investigation and a high odds ratio, finding a confounding factor will prove easier. Conversely, with a low odds ratio, the presence of a confounding factor could easily explain the results, and finding the confounding factor, if any, will prove difficult.

Some errors of interpretation commonly occur when using odds ratios or relative risks. In comparing studies of some variable that you find interesting, you should avoid segregating the studies with statistically significant associations. Identical odds ratios from two studies with different numbers of study subjects could lead to a statistically significant result with the larger study, but an insignificant finding with the smaller study. Yet, the two studies reveal the same association.

When you want to estimate a risk, you will want to explore the relationship between exposure and outcome in any available epidemiological studies. Epidemiological studies will provide the most straightforward and relevant information. The applications of epidemiology are neither restricted to disease outcomes nor confined to human populations. Epidemiology works equally well with a wide array of outcomes, including disabilities, behaviors, ecological effects, and economic impacts. Similarly, epidemiological methods apply to studies of nonhuman populations. You could study equally well the effects of water temperature on turtles.

The primary weakness of epidemiology arises from the difficulty of obtaining accurate information about exposures and outcomes. If you cannot think of a way to obtain good quality epidemiological data about the variables related to a risk, you essentially raise a question about the importance of assessing the risk. From an assessment perspective, you could estimate the risk based on theories about the exposures and outcomes. A "what if" assessment is never completely satisfying, however. As a practical matter, you will want to estimate risk based on epidemiological data. Even poor quality epidemiological data that agree with a "what-if" estimate will prove highly satisfying.

The weakness in epidemiological data comes from exposures more often than from ambiguous outcomes. Occupational studies often substantiate risks at higher exposures than the general population experiences. However, occupational exposure estimates often come from retrospective information about the typical exposures in different job classifications. If you want to estimate general population risks, these exposure estimates will prove frustrating. The process of extrapolating to lower exposures often falls into theoretical considerations.

The epidemiologist has to estimate the general exposure for a particular skill, then find the duration of employment for each worker at this task. Even this approach, which yields highly uncertain exposure estimates, is a daunting task for the investigators. Occupational studies often provide the best data available. If disease moves with the job classification, these studies can contribute greatly to the minimization of occupational risks. For purposes of estimating risks in nonoccupational populations, these studies still yield information about the right species and about similar background factors, such as socioeconomic class. Even with these crude estimates, epidemiological studies often have variable durations of exposure, magnitudes of exposure, and ages of exposure. So, they provide optimal data for modeling exposure-response relationships, if risk assessors fully use the information in the studies. Often, larger exposed populations are more available with these studies than with experimental studies.

Other complications abound. As you saw in the sections about life tables in Chapter 1, accounting for population mobility is difficult. Some persons will refuse to participate. In addition, human populations exhibit wide ranges of genetic variation that often determine susceptibility to diseases. Sometimes a study will suggest an elevated odds ratio but will contain too small a study population to yield a statistically robust conclusion. In particular, you often cannot detect small changes in relative risk as the exposure changes, eliminating the possibility of understanding the exposure-response relationship with much precision.

Epidemiology and risk assessment are overlapping disciplines, but risk estimates do not automatically follow from epidemiological studies. Frequently, epidemiologists have little interest in quantitative measures of risk. Thus, the published data will not support risk estimation. You must obtain the unpublished raw data to make much progress.

Risk assessment seeks to forecast events. Besides understanding the uncertainties in the data that lead to an exposure-response relationship, the risk assessor has to worry about changes in the future population and other significant factors. The major assumption, that the past predicts the future, will not trouble epidemiologists, nor should it.

Epidemiological studies excel in revealing the distribution of outcomes. By its very nature, the population will stratify into subpopulations whenever the investigators have noted useful characteristics, such as locations (perhaps area codes or zip codes) or socioeconomic class (perhaps family income). If you want to forecast population risk, an understanding of the distribution of risk may prove very useful. Trends in dietary practices from migration, rising income levels, varying years of education, changing family size, or decreasing population density might alter risk, although potency remains the same. Epidemiology studies can contribute information about the effects of duration of exposure, changes in exposure with dietary practices, and changes in exposures with population density and socioeconomic class.

In the following sections of this chapter, we will explore standard methods of epidemiology, traditional objectives of hazard and causation, recent transitions in this science, exposure-response relationships, combining data from different studies, uncertainties in population studies, the public health paradigm, and sources of additional information about epidemiology.

Epidemiological Methods

Prevalence is the number of individuals with some outcome (cases, diseases, characteristics) divided by the total number of individuals in a population. Because you divide a number of individuals by a number of individuals, prevalence is a unitless quality. *Incidence* is the number of individuals who newly acquire an outcome divided by the number of individuals in a population without this outcome during a specified time. Over long periods, incidence becomes prevalence. [Some epidemiology texts state this relationship as prevalence = (incidence x duration).] Because epidemiologists measure incidence over a time, it has units of $(\text{time})^{-1}$.

The incidence of a disease relates directly to the risk of the disease. While the prevalence of some outcome in a population will change, if the incidence changes, prevalence also can change when individuals migrate in and out of the population, when individuals with the outcome live for different durations, or when detection (either diagnosis or definition) of the outcome changes. Because many factors besides incidence alter prevalence, for risk assessment purposes an epidemiological study that observes incidence is intrinsically easier to interpret and understand than one that observes prevalence.

Epidemiologists define the incidence of death in a population as the *mortality rate*, and they multiply the number of deaths during some period of observation by a factor of one-thousand (1000), as a matter of convention. Usually, epidemiologists calculate mortality rates based on death certificates. At the time of death, physicians decide what caused the death. The International Classification of Diseases (ICD) standardizes the causes of death and gives each a numerical code. Even so, the opinions of physicians about disease pathology are subject to false positive and false negative conclusions, as with any other test. In addition, physicians sometimes fail to document a second disease in a person who died, when the second disease could have caused the death. Thus, death certificates sometimes underestimate the risk of a certain fatal disease.

You can calculate crude mortality rates easily. You may find these summary statistics useful in searching for patterns of disease. If a population is heterogeneous, however, crude mortality rates can suppress much useful information. For example, a crude mortality rate might miss an exposure that changed the sex ratio by causing some premature mortality only among males. If the available data do not indicate sex, epidemiologists have no alternative to crude mortality rates. Thus, you will find crude mortality rates difficult to compare, because many differences between populations can influence them. For example, an older population may have a higher mortality rate, although exposure to an infectious agent is lower than in a younger population with a greater prevalence of infection.

In epidemiology, a proportion is the number of cases with a specific characteristic divided by the number of persons with the same characteristic. Thus, the *proportionate mortality ratio* (PMR) is the proportion of deaths attributable to a specific cause, usually expressed as a percent of the total mortality rate. If five in a thousand persons infected with a virus died, compared with one in a thousand deaths in the general population, the PMR would be 500%.

Epidemiologists base specific mortality rates on homogeneous subpopulations, organized around factors that might reveal an association, such as age, sex, race, or location. Both the numerator and the denominator reflect the restriction of membership in the subpopulation. Of necessity, they obtain smaller numbers of persons (deceased or at risk). Statistically significant differences become harder to obtain. You will find calculating specific mortality rates tedious, but you will obtain more interpretable information.

To avoid distortion in comparisons between populations, epidemiologists usually adjust mortality rates to normalize factors that can change mortality rates, such as different sexes or ages. Adjusted mortality rates have less bias. Detailed directions to adjust mortality rates lie beyond the scope of this text, but the methods generally recalculate the mortality rates for a population, either by applying specific mortality rates to another, reference population (the direct method), or by applying the specific rates of a reference population, such as the entire U.S., to the composition of the study population (the indirect method). When applied to age adjustment, the indirect method generates an expected number of deaths in the study population. Epidemiologists then compare the observed deaths with the expected deaths in a ratio called the *standardized mortality ratio*, or SMR:

SMR = observed deaths/expected deaths

Distributions in Epidemiology

Epidemiology is the science of choice to explore the distributions of exposures and outcomes. All of the methods and definitions we developed in earlier chapters apply to populations. You can examine various combinations of exposure over time, space, and other population characteristics. For purposes of comparison, epidemiologists use several ways to describe distributions. Because population information often generates skewed distributions, the median or the mean, described in Chapter 2, may not portray the central tendency best. You cannot even calculate a mean with data about categorical qualities, such as sex or race. A "mean" would make no sense. Indeed, if a subpopulation mimics the distribution of the larger population, the comparison is usually of less interest. Several other measures are worth noting. The mode, the most frequently observed value, is another indicator of central tendency.

Quantile plots will help you visualize the differences between two populations. You graph the least frequent observation of one distribution against the least frequent observation of another, interpolating the values for the second distribution, where necessary, proceed to next least frequent observation, and so forth. If the two distributions coincide, a straight line will emerge across the page at an approximately 45-degree angle. (See Figure 6.1, which also illustrates the appearance when the second distribution has a different spread, skew, or location.) If the second distribution is a known probability distribution, such as the normal distribution, and the two distributions coincide, you have evidence that your experimental distribution distributes normally. Statisticians have more precise tests of distributional fits, but these methods lie beyond the sophistication of this text.

Epidemiological Surveys

In a survey, epidemiologists gather information about a population, either by soliciting the data from members of the population, such as opinions in a *poll*, or by using more objective measures, such as the levels of a metabolite in urine. While epidemiologists usually *sample* a population for a survey, sometimes they include all the population and the survey is called a *census*. The two important aspects of a survey are its objectives and the representation of the population by the survey sample.

Epidemiologists can estimate the validity of a survey in two ways. They can test the internal validity by sampling for the same variable several times, perhaps using different approaches. For example, an investigator could ask about consumption of the same food using different terminology in two different questions. If some subjects answer the two

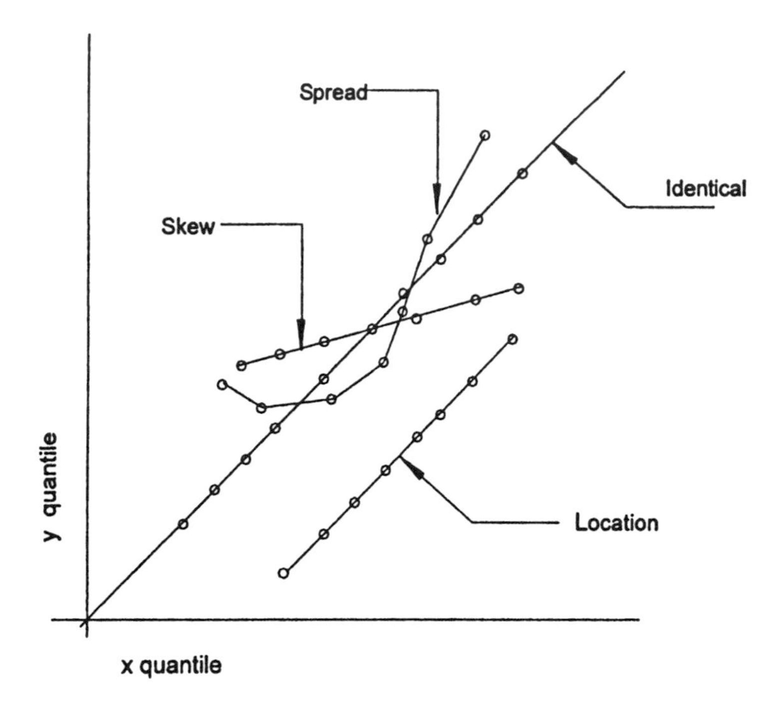

Figure 6.1 Quantile plot showing an identical pair of distributions and two pairs of distributions with location and speed differences

questions differently, internal validity is suspect. External validity means that the questionnaire yields consistent results or that it resembles another questionnaire, when applied to the same or a similar population. Validity probes the false positive and false negative results with the questionnaire. Epidemiologists call reapplying the same test to a different sample of the same population *resampling*. A survey test that does not yield similar responses after resampling will not yield valid results.

Case Reports, Case Series, and Cross-Sectional Studies

Physicians often publish case reports, which describe the symptoms, history, diagnostic features, and outcomes of one or a few patients with the same illness. When they gather many cases with similar characteristics, epidemiologists call the information a case series. Epidemiologists seldom underestimate the importance of case reports and case series. For example, case reports led to the discoveries of increased cancer risk in the daughters of diethylstilbestrol treated mothers, of acquired immune deficiency syndrome, and of the linkage between vinyl chloride and angiosarcoma of the liver.

In cross-sectional studies, epidemiologists interview subjects in-depth about outcomes as well as demographic and/or exposure characteristics, often without a prior hypothesis about possible linkages. Cross-sectional studies generate hypotheses. Thus, the investigators analyze all possible odds ratios in their data and, typically, rank the ratios by the magnitude of association. Case series and cross-sectional studies can generate useful information for risk assessments.

Case-Control Studies

Case-control studies start with observations about a population with some common characteristic or outcome, usually a disease. Epidemiologists then assemble a control population without the disease but otherwise the same characteristics (age, sex, income, occupation, and so forth). Ideally, the control population should have the same opportunities to develop the disease. One strategy to create a control population is to pick a close neighbor with similar characteristics for each case. Another approach is to choose a hospital admission with a different disease, but otherwise similar background. The primary statistics used to characterize case-control studies are odds ratios.

Case-control studies have intrinsic difficulties. Case-control studies usually involve small numbers of people. Sometimes, investigators find obtaining an appropriate control population difficult. Sometimes, the control population differs more from the background population than the case population. Sometimes investigators ask questions or apply tests differently for the cases and the controls. Case-control studies have important advantages, however. Epidemiologists can usually generate case-control findings most rapidly. Also, the studies usually cost less, and yet, they can provide definitive findings, including exposure-response information, when properly analyzed.

Cohort Studies

Cohort studies track a healthy population with a particular exposure, occupation, or characteristic, until some outcome appears. In practice, a cohort study will gather data about a population and follow it until outcomes appear (perhaps comparing it with a reference population) or gather data about two or more subpopulations and later compare the outcomes between the subpopulations. Either way, the necessity for good quality data and larger population sizes at the inception of a cohort study causes stress for the investigators. In addition, the investigators will want to follow the cohort, a resource intensive activity. Epidemiologists conventionally present the results of cohort studies as relative risks, instead of odds ratios.

Cohort studies need not be prospective. Instead, epidemiologists often take advantage of cohorts previously established for other purposes. Cohort studies can generate useful information for risk assessment from small populations with specific exposures. Exposure-response relationships from such studies can prove particularly informative. Case-control studies make efficient use of a population with a specific outcome, facilitating the exploration of many exposures. Instead, a cohort study intensively scrutinizes one exposure for all possible outcomes.

Deficiencies of Observational Studies

Case-control and cohort studies form the bulk of epidemiological investigations. In principle, they should yield the same results. So, whether prospective or retrospective, these studies have importance for society and risk assessment. Yet, epidemiologists seldom regard a single observational study as decisive. More stands at the foundation of this skeptical attitude than the normal desire of epidemiologists for replication. The findings of case-control and cohort studies have more vulnerability to chance than studies in other areas of science, and they often lack the rigor of experimental approaches where the investigators can isolate and control various potential confounding factors. Experienced investigators understand that bias, an error in the conduct of a study, can generate an exciting finding where none exists. Thus, epidemiologists are more reluctant to believe a single observational study and more likely to repeat studies. Important conclusions will often have support from ten or more studies. Thus, the interpretation of multiple studies is more important in epidemiology. We address this topic in this chapter in the section on meta-analysis.

Controlled Studies

A *controlled study* is any prospective study, or any prospective intervention study, which tests a prior hypothesis. These studies can have either case-control or cohort structure, although case-control designs are seldom used. The need to state the hypothesis explicitly before beginning the study may strike you as peculiar. Why not simply look at the data first, then think up hypotheses the study could have tested? In part, the requirement for a prior hypothesis arises from study design. If you do not know the objective of a study before you begin, how do you decide what outcomes to measure? How do you decide what population to examine? The most important rationale for a prior hypothesis, however, comes from the statistical methods you would use to analyze the data after study completion. The need to test a hypothesis and to use appropriate statistical methods forces rigorous definition of the comparison, or control, population onto the design.

The gold standard in epidemiology is a *randomized controlled trial* (also called a *controlled clinical trial*). In these studies investigators prospectively compare the outcomes of populations with different treatments, but they assign individuals to these populations at random. In analyzing the results of a randomized controlled trial, they will first explore how closely the study populations resembled each other in demographic and personal characteristics, such as age, height, weight, sex, location, and socioeconomic class. Then, they attribute the differences in outcomes between the populations to the treatments. Often, all study participants will have a common characteristic, such as a disease. Even so, the investigators probably will classify the participants according to disease severity, to insure that the results in a treatment group did not arise from prior differences in disease.

One tactic to eliminate differences between populations in randomized controlled trials is to crossover the treatments. For example, at the start of a study, one group serves as the control group and a second group receives the treatment. At the midpoint of the study, the investigators switch the groups. The first group receives the treatment and the second then serves as the control group. You can easily generalize a crossover design to more than one switch point (multiple periods of treatment and nontreatment) and more than one treatment and one control group.

Another important tactic in conducting a clinical trial is *blinding* the study. In its simplest form, the study subjects do not know whether they participate in the treatment or control group. This kind of study is *single blinded.* Investigators often devise some kind of sham treatment, such as tablets of an inert extender or injections with saline. This treatment is called a *placebo*. Surprisingly, placebos often relieve subjects' symptoms. When subjects in a treatment group receive treatment (a tablet, injection, or another kind of intervention), they experience a therapeutic setting. Subjects in a control group without a placebo will understand that they did not receive any treatment. They may embellish reports of symptoms, even produce apparently biological responses observed through apparently objective measurements, perhaps to "please" the investigators. Some studies still include such an untreated group, to establish the natural history of the study population. Particularly when the measured responses are subjective or self-reported, a placebo group can be crucial to scientific success. The difference in outcomes between a natural history group and a placebo-treated group is the placebo effect.

Louis Lasagna, a pharmacologist, carried out the most impressive of the early studies about the placebo response. He and his coworkers investigated the effects of narcotic analgesics on patients with severe postoperative pain. To the surprise of all at the time, approximately 30% of patients did not obtain relief from either placebo or analgesic, approximately 40% responded to analgesic and not placebo, and approximately 30% responded to both. In addition, many patients responded inconsistently to the placebo.

Consider the difficulties in this study. First, the severity of pain varied among patients. So, the dose of morphine to relieve pain varied, even among placebo non-responders. Second, the subjective estimation of pain relief varied among patients. Some inconsistency in response must have related to these variables. However, the tissue damage after surgery is constant over a short time, and thus, the pain was more chronic in nature. Further, by allowing repeated doses, the investigators reduced some sources of variation.

Eventually the medical community learned that very subtle influences can change the outcomes of clinical studies. Investigators administering a treatment can unintentionally communicate their expectations to study subjects in nonverbal ways. These discoveries led to the development of double blind protocols, in which neither the subjects nor the field investigators know who receives treatment and who receives placebo. In a drug study, for example, the primary investigators make the drug and placebo appear as similar as possible. The field investigators administer these apparent medications according to a set protocol without knowledge of which is which.

In pain studies, pharmacologists eventually discovered the natural endorphin system in the nervous system. Some nerve terminals can release endorphin, which can act on other nerves to relieve pain. Narcotic analgesics mimic endorphin. In situations like the one that Lasagna studied, the therapeutic setting apparently facilitates the release of endogenous endorphin in some patients.

These discoveries led to an important, if still controversial, idea of enriching the populations in clinical studies. Investigators select subjects who respond to prior treatment with a prototype medication but do not respond to a placebo. Similarly, some studies incorporate prior *run-in trials* to select subjects who do not respond to a placebo. Run-in trials are common, for example, in studies of psychopharmaceuticals, like antidepressants.

Hazard and Causation

Perhaps because epidemiology originated in studies of infectious diseases, epidemiologists direct most of their attention to identifying the infectious agent and understanding disease etiology. Is the disease caused by a specific virus? How do events progress during the disease? Because correlations and associations are potentially misleading indications, epidemiologists pay close attention to the question of causality. In addition, new applications of epidemiological methods to chronic diseases have created new problems in understanding identity and etiology.

Two British epidemiologists, Bradford Hill and Richard Doll, studied the incidence of lung cancer in British physicians who smoked cigarettes. Hill and Doll developed new criteria for causation when an exposure contributes to the risk of disease but does not act as the only cause. In 1962, the U.S. Surgeon General assembled an advisory committee to study the effects of smoking on health. The committee reported that statistical data alone does not prove causality. It listed five criteria that help to determine a causal relationship. Each of the criteria can have exceptions. Later, Hill expanded the list, now known as the *Bradford Hill criteria* to nine (strength, consistency, specificity, temporality, biological gradient, plausibility, coherence, experiment, and analogy), which differ in importance, as follows:

Strength of a correlation relates to the magnitude of relative risk. A high relative risk (>10) supports the possibility of a causal relationship. The higher the relative risk, the less likely that chance alone could generate an association. Similarly, a confounding factor has to have a higher relative risk than the suspect variable to generate a secondary association. If a confounding factor has a very high relative risk, the pattern of the studies should reveal the causal factor, which must be prominent.

Robert Koch won the Nobel Prize for Medicine and Physiology in 1905 for his many discoveries in microbiology, including the cause of tuberculosis. In earlier work with anthrax, Koch developed three postulates to prove that a pathogen causes a disease.

(1) The suspect pathogen associates with the disease.
(2) The pathogen will grow outside the host in a pure culture after isolation from a case of the disease.
(3) Inoculation of a susceptible animal with the cultured pathogen reproduces the disease, and organisms isolated from the infected animal appear identical to the suspect pathogen from the case.

Koch's first postulate merely establishes a correlation, although he meant it in an all-or-none fashion, such that the pathogen should occur in all cases of disease and in none of the controls. Most scientists believe that the first postulate is met when the pathogen generally is found in cases and generally is not found in controls. The second postulate establishes the identify of the pathogen. The third postulate tests the hypothesis that the pathogen causes the disease. Reproducing the disease with a pure culture of the organism establishes that it is not an opportunistic by-stander to a different, causative pathogen.

Modern medicine makes some exceptions to Koch's postulates. For example, association of antibodies against an organism often serves as evidence for the first postulate. Many scientists believe that protection against a disease by inoculation with an attenuated strain of the organism (one that lacks genes for virulence) meets the third postulate. A prospective intervention study that results in

In contrast, many epidemiologists think that studies with a relative risk of two or less merit little attention. Obviously, this idea is just a rule of thumb. A relative risk below two in a large population may still suggest consequences for many people. The rule of thumb arises from attribution of risk to an affected individual in the population. If the relative risk is less than two, it is more likely than not that any affected individual acquired the outcome in question from some other background factor, not the relationship under scrutiny. Hypothesis-testing studies of relative risks below two need large study populations and an excellent quality of data to persuade skeptical epidemiologists.

Specificity of a correlation requires a knowledge of which variables are independent and dependent. In most studies, we hope that exposure is the independent variable and that the disease outcome depends on exposure. Specificity refers to the uniqueness of the linkage between the two factors. For example, acquired immune deficiency syndrome (AIDS) occurs in association with intravenous drug use, but it also associates with other exposures, such as nitrous oxide inhalation or a large number of sex partners. Intravenous drug use also associates with other diseases, such as hepatitis. Even so, based on an association alone, intravenous drug use could still be the cause of AIDS. Thus, specificity contributes to a conclusion that an exposure is a causative factor for an outcome, but a lack of specificity does not detract.

Temporality in a correlation relates to the sequence of exposure and outcome and to the timing of this sequence. Cause must come before effect. If the car wreck occurs after the broken leg, we reject the notion that the accident caused the fracture. If a disease occurs before an exposure, the idea that exposure led to the disease makes no sense.

diminished prevalence of the pathogen and decreases disease prevalence, sometimes will substitute for the third postulate. For a chemical substance or form of radiation, meeting the second postulate may prove impossible. In the strictest sense, Koch's postulates may only apply to infectious diseases. Still, for any agent and disease the underlying logic remains of demonstrating (1) association, (2) independent manipulation of the agent, and (3) prospective replication of disease or its absence.

Latency complicates the temporality criterion. If a disease requires an incubation period, an exposure that occurs just before the end of the latency period still violates temporality. A different concern about timing occurs when the outcome follows the disease by such a long time that you would not think a relationship reasonable. A violation of temporality is the only Bradford Hill criterion that completely eliminates the possibility of a causal relationship. Lacking evidence for any of the other criteria, we can think of reasons why a relationship still might exist. Exceptions to the temporality criterion can still occur with poor quality data, such that disease and exposure get extensively misclassified. Even so, documented violations of temporality in a potential relationship usually are fatal to notions of causation.

Coherence of a correlation indicates a consistency of findings with the etiology and symptoms of a disease, including applications of other scientific disciplines, such as biochemistry, microbiology, anatomy, physiology, and pathology. For example, epidemiologists will map disease occurrence and study the coherence of the spatial distribution with the characteristics of disease transmission. With injuries that relate to chemical exposures, toxicologists will question whether the epidemiological manifestations of the substance coincide with its *mechanism of action,* the detailed description of molecular events that link exposure to outcome.

A *biological gradient* in a correlation corresponds to the exposure-response relationships that we have explored in this and previous chapters. Hill's idea was that disease prevalence (even better, disease incidence) generally changes in concert with increasing exposure. (Hill also implied that the severity of a disease might relate to exposure.) A biological gradient does not require a monotonic or consistent increase with each exposure. In addition, a biological gradient does not connote a linear proportionality of exposure with outcome. The absence of a relationship between exposure and prevalence (or severity) does not count against causality because a biological gradient may not change significantly when two exposures are close together, epidemiologists may not understand how exposure changes disease, or the quality of exposure data may be poor. A confounding factor that explains a correlation between an exposure and an outcome also will generate a biological gradient. For example, if industrialization leads to decreasing nest sites for storks and decreasing birth rates, discovering a biological gradient between the number of stork nests and the number of babies does not strengthen a causal relationship between storks and babies. However, the absence of a biological gradient would have weakened it. Neither the presence nor the absence of a biological gradient is of great importance.

Biological plausibility means that information about the mechanism of action is consistent with our belief that a correlation is based on a causative relationship. Simply because epidemiological observations often occur before mechanistic studies, many relationships lack biological plausibility. Thus, the absence of biological plausibility does not contradict causality. Epidemiology is notoriously weak in this area, however, and you should examine glib assertions of biological plausibility with skepticism. For example, the assertion that reducing agents chemically decrease levels of free radicals after ultraviolet irradiation of DNA seems consistent with a correlation between blood levels of natural reducing agents in blood and damage from exposure to sunlight, until we discover that chemical reaction requires molar levels of strong reducing agents, whereas human blood contains micromolar levels of weak reducing agents. Further, the wavelength of the ultraviolet light in the chemical reaction differs from the wavelength of the sunlight causing human photochemical damage.

Assertions of biological plausibility based on several independent mechanisms are particularly suspect. Far from leading to an overdetermined plausibility, the existence of more than one potential mechanism usually means that no one understands the biology. In addition, you often will encounter mechanistic explanations of effects that do not exist. The notion is that the mechanism looks so nifty that an association ought to occur, a good motivation to test for a relationship but not to assume one. Overall, biological plausibility contributes weakly to causation.

Experimental support for a correlation requires controlled testing of a proposed cause and effect relationship. For example, if we intervene to increase the birth rate by building stork nests, and new babies do not come, we severely undercut any notion of causation. Because risk assessments often lead to important social decisions, the presence or absence of experimental support is a critical factor. Often, regulatory decisions lead to costly interventions, without validation of the proposed benefits after the regulation is in force. The assessor should look carefully for hypothesis-testing studies in these circumstances. In ecological studies, investigators often can use the same species in laboratory experiments that they observed in the field.

In epidemiology, a prospective study is a powerful tool to obtain experimental support for a causal relationship. Corroborating intervention studies similarly generate strong support. Replication of an association in laboratory animals also supports confidence in causation. For example, toxicologists may reproduce an association between a chemical substance and human cancer in a controlled experiment with rodents. Overall, experimental support is a strong criterion of causality. Experimental support overlaps with biological plausibility. The difference is that biological plausibility involves theories that support an association, whereas experimental support concerns direct testing of the relationship. Hill correctly separated these two criteria.

Analogy about a correlation also relates closely to the criterion of biological plausibility. In general, any similarity between one association and another adds confidence to the idea of causation in both, but analogy is a weak criterion. For example, if two chemical substances have structural similarities, and if both lead to similar outcomes, such as peripheral nerve toxicity, the analogy contributes support to conclusions of causal relationships for both substances. Likewise, if two chemically unrelated substances both emit the same kind of ionizing radiation and generate similar outcomes, the analogy becomes stronger. If humans metabolize a chemical substance to a previously studied substance or utilize a similar metabolic pathway, analogy supports an inference of causation. The idea that some chemical substances cause cancer in humans lends support to the possibility of additional substances causing cancer, but only a little. At a minimum, risk assessors should inquire whether the organ sites and pathologies of the cancers are similar in the two associations.

Association clearly is not causation, but causation requires association. Some epidemiologists regard causation as a special form of association with statistical trends for one or more of Hill's criteria. We disagree with this notion. Instead, we regard causation as a scientific hypothesis about a mechanism of disease. Thus, you can disprove causation, but you cannot prove it. The most that you can do is summarize how you have tried to disprove causation and failed.

The evaluation of a potential relationship for causation is a crucial activity in epidemiology, and presently, epidemiology concerns hazard more than risk. Epidemiologists will properly spend extensive time evaluating the quality of a study, as part of their overall interpretation of an association. They look in depth at the analysis of data, ascertainment of outcomes, conduct, documentation, design, and exposure information. Neither hazard nor data quality is such an important topic for risk assessors. Analysis of the exposure-response relationship is more important for them. The choice of an appropriate risk model, one that incorporates as much biological information as possible, also is a crucial activity in risk assessment. Traditionally, risk models merit little attention from epidemiologists. Thus, epidemiology has much to contribute to risk, but neither hazard nor causation is necessarily a crucial topic for a risk assessment.

Epidemiologists rely on certain authoritative sources which often use groups of experts to describe hazards. For example, in cancer biology the primary authoritative source is the International Agency for Research on Cancer (IARC). IARC reviews human, animal, and mechanistic data within a classification system that differs from (but overlaps) the Bradford Hill criteria and that periodically undergoes updating. IARC publishes monographs about the classification of substances and updates the classifications whenever their system changes. The monographs place primary emphasis on epidemiological studies. In essence, IARC designates any substance with two or more good quality epidemiology studies that show excess cancer prevalence after exposure, as a known human carcinogen.

Other authoritative sources of information exist. For example, you may want to consult risk assessments or hazard evaluations performed by governmental agencies for information about the epidemiology of chemical substances and radiation.

Epidemiology in Transition

Everyday seems to bring new press reports about awful risks, as revealed by the latest epidemiology study. Over time, the inconsistency of these reports induces much skepticism in the public mind. One day, dietary fat is the scourge of humanity; the next day, a lack of dietary fat causes depression and suicide. One day, electromagnetic forces from wiring are causing widespread leukemia and brain cancer; the next day, electromagnetic forces get a clean bill of health. Somehow the overall impression from these reports is that human health grows worse in the U.S.

Risk assessors know better! Since World War II, life expectancy has increased in almost every country in the world. In some countries, life expectancy has increased by approximately twenty-five years, the same as the total life span of people living two millennia ago. World-wide, most people live to age sixty-five. Unfortunately, this information is not considered reportable, whereas oversimplifications and unqualified results from the latest epidemiology study are hot news to the press.

An incredible advance in medicine in modern industrial societies over the past two centuries succeeded in eliminating many microbial diseases. Research discoveries translated into therapies, interventions, and other public health measures. Clean drinking water and adequate removal of sewage had dramatic effects. Some diseases, such as yellow fever and polio, occur much less frequently. Through a combination of quarantine and vaccine measures, smallpox no longer exists as a disease entity. Newly arising infectious diseases, such as AIDS, although severe, have not replaced the overall impact on mortality from the older diseases. As a consequence, average age increased, and the major causes of death shifted to chronic, noninfectious diseases. Right now, several third world countries are passing through the same transition. As each population shifts, better documented causes of death become available. Epidemiological studies shows that risky exposures other than infectious diseases impact life span, such as cigarette smoking, which strongly associates with lung cancer, or sun tanning, which strongly associates with skin cancer.

This epidemiological transition imposes new difficulties on professional practice. Relative risks are lower, and causes are more difficult to discover. The publicity surrounding a new study requires more circumspection, and epidemiologists have new responsibilities to investigate potential causes in more depth.

The Impact of Information Technology

Widespread access to inexpensive information tools, such as personal computers and the Internet, has changed our society. Nowhere have the applications of this new technology caused more dramatic changes than in epidemiology. Information technology will support widespread availability of the data necessary to train new epidemiologists, to obtain independent analyses of reported studies, and to develop epidemiological methods. Ready availability of computational power frees epidemiology from burdensome and tedious calculations. Statistical programs to format and analyze epidemiology studies are available at low costs, sometimes even for free. Epidemiologists even publish descriptions of work underway to help others avoid unnecessarily duplicating studies. Some epidemiologists oppose access to study data, but overall these trends support a vigorous practice.

The Effect of Specialization

Besides general skills in observational studies and statistics, epidemiology demands an understanding of the outcomes observed. Even for medical endpoints, such as death, cancer, or neurotoxic effects, the evaluation of observations requires specialized expertise in nosology, oncology, or neurology. To date, epidemiology has avoided the tendency to fragment into many subdisciplines, although it has a general split between field work and methods development. How much longer the profession can avoid these centrifugal forces is not clear, but at present specialists in different areas can still discuss and appreciate the accomplishments of others, maintaining a robust profession.

The Perspective of Risk

Historically, risk assessors have made little use of epidemiological data. Usually, analysts juxtapose human risk estimates derived from animal data with estimates from human data to see if a contradiction exists. Similarly, epidemiologists seldom use risk assessment approaches. Estimates of the risk associated with an infectious agent or a chemical substance are not the same as attributions of risk. In the latter, epidemiologists partition risk from various factors to the increased prevalence of disease or death in an exposed population. While risk assessors can convert these estimates into potency estimates, they are not the same. Risk factors generally partition the observed cases of disease, assuming equivalent exposure throughout a population. So, conversion yields a point estimate (prevalence/exposure). Often, the stated risk factors add to significantly more, or fewer, cases than those observed. Interactions between factors could account for these differences, but inhomogeneous exposures and nonlinear dose-response relationships will generate the same observations.

Epidemiologists and risk assessors also have different attitudes towards causation and hazard. (See the section on the Public Health Paradigm in this chapter.) Risk assessment does not require precision in either the specification of a hazard or the attribution of its cause. Instead, risk assessment treats these parameters as sources of uncertainty. The current practice of epidemiology does not focus much attention on dose-response relationships. Of the Bradford Hill criteria, biological gradient is most relevant to risk assessment but contributes only weakly to causation. A simple trend within one or more studies is sufficient; quantitative agreement in potency estimates is not required. The strength of a correlation is a persuasive criterion of causation but merely a happenstance of exposure magnitude in risk assessment. Experimental support for a correlation also is a powerful criterion of causation, but risk assessors seldom get around to testing hypotheses. Biological plausibility contributes only weakly to causation, but a mechanism of action often is a vital precursor for model construction in risk assessment.

Risk assessment is important in governmental regulation of hazards. When epidemiologic data support the conclusions of a risk assessment, obviously risk management can proceed with greater certainty. Sometimes, only epidemiologic data support major regulatory actions. Some recent examples are inhaled particulates and respiratory disease, ozone and respiratory distress of asthmatics, benzene and leukemia, and arsenic and cancer. Because remedial action is the equivalent of an intervention study, and the consumers of risk assessments often provide support for epidemiological studies, risk assessment also is of great interest to epidemiologists. Their science will progressively absorb more risk assessment methods, a trend already underway. The risk assessment void in epidemiology can only be temporary. We view it as an opportunity.

Cancer Epidemiology

Cancer is a generic term for a group of diseases characterized by a fatal, uncontrolled growth of aberrant cells derived from normal body tissues at inappropriate body locations. Each disease has its own tissue of origin, medical characteristics, and progression.

During both development of the body and replacement of lost cells in the adult, stem cells produce two kinds of progeny: (1) new stem cells and (2) mature, differentiated cells that acquire the properties and functions of a particular class. As cells differentiate more, they lose the capacity to divide and produce progeny. For example, adults constantly lose and replace skin cells and red blood cells. The stem cells and signals that produce skin cells clearly differ from those that produce red blood cells. The overall process of differentiation is tightly controlled.

Intercellular signals notify stem cells to produce or to cease producing differentiated cells, as required, so that the body does not overproduce or underproduce replacement cells. Biological signals also control the location of the cells, such that skin cells do not grow in the blood compartment, and blood cells do not appear on the surface of the skin. Similarly, animals tightly control the pool sizes of different kinds of stem cells. During development, animals increase their overall numbers of differentiated cells to achieve adult body size. In addition, they need to create specific kinds of cells at appropriate times and in patterns for the correct sizes of organs and tissues.

Some noncancerous diseases of growth control exist. For example, skin cells pile up inappropriately to produce psoriasis, and patients overproduce red blood cells in polycythemia. A tumor is a visibly inappropriate accumulation of cells at a normal location. In addition, diseases of cell location exist when cells of one tissue grow at an inappropriate site, such as an extra nipple. Some tumors, such as brain tumors, kill because of excess cell proliferation in a closed compartment. Mostly, tumors disfigure cosmetically but do not kill. The essential property of most fatal cancers is loss of normal growth control at the original site, as well as metastasis, or uncontrolled growth of tumor cells at distant sites.

The number of human carcinogenic diseases differs from authority to authority, usually varying between sixty and one hundred. The gold standard for diagnosis also varies between the diseases and is usually multifactorial, requiring knowledge of clinical symptoms, biochemical tests, and histopathological examination of biopsy tissue. Cancer cells usually have a distinctive appearance under the microscope.

Morphologically, most cancers display a common progression from (1) apparently healthy cells to (2) dysplastic cells to (3) carcinoma cells localized at one site to (4) a tumor that invades surrounding membranes to (5) detached cells that spread through lymph nodes and create secondary tumors to (6) metastatic cells that aggressively invade many tissues. Some kinds of cancers seem to skip one or more of these stages. For example, breast cancer has usually reached step six when it is first diagnosed. Most likely, intermediate stages do occur but are not visible to pathologists. Recent studies with molecular markers of carcinogenesis, particularly with gastrointestinal and brain cancers, show that some genetic changes, which predispose the cells to cancer, take place while the tissues still appear normal.

Genetic alterations of cells lead to heritable losses of both location and growth control in cancer cells, which retain these properties when grown in tissue culture or passed through new host animals. Unlike germ line mutations, *somatic mutations* do not pass between generations. Somatic mutations occur constantly from background processes. The mutational process has two requirements: (1) alteration and (2) fixation. In addition, two kinds of genetic changes can occur: (1) loss of tumor suppressor genes and (2) changes in the properties of normal genes, such that the altered phenotype predisposes to cancer. The latter are called "oncogenes." Unlike the expectations of many cancer biologists, the advancing genetic changes do not occur in a consistent pattern and often do not correlate closely with morphological progression. Instead of mutation A followed by mutation B followed by mutation C, the mutations occur in a random pattern. For example, mutation A, B, or C precedes a mutation in any other of the three genes, A, B, C, or even in a new gene, D. Often, six or seven mutations may accumulate in a diagnosed tumor. Some mutations are more devastating, however, and the loss of a tumor suppressor gene usually predisposes more than an oncogene mutation.

Many different processes lead to genetic alterations, including radiation, chemical adducts attached to DNA, the genetic material, insertion of virus genes into DNA, normal oxidative processes, some hormones, and background errors in the normal replicative machinery. Animal cells have a variety of defense mechanisms against these insults, including DNA repair processes. Some mutations have lethal consequences, and the mutated cells die. Some mutated cells simply do not replicate; so, no progeny cells can result to form a cancer.

Some toxic insults cause normally quiescent cells, which already have mutations from a background process, to divide. These cells may fix the mutations. These cells will accumulate in number, and they already have mutations that predispose to cancer. The target population for subsequent changes increases. Thus, all processes that lead to cancer cells require cell division. High rates of cell death lead to high cell division rates through efforts to repair tissues. In addition, any process that increases the rate of mutagenic alterations of DNA, or decreases DNA repair, will increase the incidence of cancer cells among dividing cells. Because of the requirement for loss of both growth control and location control, we expect that a minimum of two mutations must occur before an initial cancer cell arises. Because mammalian cells carry two copies of every gene, located on different copies of the same chromosome, even the loss of tumor suppression requires two separate mutations.

Epidemiological studies have shown that incidence of any cancer varies with age and other demographic characteristics. The incidence of most cancers increases with age, as predicted from the need to accumulate several mutations within one cell before the cell exhibits carcinogenic properties. Malcomb Pike and his coworkers have shown that breast cancer depends on the lifetime supply of the female hormone, estrogen. The incidence increases in proportion to age, starting from a low rate in puberty and rising until menopause, then declining again. Some kinds of cells in breast tissues proliferate in response to higher estrogen levels which occur during their monthly rise and fall in the menstrual cycle. Thus, the incidence of breast cancer relates to hormonal control of the many cycles of cell proliferation during the portion of life with active estrogen secretion.

Some pediatric cancers, such as childhood leukemia and some kinds of testicular cancer, peak in incidence early in life. We expect this phenomenon because some kinds of cells proliferate rapidly during early developmental stages, and proliferation ceases when tissue formation is complete. Cell replication carries an intrinsic risk of somatic mutation.

Genetic predispositions to cancer in families fits well with this picture of somatic mutation. For example, Li-Fraumeni syndrome is a germ line mutation of a tumor suppressor gene. Families with this mutation express higher incidences of many kinds of cancers. When passed from one parent, only one defective gene appears in a child. However, half the normal allotment of suppressor genes still predisposes to cancer in general.

Some infectious diseases are associated with cancers. Hepatitis B virus (HBV) associates with liver cancer. Intervention by vaccination against HBV reduces the incidence of liver cancer. Liver flukes also are associated with liver cancer. Whether these agents cause cancer by increasing mutation rates, increasing cell turnover, or both, is not well understood. Many viruses insert their genes into the chromosomes of host cells, which may delete genes from the host. If the host gene codes for a tumor suppressor, the progeny will have a predisposition to cancer. In addition, many viruses kill host cells, leading to many rounds of replication by uninfected, normal cells as replacements. Waste products from the invading organism or from the inflammation at the site of infection may produce DNA adducts.

Aflatoxin B in food also is associated with human liver cancer. *Aspergillus flavus*, a mold that grows on stored food, secretes aflatoxins in response to stress, which cause severe, acute liver cell death. Metabolites of aflatoxin B form covalent adducts with DNA that are mutagenic. Thus, aflatoxin B has both of the properties that lead to cancer induction, increased mutation rate and increased cell turnover from repair of the damaged liver. The population in the south of China has a high prevalence of liver cancer associated with excess aflatoxin ingestion. Most, if not all, deaths from liver cancer in epidemiology studies of this population occurred in people who had chronic HBV infections. The relationship between HBV infection and aflatoxin B in inducing human liver cancer is not well understood. Yet, these studies form the basis for regulatory exposure limits for aflatoxin in countries like the U.S., where HBV infection is unusual (WHO, 1997). Bladder cancer is associated with beta-napthylamine in certain dyes that are occupational hazards. Metabolites of beta-napthylamine form covalent adducts with DNA that lead to mutations.

Similarly, ionizing radiation has sufficient energy to disrupt covalent bonds in DNA, and different sources of ionizing radiation are associated with increased cancer rates. Because external radiation sources often lack much specificity for target tissues, radiation sources are associated with many kinds of cancers. Higher prevalence is seen in groups exposed to more radiation, such as uranium miners. Approximately 0.01 of the much studied Japanese atomic bomb survivors died of cancers related to radiation, including leukemia and solid cancers of the bladder, breast, colon, lung, stomach, and thyroid.

Nonionizing radiation in the ultraviolet portion of the energy spectrum induces lesions in DNA that correlate with skin cancer.

Exposures to radiation sources, chemical substances, and infectious diseases vary with geographical location and over time. In addition, cancer prevalence genes congregate locally in particular populations through families. Thus, we expect cancer to exhibit geographic and temporal variation. Often, trends signify specific causes. High use of chewing tobacco correlates with a high incidence of oral cancer in India. Currently, skin cancer is highly prevalent in our society. Most of the melanoma, basal cell carcinoma, and squamous cell carcinoma cases in the U.S. over the last few decades correlate with preventable exposure to sunlight. Cancers associated with AIDS, such as Kaposi's sarcoma, have become common as the AIDS epidemic spreads.

The age-specific incidence of different kinds of cancer will change with time and will concentrate in regions. Thus, cancer trends should be distinguished from changes in prevalence caused by increased age in a population. Also, decreases in any competing cause of death will increase cancer prevalence as will demographic age bulges caused by changes in birth rates. For example, the baby boomer generation will increase cancer prevalence as it ages. Similarly, improved diagnostic techniques, particularly earlier diagnosis, will lead to short-term increases in cancer prevalence. For a while, the cases previously diagnosed during a short interval will overlap with cases diagnosed earlier. When this phenomenon occurs, prevalence eventually falls back to the rate seen before the introduction of the new diagnostic procedure.

In the U.S. we have excellent statistical data about the age-specific incidence of many cancers due to the efforts of several state and local cancer registries, the American Cancer Society, and the National Cancer Institute. (See Figure 6.2.)

Because life span has increased in the U.S., cancer prevalence has generally increased. However, overall cancer incidence has fallen. This statistic is the product of many trends since major changes have occurred in the incidence of several cancers. For example, lung cancer incidence has increased, while stomach cancer incidence has fallen.

Not all cancer cases die; cancer mortality depends on both incidence and survival. Some cancers, for example, some kinds of skin cancer, are highly treatable. In addition, some cancers progress slowly, and competing causes of death reduce the death rate from the cancer. Mortality declines as treatment improves. The last half of the century has witnessed incredible progress in treatment. Fifty years ago, pediatric leukemia was usually fatal. Today, more than nine out of ten cases of pediatric leukemia survive, and more than forty thousand of these children are now adults—enough to populate a small city. Five-year survival is the gold standard for effective cancer treatment. The fatality ratio (deaths/cases) for any kind of cancer is the probability of surviving for five years after diagnosis.

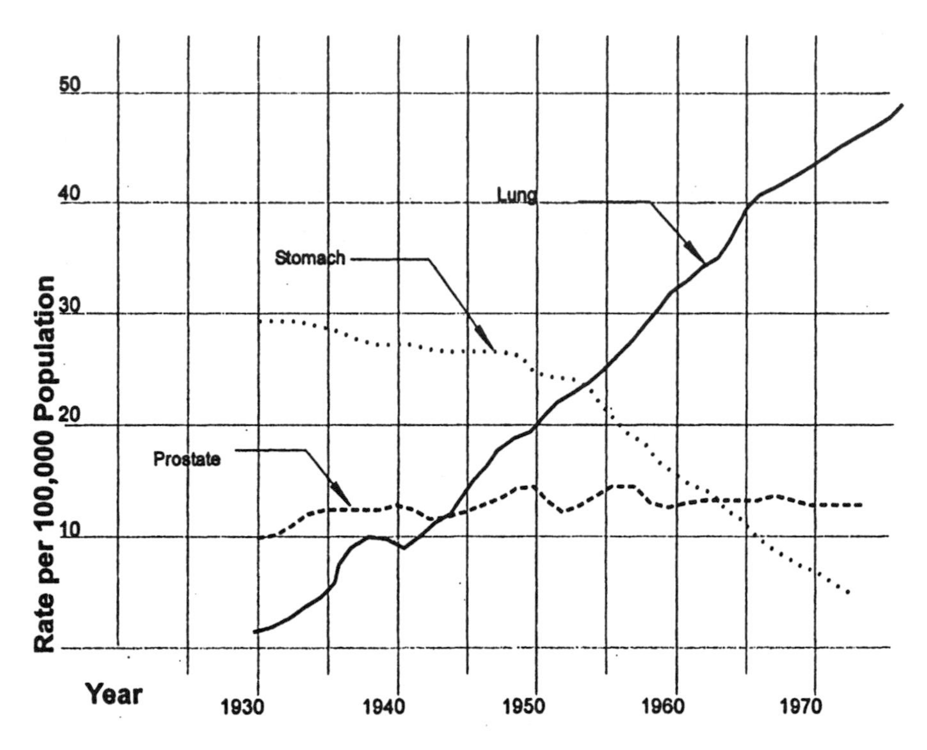

Figure 6.2 Age adjusted cancer death rates of U.S. men, 1930 to 1970 (National Cancer Institutes data)

The International Agency for Research on Cancer has listed more than thirty known human carcinogens. Some of them are iatrogenic, that is, physician-associated. Use of diethylstilbestrol, an estrogen substitute, to sustain pregnancy, led to increased breast cancer in the treated women and to cervical carcinoma in their daughters. Use of alkylating agents in cancer chemotherapy correlates with leukemia. Use of tamoxifen in breast cancer treatment correlates with endometrial cancer.

However, the mechanisms behind some well known associations with cancers remain unknown. Some authorities attribute as much as one-third of cancer cases to diet. Aflatoxin and certain food preservation techniques, such as pickling meats, contribute only a small portion of this association, and they are the only known causes. At present, diets with higher fruit, vegetable, and grain content are known to correlate with lower incidence of several cancers. Speculations abound about other causes, such as dietary fat content or the presence of specific fats, and are usually accompanied by poor quality data. Two common problems occur throughout all nutritional studies of cancer: (1) an increase in other substances replaces each substance diminished in the diet and (2) overall food consumption varies with the quality of the diet. Thus, something as basic as total calories may eventually explain the association between diet and cancer.

Similarly, lower socioeconomic status (SES) correlates with higher overall mortality and with a higher incidence of many cancers. Thus, age effects do not explain the correlations with cancers. Any reasonable quality epidemiological study will account for SES, usually through several indicator variables, such as family income. Lower family income correlates positively with other cancer-related variables, such as cigarette smoking and poor quality diet. In contrast, higher SES correlates with a higher incidence of breast and testicular cancers. Thus, SES is a widespread and poorly understood confounder of all epidemiological studies of cancer.

An emphasis on external causes of cancer has led to a belief that all cancer is preventable and a misperception that environment causes most cancers. The former is possible but not likely. The latter notion clearly contradicts the factual evidence. No credible authorities attribute more than 5% of all cancers to environmental causes; most experts suggest no more than 2%.

These notions are important because they often distort the regulatory agenda. Cancer assumes disproportionate importance in regulation. In addition, these ideas point to a major difficulty in analyzing epidemiological data for risk assessment purposes: whether the excess of cancer associated with an exposure is absolute or relative. An absolute increase would occur independently from the background incidence, whereas a relative increase interacts with the background rate. Under the relative risk idea, a population with a low background incidence of a cancer will exhibit a smaller incidence after the same exposure as a population with a higher background incidence. Under the absolute risk idea, an increase in the incidence depends only on exposure.

Exposure-Response Relationships in Epidemiology

The simplest circumstance in epidemiology is a comparison of an exposed population to an unexposed population. You have little choice but to draw a straight line between the two points and assume that prevalence (or incidence) is proportional to exposure. If the control group had no exposure and no apparent outcomes, the result of this mathematical procedure suggests a linear exposure-response relationship without a threshold. However, individuals within the exposed population were initially exposed at different ages; they experienced different magnitudes and different durations of exposure. Even if exposure is based only on qualitative characteristics, such as job classification, placing all of the exposed individuals into one group loses much information.

At the most complex level of exposure-response analysis (the opposite of lumping all of the exposed study subjects into one group), each study subject becomes a point of comparison. The exposure and outcome of less than one individual makes little sense. Thus, exposure of the individual ultimately is the proper basis of an exposure-response model, if we want to extract the most information from an epidemiology study. To analyze data about each individual in a study, however, we will have to have some models that we can compare.

Intermediate approaches are possible by classifying the subjects into several groups, for example, high, medium, and low exposures. The more groups, the better the discrimination of the exposure-response relationship, but the greater the range of variation about the shape. The same number of subjects gets divided into more, but smaller, units of observation. Even this approach has the liability that the chosen number of groups or the classification procedure may generate a result that the data, on closer inspection, turn out not to support. (This situation partially explains why qualified independent investigators should have access to the detailed results of epidemiology studies. Re-analysis may prove crucial.) As a routine matter, most epidemiology studies do not apply models to observational data about the individual subjects. Many exposure-response models assume an in-depth understanding of the disease relationship, and mechanistic models provide the most informative approach. The assessor should beware of studies that report shapes for their exposure-response relationships that actually portray the shape assumed in composing the mechanistic model used to evaluate the data.

Most regulatory authorities base the standard for aflatoxin content of food on a potency value obtained from one epidemiology study of liver cancer induction (Yeh et al., 1989). This study had many confounding factors that perplexed the estimation effort. For example, all of the liver cancer cases occurred among people infected with hepatitis B virus (HBV). HBV infection is chronic, and it also causes liver cancer. Conventional risk assessment models do not work well with epidemiological data about populations experiencing two exposures that correlate with the same outcomes (Bowers et al., 1993). Most assessments have to use a series of dubious assumptions, such as additivity between aflatoxin and HBV.

The study populations (Yeh et al., 1989) suffered from several potential causes of liver cancer, including aflatoxins and chronic infection with hepatitis B virus. But, allocating risk of liver cancer among several apparently independent factors has proven quite a problem. Conventional multifactorial statistics did not work with this population (Wu-Williams et al., 1992). In a multifactorial approach, statisticians assume interactions between different factors. When all plausible assumptions lead to the same result, the results are thought to be indicative. However, in this case, different, plausible assumptions generated very different results. The assumption of a linear interaction between (equal contributions of) hepatitis B infection and aflatoxin intake is clearly wrong, both in fit to data and in biological basis. Yet, the available regulatory analyses have not estimated nonlinear relationships.

In contrast to the situation with aflatoxin, epidemiologists have obtained good success with radon. (For more background information about the risks of radon, see the case studies in Chapter 13.) The circumstances are fairly straightforward. Health experts believe that both radon and cigarette smoking associate with lung cancer. Miners have higher exposures to radon, because the movement of rock releases radon and underground containment traps this gas. Because uranium breaks down to radon, radon levels are particularly high within uranium mines. If we compare the cases of lung cancer in a mining cohort to nonminers, the following three questions arise:

(1) Do radon and cigarette smoke interact in any way? If not, the analysis of the data seems more straightforward. We can account for each lung cancer case as an outcome of exposure to radon *or* to cigarette smoke. We could apply the "OR" logic described in Chapter 1, which suggests additive risk. However, if we work with grouped data, this approach becomes difficult. We could develop a cohort of miners exposed to one average level of radon, and a cohort of nonminers exposed to a lower level of radon. If the two cohorts have the same proportion of smokers, we might attribute the increased relative risk of lung cancer to radon alone. (The two cohorts may not have the same proportion of smokers. For example, a greater proportion of miners might smoke. If so, we force onto the study the bizarre conclusion that cigarette smoking "causes" mining. This erroneous reasoning is another example of why association is not the same as causation.) We could reach outside the study data and obtain potencies for radon and for cigarette smoke. After we apply them, if we have some extra lung cancer cases, we could attribute them to measurement uncertainties, mis-estimation of radon levels, for example. If we get lucky, we will have few "extra" cases. But if we only have a potency for cigarette smoke, the situation becomes more difficult. We would depend on the precision of the potency for cigarette smoke, because our estimate of radon potency would depend on the lung cancer risk left over, after we subtract the cigarette smoke. Uncertainty about the potency of cigarette smoke would overwhelm our analysis. The alternative approach of staying within the comparison of the two cohorts would impose a different approach on the analysis of data. We would look at the risk of lung cancer among miners who do not smoke cigarettes, compared to miners who do smoke, and nonminers who do smoke, compared to nonminers who do not smoke. Notice that we now have four exposure groups.

(2) Do radon and cigarette smoke interact? We could approach the data from a different perspective and ask what the comparison of the two cohorts tells us about the joint effects of radon and cigarette smoke. The most radical conclusion might be that radon exposure has no risk of inducing lung cancer. Instead, radon simply increases the risk of lung cancer among cigarette smokers. (Alternatively, we could assume that cigarette smoking does not induce lung cancer, cigarette smoking simply increases the risk of lung cancer among those exposed to radon.) We would look at the extra cases from our earlier analysis and attribute them to a joint effect of radon and cigarette smoke using "AND" logic. We could look for a multiplier of the multiplication product of the two potencies that would yield the number of extra cases observed. The difficulty of assuming single potencies for mining or not, and for cigarette smok-

ing or not, would soon become apparent. The range of risks would overwhelm our analysis. In effect, the uncertainties in the potencies could rationalize any number of extra cases.

(3) What is the nature of the interaction between radon and cigarette smoke? To understand the interaction within the study, we will need to compare quantitative exposures to cigarette smoke and radon. Our efforts to put the data into four groups having failed, we need to account for the magnitudes of the exposures. Perhaps one miner smokes ten cigarettes a day, another 50 cigarettes a day. Perhaps one subject only mined for one year, another for a lifetime. As we proceed to refine our analysis, we cannot divide the cohort more finely than down to the individual miner.

Once we begin a quantitative analysis, we will encounter new difficulties. For example, cancer typically has a latency. From the time of exposure, we expect a delay in the appearance of lung cancer. We will need to account for miners who have not been exposed for sufficient time for lung cancer to occur. Similarly, radon and cigarette smoke have other adverse health effects. Some miners, exposed long ago, may have died of a competing effect of exposure. Accounting for these effects is difficult. The difficulty is an inevitable consequence of the question we have asked about risk. Fortunately, Moolgavkar and his coworkers have successfully analyzed the risk of lung cancer risk from radon exposure (Moolgavkar et al., 1993). They found a slightly more than additive interaction between radon and cigarette smoke, but a less than multiplicative interaction, in a cohort of uranium miners.

Epidemiological analyses best use data about each individual. No one ever gets the perfect model, but any study can test some simple, informative exposure-response relationships, such as the best fitting powers of exposure and outcome, to explain the data; for example, $exposure^2$ (exposure squared) correlates with outcome given the study population. Some mechanistic models are more interesting than possible models. Modeling uncertainty becomes more important in devising epidemiological disease models, and the risk analyst needs to pay more attention to the questions asked and the biological process involved. Even if an explanatory model fits the data better than other models, it is only correct until someone comes up with a new model that fits the data even better, or fits the data almost as well and is more explanatory, or until data from a new, better quality study fails to fit.

Meta-Analysis

Epidemiologists often have two or more studies that suggest an association between exposure to a substance (or process) and an outcome. If all of the studies lead to a similar conclusion about the potency of the substance, the studies present no difficulty for risk assessors. However, uniform agreement between multiple studies is a rare situation. Most often, the studies conflict. Sometimes, they reach diametrically opposite conclusions; for instance, the sign accompanying the risk changes from positive to negative, and the substance appears to have antirisk properties, perhaps protecting against risks from other substances operating in the background.

Epidemiologists reach overall conclusions about multiple studies in various ways, including by comparing them in tables or graphs. Epidemiologists have worked out standardized ways of displaying data that emphasize confidence limits on correlations, which help to visualize study size as an explanation for variable findings (Walker et al., 1988). These approaches are most useful in deciding whether the data consistently support one conclusion or whether epidemiologists can eliminate some unusual results from outliers. The risk assessor also can question whether the epidemiologist has summarized all of the relevant studies by conducting independent literature searches.

By presenting and analyzing a group of studies, review articles usually reach an informed judgement. In this way, we rely on expert opinion of the reviewer to integrate the data about multiple studies of the same substance. Reviewers weigh many kinds of evidence in reaching expert opinions about multiple studies, including quality, experience of the investigators, statistical power, elegance of methods, and the citation index of the journal publishing them (giving weight to the quality of the journal's peer reviewers and editors). They weigh the evidence to reach conclusions about the hypotheses (even if the hypotheses are not stated) for all of the relevant studies. However, consistency of findings within a group of studies is not always sufficient to reach conclusions. Often scientists will weigh only evidence that supports a particular hypothesis, either intentionally or accidentally. (This approach is called *strength of the evidence.*)

The subject of scientific inference lies beyond the scope of this book. A recent book by Howson and Urbach (1989) illustrates some of the difficulties of scientific inference. In many areas of science, including the basis for regulatory decisions that identify "known human carcinogens," one positive study will outweigh many negative studies. The underlying logic is very similar to the approach in organic chemistry of describing a route of synthesis, even if the chemical reaction often fails in the laboratory. In contrast, risk assessors have to turn a hypothesis on its head to make sense of a model. In addition to asking what evidence supports a model, they have to ask what evidence fails to support the model, and what evidence supports alternative models.

In comparing and choosing different epidemiology studies, epidemiologists also can combine their data, as if one investigator conducted one large study, with different portions of the cohort reported by different subinvestigators. Mathematically combining data from different studies through mathematical procedures is called *meta-analysis* (Greenland, 1987). It has some advantages. Because sample size increases, statistical power increases and the confidence limits decrease. Because the procedure is mathematical, it appears more value neutral. Indeed, combining studies at the level of the individual study subject may give a meta-analysis these advantages. However, most meta-analyses cannot achieve this kind of synthesis. Instead, by combining summary results from many individual studies, they create many mathematical problems, some of them intractable. In addition, all meta-analyses have some common problems: they cannot account for unquantified differences, variable methods, and discrepant measurement techniques between individual studies.

So, meta-analysis is not a substitute for the quantitative analysis of each study. Finally, the idea that a mathematical procedure eliminates personal judgement is a fallacy. Because many ways exist to combine data, different approaches to meta-analysis can lead to diametrically opposite conclusions. A meta-analysis requires judgement in the choice of integration techniques and in the summary of quantitative variables.

Keeping these conditions in mind, the steps in a meta-analysis are: review of the literature; evaluation of the existing literature; comparison of differences in outcome ascertainment, exposure methods, and statistical techniques; selection of a model to integrate the data; description of bias, and expression of the results, indicating variation and sensitivity. The final meta-analysis lends itself to risk assessment; extracting a range of potency values from the meta-analysis is simple. The notion of using a range of potency is not difficult. However, many risk assessments, particularly those where carcinogenicity is suspected, do not reduce to a single risk figure or to a characterization that anyone can cite with confidence. The exact relationship of animal tests to human outcomes, even the predictive value of high exposure occupational epidemiology studies to the risks of environmental exposures, are open to question, especially if different studies are to be combined. But, choosing the best model to integrate data in a meta-analysis is somewhat similar to choosing the most appropriate mathematical model to extrapolate from high to low dose. Additionally, exposure involves multiple pathways, but risk assessments usually substantially simplify these pathways.

We cannot measure systematic errors. So, the assessment of risk from multiple studies becomes a matter for subjective judgement. But, just as the confidence that you can place in conclusions drawn from samples depends in part on sample size, your confidence in a risk estimate increases when it draws on the results of multiple independent studies.

The International Classification of Diseases, Injuries, and Causes of Death (ICD), published by the World Health Organization, assigns a three character alphanumeric code to every major condition. Often the ICD adds a fourth character for more exact specifications. For example, ICD C92 is "myeloid leukemia," which may additionally be specified as C92.0 ("acute") or C92.1 ("chronic"). Broader groupings are readily formed—for example, ICD C81-C96 consists of all malignant neoplasia of lymphatic and hematopoietic tissues. Physicians use this system to code death certificates, and epidemiologists rely on it. It determines the presentation of results in the registrar general's reports and in the diagnostic registers of most hospitals.

The ICD undergoes periodic revision to keep pace with medical usage. The ninth revision came into general use in 1979, and has now been superseded by the 10th revision for many applications. When the classification alters from one revision to the next, disease rates may not be directly comparable before and after the change. For example, the eighth revision included separate categories for gastric ulcer and peptic ulcer of unspecified sites, whereas in the seventh revision this distinction was not made. A meta-analysis involving studies of ulcers before and after the eighth edition needs to aggregate these categories before comparing or integrating data.

Measurement error can differ between studies we wish to combine, even if all of the studies appear to use the same analytical method. Errors in measuring exposure are an important source of bias in epidemiological studies. In combining studies, we need to assess the quality of measurements. Sometimes a reliable standard is available and is used by both studies, making comparison and integration more reliable.

We first need to evaluate a study to identify any major potential for bias. Almost all epidemiological studies are subject to bias of one sort or another. This does not mean that they are scientifically unacceptable and should be disregarded by us. However, it is important to assess the probable impact of biases and to allow for them when drawing conclusions. In what direction is each bias likely to have affected outcome, and by how much? When looking for possible biases in studies we wish to combine, three aspects are particularly worth considering:

(1) How were subjects selected for investigation, and how representative were they of the target population with regard to the study question?

(2) What was the response rate, and might responders and nonresponders have differed in important ways? As with the choice of the study sample, it matters only if respondents are atypical in relation to the study question.

(3) How accurately were exposure and outcome variables measured? Here the scope for bias will depend on the study question and on the pattern of measurement error. Random errors in assessing intelligence quotient (IQ) will produce no bias at all if the aim is simply to estimate the mean score for a population. On the other hand, in a study of the association between low IQ and environmental exposure to lead, random measurement errors would tend to obscure any relation—that is, to bias estimates of relative risk towards one. If the errors in measurement were nonrandom, the bias would be different again. For example, if IQs were selectively under-recorded in subjects with high lead exposure, the effect would be to exaggerate risk estimates.

There is no simple formula for assessing biases. Each must be considered on its own merits in the context of the study question.

The Public Health Paradigm

During the last two decades, government and automobile companies have cooperated to achieve a significant reduction in automobile fatalities. This accomplishment primarily resulted from a different approach, the public health paradigm, and not from an application of the risk analysis paradigm. Epidemiology is closely associated with the public health paradigm, but epidemiology and public health are not the same. The risk analysis paradigm involves risk assessment, risk management, and risk communication. Instead, the public health paradigm involves four steps::

(1) Define the problem. In public health, reasoning proceeds from survey data about mortality and morbidity. The public health system documented that injuries and deaths from automobile accidents contributed substantially to reduced life span in the U.S.. Accumulating accurate data—life table data—is a problem. Many persons contribute to the reporting and tabulation of deaths, a resource intensive activity. This step involves many epidemiological techniques.

(2) Elicit risk factors. Another difficult problem in public health is the attribution of potential causes. Here the association between epidemiology and public health is more loose. Many public health officials are careful to avoid the notion of causation. If the search for causation slows down a reduction in prevalence of some disease, then a concise understanding of mechanism is not necessary and may prove undesirable. A relative risk that does not look significant in an epidemiology study can imply many deaths in public health, if a large population is exposed to the risk factor.

(3) Intervene. Public health officials sometimes conduct intervention trials, but they more often simply intervene when the potential reduction in mortality and morbidity looks strong enough. The latter action sometimes makes epidemiologists shudder, because hasty intervention usually fails to accumulate data of sufficient quality to show that the intervention worked. Specifying adequate control populations often is scientifically difficult.

(4) Advocate public policies based on the intervention results. Here the public health paradigm diverges most strongly from the risk analysis paradigm. From the perspective of public health workers, scientific niceties like adequate controls and statistical significance are not important if the policies improve mortality and morbidity of a population. Doing the right thing for the wrong reason is acceptable.

In the U.S and most European countries, the law accords great social power to public health experts. It may astonish many, but local and state public health authorities have the ability to confiscate and destroy property, confine people against their wills, and subject them to medical treatments, all without due process. This power arises from the need to protect the larger public from rapidly spreading infectious diseases. With this power comes great responsibility to use it to prevent epidemics. Public health agencies understand that the discovery of even a few abuses for reasons of personal revenge or economic gain might jeopardize the authority to impose quarantine.

John Snow, a London physician who lived during the early 19th century, published a pamphlet, *On the Mode of Communication of Cholera*, proposing that cholera was an infectious disease mostly transmitted through contaminated water. Snow later proved his hypothesis by mapping the location of cholera cases during an epidemic. The map revealed that more cases of cholera occurred among customers of a company that obtained sewage-polluted water from the lower Thames than among customers of a company that obtained water from the upper Thames before the river became contaminated. Snow recommended that officials remove the handle of the water pump at the intersection of Cambridge and Broad Streets, where the prevalence of cholera concentrated. Afterward, the epidemic ceased. This incident illustrates both the power of hypothesis validation and the public health paradigm. Given this inspiration, resisting the temptation to reach a premature conclusion about a disease process becomes difficult. The problem with the public health paradigm is that it has no stopping point; it has no index to suggest when officials have gone far enough.

Epidemiological Uncertainties

Overall, epidemiologists do an excellent job of describing the uncertainties in their studies. More recently, as epidemiologists have applied their methods to chronic diseases and nonmicrobial toxicities, the use of these studies in regulatory, public health, and public information venues has sometimes become problematic. A primary basis for the problem is the relatively low risks of these substances and processes. While a low relative risk applied to a large population suggests a large number of cases, a low relative risk also is difficult to study, and causation becomes very difficult to ascertain. Controversy and confusion soon follow, particularly within the press, because reporters are not trained in epidemiology.

In addition, sometimes epidemiologists reach erroneous conclusions. A good example is the idea that electromagnetic force (EMF) increases cancer risks. Persons in proximity to power lines experience EMF because electricity passing through wires generates a surrounding force field. Investigators experienced great difficulty in measuring such exposure. Some studies measured exposure through surrogate means, such as employment classifications (job titles) and attributed an average EMF to all persons who worked in this category. Similarly, on-site measurements of field strengths do not reveal how often a person moves through the field.

A 1993 study by Feychting and Ahlbom seemed to overcome the problem of exposure measurement. They found an electric utility with records of the amount of electricity in power lines at different times. So, they could estimate the magnetic field at any distance from the wire at any time. They studied cancer among all subjects within 300 meters of a high-voltage transmission system over a 25-year period, in comparison to EMF at the time of diagnosis, taking care to eliminate confounding variables and bias as much as possible. They found a statistically significant association between EMF and childhood leukemia. However, the study now serves as an outstanding example of the multiple comparisons fallacy.

Feychting and Ahlbom did not specify an hypothesis before conducting the study. While their publication summarized their methods, the investigators actually studied twelve cancer rates using many exposure scales. Their approach calculated 800 relative risks. Random variation alone predicts that 40 of the relative risks would exceed statistical significance at a 95% confidence level (800 x 0.05 = 40). Similarly, we would expect 40 of the relative risks to decline. A report that recommended living close to a power line to avoid a disease, based on these data, would seem equally suspect. The reported association did not allow for much latency between exposure and childhood leukemia. The unreported supporting data revealed that the association disappeared after five or ten years latency. Similarly, the study revealed a positive association only for children living in single-family homes, not in apartments. In addition, other studies have reported inconclusive results.

You can see how suspect a risk assessment based on only one "positive" study might become, and how difficult a meta-analysis of summary data from similar publications might be. In this study, however, the odds ratio for childhood leukemia and EMF was high, approximately four. Imagine the inherent difficulty, if the odds ratio had been 1.4.

Sources of Additional Information about Epidemiology

Extensive information about epidemiology is available, both in general and about specific substances or processes. A good starting point to search the Internet for information about epidemiology and public health is the Hardin Meta Directory of Internet Health Sources. It has several mirror sites in the U.S. Go to www.arcade.uiowa.edu/hardin-www/md.html. Similarly, the World Wide Web Virtual Library has a section about epidemiology at www.epibiostat.ucsf.edu/epidem/epidem.html, and Yahoo maintains a list of sites at Health:Medicine:Epidemiology. See www.yahoo.com/Health/Medicine/Epidemiology. The primary professional organization for epidemiologists is the Society for Epidemiologic Research (SER). You can take a look at SER's newsletters at phweb.sph.jhu.edu/pubs/jepi/aprnews.htm. Many good epidemiology journals exist. SER sponsors the *American Journal of Epidemiology*. The British Medical Journal maintains a marvelous Internet site about epidemiology, virtually a short course for the uninitiated. See Epidemiology for the Uninitiated by D. Coggon, G. Rose, and D. J. P. Barker, *British Medical*

Journal, Fourth Edition, (BMJ Publishing Group, 1997), or visit the BMJ site at www.bmj.com. The University of Pittsburgh sponsors a similar Internet site, called the Epidemiology Supercourse at www.pitt.edu/~super1.

For more information about causation and the use of epidemiology in risk assessment, take a look at the "London Principles" developed by Federal Focus: *Principles for Evaluating Epidemiological Data in Regulatory Risk Assessment* (1996). You also can learn more about exposure-response relationships in Modeling Epidemiologic Studies of Occupational Cohorts for the Quantitative Assessment of Carcinogenic Hazards by L. Stayner, R. Smith, J. Bailer, E.G. Luebeck, and S.H. Moolgavkar, found in *Am. J. Ind. Med.* 27: 155-70 (1995).

Some government sites also provide extensive epidemiological data. The Centers for Disease Control (CDC) epidemiology home page, www.cdc.gov/epo/epi/epiinfo.htm, is a good place to start. CDC publishes *Morbidity & Mortality Weekly Report* a must read newsletter for epidemiologists and physicians with public health reporting responsibilities. Go to www.cdc.gov/epo/mmwr/mmwr.html to get started. The National Center of Health Statistics is part of CDC, and it publishes very useful lifetable compendia at www.cdc.gov/nchswww. More information about John Snow is available from CDC's Division of Bacterial and Mycotic Diseases at www.cdc.gov/ncidod/dbmd/snowinfo.htm, including a link to Snow's original map on the location of 578 deaths from cholera. Extensive information about the epidemiology of cancer is available from the National Cancer Institute. A good place to begin is the SEER homepage at www-seer.ims.nci.nih.gov. The World Health Organization (WHO), which has a homepage at www.who.org, has a list of authoritative sources, including IARC. WHO also supports an email listserver, *Weekly Epidemiology Report,* with a website at www.who.ch/wer/wer_home.htm.

Several classic epidemiology text books have been published, including:

Rothman, K. J. 1986. *Modern Epidemiology.* Boston: Little Brown and Co.

Mausner, J.S. and S. Kramer. 1985. *Epidemiology: An Introductory Text.* Philadelphia: WB Saunders Company.

Lilienfeld, A.M. and D.E. Lilienfeld. 1980. *Foundations of Epidemiology.* New York: Oxford UP.

Macmahon, B. and T. F. Pugh. 1970. *Epidemiology: Principles and Methods.* Boston: Little Brown and Co..

We usually recommend the first of these texts to students who want to acquire more in-depth knowledge of the subject.

References

Bowers, J. C., B. Brown, J. Springer, L. Tollefson, R. Lorentzen, and S. Henry. 1993. Risk Assessment for Aflatoxin: An Evaluation Based on the Multistage Model. *Risk Analysis* 13: 637–642.

Feychting, M. and A. Ahlborn. 1992. Magnetic fields and cancer in children residing near Swedish high-voltage power lines. Amer. J. Epidemiol. 138: 467–481.

Greenland, S. 1987. Quantitative Methods in the Review of Epidemiological Literature. *Epidemiol. Reviews* 9: 1–30.

Howson, C. and P. Urbach. 1989. *Scientific Reasoning: The Bayesian Approach*. La Salle, IL: Open Court Publishing.

Moolgavkar, S. H., E. G. Luebeck, D. Krewski, and J. M. Zielinski. 1993. Radon, Cigarette Smoke, and Lung Cancer: A Re-Analysis of the Colorado Plateau Uranium Miners' Data. *Epidemiology* 4: 204–217.

Walker, A. M., J. M. Martin-Moreno, and F. R. Artalejo. 1988. Odd Man Out: A Graphical Approach to Meta-Analysis. *Amer. J. Public Health* 78: 961–968.

WHO. 1997. *Evaluation of Certain Food Additives and Contaminant: Forty-ninth report of the Joint FAO/WHO Expert Committee on Food Additives*. Geneva.

Wu-Williams, A. H., L. Zeise, and D. Thomas. 1992. Risk Assessment for Aflatoxin B_1: A Modeling Approach. *Risk Analysis* 12: 559–567.

Yeh, F. S., M. C. Yu, C-C. Mo, S. Luo, M. J. Tong, and B. J. Henderson. 1989. Hepatits B Virus, Aflatoxin, and Hepatocellular Carcinoma in Southern Guangxi, China. *Cancer Research* 49: 2506–2509.

Chapter Seven

Toxicology

Toxicology

Toxicology and epidemiology are closely intertwined subjects. From the perspective of this book, the two disciplines are alternative, sometimes complementary, ways to establish a potency estimate. Risk assessors multiply potency by exposure (or dose) to calculate risk.

Toxicology is the study of poisons. Unlike common belief, toxicology neither is limited to animal studies nor is it the opposite of epidemiology. Some epidemiologists study toxic substances; therefore they simultaneously engage in toxicology. Some toxicologists study populations; therefore they simultaneously engage in epidemiology. The two disciplines often overlap.

Suppose, for example, you wanted to study the potential effects of the effluent from a manufacturing facility on nearby residents. Because of a sparse human population, however, an epidemiologist might want to study wild mice living near the facility, and compare them with mice living farther away from the effluent. An epidemiologist might develop life tables that compared the two mouse populations. A toxicologist also might study the effects of the effluent on mice but probably would conduct controlled experiments with mice exposed to concentrated effluent in a laboratory setting.

Toxicology is a biological science. The scope of toxicology in standard textbooks generally covers radiation, chemical substances, and toxins such as snake venoms. It usually does not include the pathological effects of microorganisms but does address the toxic effects of microbial excretion products, such as botulinum toxin. It usually does not include the adverse effects of an immune response, but immunotoxicology does study the effects of radiation, chemical substances, and microbial excretion products on immunity.

While toxicology extends to radiation and biological toxins, most scientists think of it as a subdiscipline of chemical biology. Unlike biochemistry, which generally looks at the chemistry of biological processes, chemical biology is the study of biological responses to chemical substances. It covers toxicology, chemotherapy, and pharmacology. Most of the substances studied are *xenobiotics*, chemical substances that are foreign to the body. However, chemical biology also studies the outcomes of exposures to naturally arising substances such as hormones, particularly at higher than endogenous levels. The concepts remain much the same for radiation and microbial excretion products, or poisonous mushrooms.

Hormones are chemical substances produced by one organ in an intact animal that circulate to distant organs when released and regulate the activity in the distant organ. For example, insulin originates in the pancreas and circulates through the blood to many tissues, particularly muscles, to regulate the uptake of glucose, a primary source of energy. Similarly, testosterone originates in the male testes and circulates to many tissues, producing male characteristics.

Toxicologists directly study the mortality, morbidity, and pathology resulting from an exposure to xenobiotics, such as drugs, hormones, or environmental pollutants. One way that toxicologists summarize data about many substances, as obtained under the same set of circumstances, is to tabulate lists of the exposures or doses that induce the same prevalence of some biological effect (ED). When the effect is death, toxicologists call the dose the lethal dose (LD). Toxicologists often look for explanations of relative EDs or LDs in the structures of the chemicals. We call these explanations structure-activity relationships (SARs). For example, potencies sometimes correlate with lipid solubility within a structurally related group of chemicals. Chemists can infer lipid solubility from structures.

Dose-response relationships have primacy in toxicology, particularly for lethal effects. In one of the earliest observations in toxicology, Paracelsus noted that the dose makes the poison. Thus, even water is lethal if the dose is high enough. Toxicologists usually think about doses, not exposures. In part, this convention arises from their frequent use of intravenous, intramuscular, and intraperitoneal routes of administration. These routes do directly equate the available amount of an administered substance with a dose. However, many chemical biologists also assume that the oral route of administration is a dose. The underlying assumption is that subjects absorb 100% of the administered substance, a dubious and generally false assumption.

Technically, a 50% response is usually the most accurate point in an experiment, whether measuring the magnitude of response in one subject or the prevalence of response in a population. Thus, chemical biologists often denote LD and ED with precision by defining the response quantitatively, for example, as LD_{50} and ED_{50}. However, this prevalence can change. For example, the 20% response is denoted LD_{20} or ED_{20}.

Chemical structure is another primary basis for toxicological effects of chemical substances. Minor changes in chemical structure change the resulting biological responses. Besides altering the exposure-response relationship for the same outcome, even minor modifications in chemical structure can shift the qualitative outcome, for example, from cancer to neurotoxicity. For some drugs, only the optical isomer that bends light in one direction (a stereoisomer) produces an effect, although the other isomer has the same general structure.

Chemotherapy principally involves the chemotherapeutic index (CI), the ratio of two toxic potencies of a substance: the toxicity for a host species to the toxicity for a foreign organism, such as a microbe or tumor, at equivalent prevalence of response. If it takes less of a substance, like penicillin, to kill bacteria than it does to kill the people infected with the bacteria, the substance has a potentially useful chemotherapeutic index; the larger the ratio, the better. Chemotherapists need toxicological expertise about two different species: the host and the invading organism:

$$CI = LD_{30,HOST} / LD_{30,ORGANISM}$$

Pharmacology involves the therapeutic index, also a ratio of two potencies of a substance: toxicity for an organism to desired effects in the same organism. If it takes less of a substance, such as aspirin, to obtain a desirable effect such as a reduction in temperature than it takes to induce a toxic effect such as tinnitus (ringing in the ears), the substance has a potentially useful therapeutic index. The larger the therapeutic index, the more favorable the substance is likely to be for the organism.

$$TI = LD_{50} / ED_{50}$$

Duration of exposure also is an important variable in the magnitude of a toxic effect in an animal or the prevalence of toxicity in a population. Thus, neither these indices nor structure-activity relationships are considered valid if the EDs or LDs come from experiments of very different duration. At present, toxicologists use several conventional descriptors for the durations of laboratory experiments. *Acute* exposures for rodents, such as mice and rats, imply a brief, usually single dose administration, or a human duration of less than a day. (An acute exposure for a rodent approximately corresponds to a human exposure of approximately a month or less.) *Chronic* exposures are human exposures of approximately ten years to lifetime duration. Lifetime exposure for laboratory rodent species is approximately two years or until death, sometimes more but usually less in some inbred strains. By convention, chronic exposure experiments usually begin at a rodent age roughly equivalent to human early adolescence, for example, with six to eight-week-old mice. The basis for this convention is that a constant applied exposure rate will not generate wildly different delivered doses over the duration of the experiment. Usually a chronic exposure or dose experiment in the laboratory means a near lifetime duration, but in principle a toxicologist could refer to any period longer than four months as a chronic exposure. Durations between acute and chronic are *subchronic*, from a week to a few years for human exposures or from several applications to approximately ninety days in laboratory rodents.

Most toxicology studies yield observations that are "time-sliced." Separate acute, subchronic, and chronic data about substances do not reveal the effects of duration in a continuous, coherent way. Instead, comparisons of these durations yield an impression of the effect of duration. In contrast, epidemiological data typically cover a range of duration and provide a better source of information about the effects of time of exposure. Risk assessors have a strong interest in the exposure-time-response relationship for a substance. Over many structures, the acute exposures to achieve an effect will correlate closely with the subchronic or chronic exposures for equivalent effects, particularly for lethality, and usually even for different effects. Thus, acute toxicity experiments are highly informative, as they are predictive of subchronic and chronic exposure-response relationships.

The frequency of application in laboratory experiments is another important variable. When an intravenous, intraperitoneal, or intragastic route is used, technical personnel seldom can sustain daily dosing for more than two years, the life span of many rodent species. Instead, by convention, animals receive exposures once a day for five, sometimes six, days a week. The pattern of exposure is not continuous but frequently repeated brief exposures. The same pattern usually occurs with inhalation or dermal exposures, but the period of daily exposure varies. For example, the animal may inhale a substance in air for four hours a day, five days a week. For ingestion exposures, however, chemical substances often get incorporated into the animals' feed or water. Even this approach does not necessarily generate a constant delivered dose. Mice, for example, are nocturnal animals, and they mostly eat and drink during several episodes in the night. In this regard, toxicological exposures more generally resemble human exposures, which also are intermittent (see Table 7.1).

Table 7.1 Toxicological Exposures

Animal	BW (kg)	mg/kg.d ingested with one ppm of substance in food	mg/kg.d ingested with one ppm of substance in water
Mouse	0.02	.150 (dry)	0.25
Rat	0.4	.050 (dry)	0.1
Dog	10	.075 (moist)	0.05
Human	60	.025 (moist)	0.03

Duration of exposure usually corresponds to duration of observation or duration of effect in toxicology experiments, but not always. For example, a toxicologist may observe animals for several months after giving them a single interperitoneal dose of a substance. In this case, subchronic effects may result from an acute exposure. A useful toxicological data set collected all of the instances of carcinogenic effects after a single exposure to a chemical substance. Carcinogenic effects have long latency and require a long duration of observation for expression, usually a lifetime.

Toxicology has many important practical applications. All three branches of chemical biology require an in-depth knowledge of toxicology. This field relates to protection from adverse effects, alleviation of infections and cancers, and medication of diseases. All three branches are deeply concerned with the *mechanism of action*, a detailed description of the chain of molecular, biochemical, and biological events that occur between exposure and biological response. Toxicologists often may not understand the complete mechanism but may know about critical components of the mechanism that are prominent in its exposure-time-response relationship.

Some toxicologists have recently described such partial knowledge of a mechanism as a *mode of action*. For example, nicotine, a biologically active substance, stimulates contraction of muscle fibers, an activity that directly relates to its lethal effect. At high exposures nicotine induces convulsions and death through a failure of respiration. We can think of muscle contraction as the mode of nicotine. However, nicotine induces complex outcomes, often paradoxical biological effects, which have both simulating and paralyzing effects on nerve fibers that respond to acetylcholine, a natural neurotransmitter. A full description of the mechanism of nicotine poisoning includes paralysis of the respiratory centers in the brain. Thus the distinction between mechanism and mode is sometimes not clear, and the latter term is not used universally.

SARs are based on the proven principle that similar chemical structures tend to have similar mechanisms. The SAR relationship for any specific effect is not all-or-none; it is probabilistic. For example, pharmacologists have isolated, synthesized, or studied many structural analogues of acetylcholine besides nicotine for potential medical uses, such as substances that block acetylcholine at neuromuscular junctions and induce relaxation. Thus, any SAR leads to predictions of qualitative effects and also quantitative exposure-response relationships. Similarly, as knowledge of the mechanism of a substance increases, so does the ability to use the substance beneficially, to avoid adverse effects, and to develop even better agents. In this context, scientists can discuss a mechanism of toxicity, a mechanism of lethality (usually in chemotherapy), or a mechanism of action (usually in pharmacology).

One major classification of mechanisms and substances in toxicology involves interactions with biological *receptors* to produce different effects. In this context, receptor binding means that a tight, specific, but noncovalent attachment of a substance (or a metabolite of the substance) occurs with a biological macromolecule, the receptor. Receptors may occur on the surface or within cells. Binding to the receptor mediates the observed biological effect. Chemical biologists sometimes refer to lower molecular weight chemicals as ligands (L), particularly when they observe the ligand-receptor (LR) interaction outside an intact animal. Binding reactions are highly specific to the structure of the ligand. This high structural specificity connotes activity at extremely low concentrations. Such structural specificity is one reason that chemical biologists are so contemptuous of advocacy-based attacks on atomic elements such as chlorine. The same chlorine substituent in molecules with different structures may range in activity from inert to cancer to liver inflammation.

Much of the early history of pharmacology involved the study of hormones or chemical analogues of hormones. Looked at the other way around, we define the part of a cell that interacts with a hormone to produce an effect as the receptor for the hormone. Receptors for a hormone often occur only on certain organs in the body. During the early development of receptor theory, little was known about the biochemical nature of receptors. Yet the intellectual concept turned out to have extensive utility in medicine, and scientists accepted it long before they explained the physical and biochemical structures of the receptor molecule.

In the earliest discoveries, Paul Ehrlich noticed that certain dyes stain some kinds of cells selectively, implying that some component only found in these cells adhered to the dye. Ehrlich called this component the receptor. Ehrlich went on to develop a theory of drug action based on this implication, and he successfully pursued drug discovery using the theory. Soon after, Langley applied nicotine to small regions of muscle surfaces and noticed that twitches occurred after application to certain areas but not others (Langley, 1905). He suggested that a hypothetical receptor substance existed only on the surface of the responding cells and only in the areas that responded to nicotine.

Subsequent investigators discovered substances that selectively inhibited (or antagonized) the biological effects of hormones and other biologically active substances, called agonists. Nicotine is an agonist of acetylcholine. Many of these substances had chemical structures similar to the inhibited hormones. Chemical biologists accepted that an inhibiting substance (usually called an "antagonist" in the pharmacological literature) attached to the same receptor as the hormone but lacked some chemical feature required to initiate or sustain biological action. An antagonist blocked drug action by denying access of the hormone to its receptor. The accumulation of many similar experiments with different kinds of hormones, agonists, and antagonists gave general credibility to the concept of receptors.

In other instances, with drugs such as morphine or aspirin, pharmacologists found that similar principles applied, although they had no knowledge of the mechanism or receptors involved. Pharmacologists could synthesize structurally related chemical substances that blocked or mimicked the actions of these drugs, eventually creating a specific structure-activity relationship for each class that suggested some features of the apparent receptor. For example, pharmacologists discovered two major divisions of the autonomic nervous system. Both divisions responded to endogenously released acetylcholine, but one division responded only to nicotine, whereas the other responded to muscarine, a substance isolated from mushrooms. It turned out that acetylcholine has multiple receptors, including receptors specific for nicotine or for muscarine, but not both. These classifications gave pharmacologists a great advantage in studying new drugs. If a known antagonist selectively blocked the activity of a new drug, they could predict that the new drug would have all of the actions associated with binding to the same receptor. For example, a nicotinic antagonist might block the drug's action, whereas a muscarinic antagonist would not. The toxicity of the drug would resemble the toxicity of nicotine, but not muscarine.

In a major scientific advance, A.J. Clark (1933) first showed that typical equilibrium data between drug concentration (or dose) and the magnitude of biological effect coincided with the hyperbolic relationship expected for the formation of a drug-receptor complex according to the law of mass action. Clark's idea was confirmed and extended through many experiments with structural analogues of drugs and hormones. Pharmacologists could often measure the apparent affinity and equilibrium binding constants for drug-receptor complexes, usually without any knowledge of the biochemical structure of the receptor molecules involved, by carefully measuring exposure-response relationships. Clark's discoveries linked quantitative measurements of potency, structure-activity relationships, and notions of receptors as cellular macromolecules that transferred energy from the binding ligand to a biological system. This large body of information led to the idea that receptor-mediated effects often had hyperbolic exposure-response relationships. An exposure-response relationship is the net result of both pharmacokinetic and receptor binding processes.

The logarithm of ligand concentration plotted versus the magnitude of an effect usually produces a sigmoidal curve, as suggested by the hyperbolic equations. On examination, these sigmoidal curves have the characteristics of normal distributions, suggesting an underlying lognormal tolerance distribution of sensitivity in the populations (of animals, cells, or enzymes). Consequently, toxicologists often plot their data as the probit of response versus the logarithm of exposure, so that the data will fit a straight line. The resemblance in equations for enzyme kinetics and exposure-response relationships led to the belief that receptors probably were proteins, just like enzymes. In a few instances, such as the drug methotrexate, its receptor, dihydrofolic acid reductase, actually is an enzyme. Subsequent discoveries have proved that receptors generally are proteins. Because genes code for proteins, genetic analysis of receptors has become an important method of understanding toxic reactions.

Classical receptor theory led to major medical discoveries. Quantitative classifications of receptors simplified the presentation of pharmacological classes of drugs, instead of a random list of medications. For example, adrenal steroids (like hydrocortisone) are distinct from other steroids, such as mineralocorticoid steroids or sex steroids. Pharmacologists used this knowledge to develop important new drugs. Biologists mapped neuronal pathways in the brain by staining for the presence of specific receptors, such as dopamine receptors or acetylcholine receptors. Today, mapping the genes that code for receptors has generally upheld receptor theory and provided more details to the classification.

Despite its success, classical receptor theory did not explain all pharmacological phenomena satisfactorily. Interest by many scientists in phenomena that the classical theory could not explain led to a general receptor theory (see The Toxicology of Carcinogens section in this chapter). General receptor theory essentially subsumes the ideas of the classical theory, leaving it as a special case. The easiest way to think about general receptor theory is that the midpoint of an exposure-response relationship does not necessarily imply occupation of half the available receptors. Instead, achieving this extent of effect may require occupation of most of the receptors or only a small portion. The latter condition, called receptor reserve, occurs commonly. The relationship between receptor occupation and extent of effect is often nonlinear.

One of the earliest questions that toxicologists ask in elucidating the mechanism of toxicity of a substance is whether it operates by binding to some specific class of receptors, perhaps by mimicking or inhibiting the action of some endogenous hormone. You may want to consult a textbook of pharmacology to gain an idea of the extensive number of receptors operating in a typical mammal. The classic example is dioxin, an extremely potent and very stable chemical structure, which binds to receptors within cells. The dioxin-receptor complex migrates to the nucleus within cells, changing gene expression, which is the transcription of several proteins.

The prominent toxic effects of many substances do not, however, arise from binding to a specific protein receptor. Nonreceptor mechanisms are the other major category of responses in toxicology. Either the substance interacts with many receptor systems at approximately the same delivered dose of the substance, or it interacts nonspecifically with many macromolecules in the body. For example, ethanol interacts with a bewildering variety of receptors. One hypothesis even suggests that ethanol depresses neuronal activity by altering the structure of neuronal membranes, not invoking an interaction with receptors at all. As another example, many carcinogenic substances (or their metabolites) generate DNA adducts. These adducts usually do not occur at selective locations. Instead, the adducts cause a variety of biological effects, one of which is the induction of mutations. As discussed in Chapter 6, some mutations lead to cancers.

Even when a substance does not interact with a specific receptor, it still may produce biological effects specific to an organ system. For example, only the kidney may prove susceptible to the expression of a mutagenic change such as cancer, even if the same mutation occurs in other organs. Substances that interact with receptors often achieve organ specificity because only that organ expresses the receptor. Organ specificity may explain differences of biological effects between species. The same receptor expressed in the brain of rats may occur in the brain and peripheral nervous system of dogs. Toxicology has many specialties. Many of them relate to organ systems, such as liver, kidney, or nervous system, even when the endpoint of an organ-specific effect is death. Another relates to effects that only occur during specific developmental stages. Some relate to the kind of toxicity, such as cancer, which occurs in many organs and developmental stages. Often, the individual toxicologist has more in common with other nontoxicologists who share an interest in a biological specialty, such as neurobiology or oncology, than across toxicological subdisciplines.

Toxicologists use any and all kinds of information about poisons. They try to integrate dosimetric and pathological data. As experimental scientists, they use biological models extensively. Biological models, such as mice that serve as surrogates for humans, have advantages and disadvantages. The major advantage is that toxicologists gain better, often rigorous, control of experimental conditions. So, confounding factors and bias trouble the interpretation of toxicological results much less than we saw with epidemiology in Chapter 6. Both experimental and observational studies are subject to chance findings. Biological models simplify the design of experiments and help to eliminate chance, thus reducing the need to repeat studies many times.

When toxicologists integrate epidemiological information into this understanding, the information about biological models becomes particularly valuable. If information about human outcomes comes from human data, experimental data from laboratory animal experiments lets the toxicologist know when the animal is a biological model in the traditional sense, one that reacts to a substance in the same way as the species of interest. So, mechanism of action information obtained from controlled experiments with the animal model will apply directly to humans. Because of the difficulty in obtaining human data, the animal data can achieve otherwise unreachable insights that allow better use of the tested substance. However, mechanistic information is usually of scientific interest to toxicologists, even when its human applications remain unknown.

Conversely, toxicologists confront major disadvantages in using biological models. Because they study the effects of a substance on one species to understand the effects of the substance on another species, toxicologists always confront problems of inference, trying to decide the relevance to living people of experiments with intact mice or human cells in culture. Relevance was not such an important question in epidemiology. In this chapter we will cover a variety of biological models, from clinical studies of the appropriate species to chemical reactions. We will primarily emphasize experiments with intact animals as biological models of humans, but we also will cover other kinds of experimental data in relation to effects in intact animals.

Toxicologists also use laboratory animals as biological models in a predictive way, testing a substance in animals without human data and trying to forecast the exposure at which qualitative effects might occur in man. This use of toxicological data exacerbates the problem of scientific inference. The biological model lacks validation. For example, even if a laboratory animal produces the same response as humans with some substances, how does the investigator know that the study substance will not be an exception? Toxicologists have to extrapolate both potency as exposure, and outcome as pathological effects, between the two species simultaneously. The usual absence of detailed information about false positive and false negative results with the same model also denies the toxicologist the opportunity to make mathematically based predictions.

Federal regulatory agencies routinely base most risk assessments of chemical substances on potencies obtained from animal bioassays. Even when these agencies possess apparently adequate human data, they seldom use these data as more than checks on the potencies derived from animal experiments, particularly for carcinogenic substances. In part, this practice derives from bureaucratic inertia. (We will explore more reasons for this in The Toxicology of Carcinogens section below.) Many regulators regard rodents (particularly rats and mice) as predictive biological models of human carcinogenesis out of necessity, although this idea is controversial scientifically. Ignoring information from an animal model that an effect occurs with some substance is difficult in both maintaining political office and protecting the public health. In addition, most bioassays do not provide data suitable for more than defining breakpoints in exposure-response relationships: effectiveness levels, or an external exposure that causes no apparent adverse effects in a group of animals (the no-observed-effect-level, NOEL).

For practical, economic, ethical, and statistical reasons, the number of animals tested in predictive studies seldom exceeds four hundred per exposure, and toxicologists frequently use much fewer animals. From a risk assessment perspective, the sensitivity of detection in such experiments is poor if a large human population will have exposure to a substance. We saw the same problem in epidemiological studies, where relative risks of less than two seldom have much value. However, in predictive toxicology, investigators can use high exposure levels to compensate for small test population size. This practice imposes a different limitation on the inference of human health effects from animal data. Risk assessors usually have to interpolate high-test exposures to much lower ambient human exposures. Estimation of potency and outcome then involves assumptions that many scientists question. For example, if the exposure is so high that it forces metabolism through a minor pathway, and a different metabolite causes the toxic effect observed in the test animals, the data will be useless in predicting an effect in humans at exposure levels a thousandfold lower.

Toxicology focuses on information about poisons, and is a less methodological approach than epidemiology. Epidemiologists become expert at conducting clinical studies and observational studies. They spend a great deal of time working on appropriate techniques and methods to analyze the data from these studies. Because toxicologists focus on the properties of poisons, they take a less descriptive and more hypothesis testing approach in trying to understand the properties of poisons and ultimately their mechanism of action. Toxicologists are less bound to specific techniques and use a variety of approaches to test hypotheses, ranging from tests of animal populations to genetic analyses.

The classical experiment in toxicology involves administering a substance to groups of animals at different exposures over a single period of time and accounting for the outcomes. These experiments involve animal husbandry, pathology, statistical analysis, and analytical chemistry to define the test substance. Given the wide range of methods, however, many toxicologists have only read about these experiments.

Toxicologists can use many plants or animals as bioassay subjects. Environmental toxicologists perform direct studies with the species of interest in the laboratory and reproduce observations in the field. In addition, toxicologists use isolated organ studies, cell culture studies, enzyme studies, protein binding studies, studies of chemical reactivity, and even computational SAR studies.

Detailed descriptions of methods used in toxicology are available in handbooks and the materials and methods sections accompanying publications in the primary literature. Our purpose here is not to provide a "cookbook" approach to toxicology, but to give you a sense of the data available for risk assessment.

In an ideal world, toxicologists would begin with a complete set of pharmacokinetic data. As a practical matter, these data come later in the process. Specifically, toxicologists want to know whether the parent compound or a metabolite is the biologically active species. Often, physicians have seen effects in humans (or suspected effects in humans). The animal studies have the goal of observing biological effects of the same exposure at different exposures and durations of exposure.

We might expect that toxicologists would be bioassay experts with animals, as epidemiologists are observational study experts with humans. However, this is not so. In a typical bioassay, exposures are chosen either to invoke the toxic effect at very high exposures or to come as close to exposures that will cause toxic effects as possible, the so-called maximum tolerated dose, in order to mimic exposures seen for humans. Sometimes toxicologists will use both kinds of exposures and conduct experiments to detect the effects in a colony of rodents over different durations.

At the simplest level, a prior hypothesis about the expected prevalence of effects is possible, because of the correlations between bioassays of different duration or between bioassays with similar substances. If so, we can design the number of animals exposed to elicit a most meaningful and efficient statistical endpoint. However, this approach is unusual in toxicology. Typically, regulatory agencies specify standard group sizes for each exposure, for example six animals at exposures spaced twofold apart over a range of perhaps eight or nine different exposures. So, interpretation of the exposure-response curve becomes more of a biological than statistical exercise.

The toxicologist observes the transition from a high proportion or all of the animals exhibiting an effect to an exposure where none experience more than the background prevalence of the effect. Naturally, if small numbers of animals are used, comparisons between these two groups are not statistically robust. So we often have a qualitative description of the range of transition from no effects to effects seen in all the animals. Newer techniques use very small numbers of animals in range-finding experiments to locate this exposure range. This practice preserves the integrity of the scientific information while satisfying ethical requirements of animal experimentation. In these studies the same animal is given incrementally increasing exposures until an effect is seen or the experimental methodology interferes with observations. Then a second animal is given increasing exposures, and so forth, until an exposure range is obtained in which the animals clearly exhibit effects. Such range-finding approaches often substitute for the determination of the LD_{50}.

Again, if pharmacokinetic information is available, we will have some understanding of the active metabolite and the organs where the metabolite is found. Radioautography involves the administration of a radioactively labeled substance and the anatomical location of the radioactivity at different times on photographic plates. Thus, radioautographic information before the design of bioassay experiments is particularly valuable. With such data at hand, the toxicologist can begin with a good idea of the organs and tissues in which the substance concentrates.

Toxicologists often look for biomarkers of effect that they can substitute or use as surrogates for the biological effect of interest—indicators such as histopathological changes, alterations in enzyme activity, changes in cell number, and so forth. For example, in immunological studies, if the overt effect in the animal is susceptibility to infectious diseases, and immune depression is observed in isolated cultures, the toxicologist can probably observe a decrease in specific immune cells. Carrying this logic one step further, many toxicologists engage in cell culture experiments. Overall, a good correlation exists between toxicity for cells in culture and the toxicities of the same substances in intact animals when the following conditions are met: either the substance induces nonspecific, widespread toxicities on many kinds of cells throughout the organism and in all tissues, or the cultured cells express the same specific receptors for the active metabolite involved in the effects seen in the intact animal.

Many toxic effects depend on effects on an enzyme, either by direct inhibition of the enzyme or through an inability to realize the enzyme activity which is normally induced. Where the former is the case, toxicologists can study the enzyme inhibition in an isolated biochemical system. In the latter case, toxicologists can attempt to modify enzyme expression in cultured cells. Even when the protein that binds the toxic substance does not engage in some catalytic function, toxicologists can probably isolate a protein that binds to a substance (or its metabolite) with high affinity and specificity, like a receptor. For example, when looking at a substance that has effects like testosterone in an intact animal, toxicologists can construct cells that have the receptor for testosterone, and they can link a protein induced by testosterone to a protein that they easily detect, like firefly luciferase. So, a cell that binds a testosterone-like substance will express luciferase and glow (actually, it is phosphorescence). The testosterone receptor can be produced in large amounts in relative purity, and toxicologists can look at the ability of substances to bind to this protein or to displace bound testosterone from the protein. These protein binding methods are useful to elicit SARs.

Classically, toxicologists drew chemical structures on paper and had a good knowledge of how these graphic representations of chemical structure related to chemical properties in solution. A classic textbook by Adrian Albert has extensive descriptions of structure activity relationship techniques using simple but straight forward "paper toxicology" methods. Over the last two decades computational methods have been applied to structure activity relationships, often with great success. Many different approaches to structure activity relationships are possible. One can look at overall electronic structure, chemical reactivity, or space filling structures. However, the best successes to date have come from methods that essentially break chemical structure into pieces and ask whether the individual pieces resemble pieces in other substances known either to have effects or not to have effects. As we outlined in Chapters 1 and 2, the effort here is to obtain both sensitivity and specificity in SARs. Chemical-fragment or substructure methods have been highly successful both in predicting potency of simple biological effects like lethality and in predicting qualitative effects such as carcinogenesis, teratological or developmental effects, or neurotoxic effects.

Classic mechanism of action, for example, can involve organophosphate substances used as insecticides. These acetylcholinesterase inhibitors also have toxic effects in mammals. Their target enzyme, acetylcholinesterase, normally degrades acetylcholine after its release at nerve endings. When acetylcholinesterase is inhibited, acetylcholine cannot be degraded without this enzyme. Acetylcholine accumulates at the nerve ending and the nerve cells fire repeatedly until tetanus occurs. The action is like the stimulatory effect of nicotine plus the effects of muscarine, only more widespread. Obviously this effect is toxic to both insects and mammals; it can be lethal. Exact expression of organophosphate toxicity is complex, and the mechanism of any one inhibitor involves more than inhibition of acetylcholinesterase. For example, some organophosphates elicit nicotinic reactions, others muscarinic. Different organophosphates localize in different target organs, such as nerve terminals in the peripheral nervous system rather than those of the central nervous system. The chemical structure necessary to bind to and inhibit acetylcholinesterase in insects and in mammals is known in great detail.

Almost all toxicological methods also can be used as predictive tests. Predictive testing can be quite controversial because the toxicologist usually has to extrapolate across species and levels of emergence, moving up from chemical structure to protein binding to enzyme inhibition to effects in cell culture to effects on isolated organs to effects in the intact animal.

Despite these difficulties of interpretation, much predictive testing is done for purposes of screening potential new substances to prevent undue toxic effects from occurring when they are introduced into commerce. Data from these predictive assays are often used in risk assessments in which the assessors attempt to state the probability that an adverse effect will occur under certain conditions of use of a new substance. We review some of these applications in the section below on Safety Assessment. In the last section of this chapter, we include some sources of additional information about toxicological methods.

Weight of the Evidence in Toxicology

Like epidemiologists, toxicologists also have to develop an informed judgement about a group of studies. For toxicologists, the wide range of experimental data makes the task more difficult. One way to look at this problem is proportional response. If the inhibition of an enzyme by one substance is, for example, three times greater than the inhibition of the same enzyme by a second substance, perhaps the LD_{50} of the first substance also will be three times greater than the LD_{50} of the second substance;

$$\frac{ED_A}{ED_B} = \frac{LD_A}{LD_B}$$

Similarly, the ratio of the inhibition of an enzyme in one species to the inhibition of the same enzyme in a different species might be the same as the ratio of LD_{50}s in the two species. Obviously, the relationship between enzyme inhibition and lethality in response to two substances or in two species can be much more complicated. The relationship may involve nonlinear relationships. If so, we should not feel surprised if simple proportionality does not hold. At worst, inhibition of the enzyme may prove unrelated to lethal effects, in which case the two proportions will simply vary at random.

Toxicologists almost entirely rely on expert opinion to integrate the data about multiple studies. They seldom try to apply the kinds of mathematical relationships used in meta-analysis to integrate information (see Chapter 6). Too many factors, many unknown or unquantifiable, influence the outcomes. Toxicologists weigh many kinds of evidence in reaching expert opinions about multiple studies, including the study's quality, experience of their investigators, capabilities to reproduce results, methods, and extent to which the data support or negate hypotheses crucial to the mechanism of toxicity. It is worth recalling that weight of the evidence differs from *strength of the evidence,* which only evaluates the adequacy of data that support a particular hypothesis.

The subject of scientific inference lies beyond the scope of this book. In many areas of regulatory toxicology, including the classification of substances with particular outcomes, such as "carcinogenic substances," one positive study will outweigh many negative studies, and two positive studies are taken as an indication of reproducibility, even in the face of many negative studies. Like epidemiologists, toxicologists rely on authoritative sources in which groups of experts classify substances. Often these authorities rely on criteria similar to the Bradford Hill criteria to integrate animal bioassay and mechanistic data into a classification. EPA's current proposed *Carcinogen Risk Assessment Guideline* specifically addresses the weight of evidence procedure for the regulatory classification of carcinogenic substances. Results from long-term animal studies are supplemented with available information from short-term tests, pharmacokinetic studies, comparative metabolism studies, structure activity relationships, and other relevant toxicological studies. Regulatory toxicologists must judge the quality, adequacy, and appropriateness of these data. Similar schemes have been applied to substances with developmental, reproductive, and neurotoxic effects. Discrepancies between different classifications of the same substance for the same effect should not surprise risk assessors.

Exposure-Response Relationships in Toxicology

Exposure-response models (as described in Chapter 2) are important, helpful, and theoretically based. Following the division of toxic phenomena into receptor-mediated and nonreceptor outcomes, we will start with the first category. To most observers, the development of classical receptor theory culminated with E. J. Ariens' publications (Ariens, 1954, 1964). He expanded the theory and created an overarching algebraic system to manipulate its findings. Ariens also contributed the concept of intrinsic activity to the classical theory. Some substances apparently bind to a receptor fully, yet they elicit only a partial biological response. In Ariens' view these "partial agonists" had some, but reduced intrinsic activity in comparison to a "full agonist" (usually the parent drug or a hormone to which the partial agonist was compared). Most important, the work of A. J. Clark, Ariens, and many others supports the use of log-normal distributions to describe and interpret the results of toxicology studies, particularly the results of animal bioassays.

Both Clark and Ariens used the assumption that receptor occupancy was proportional to the percent of maximum effect to obtain a simple conceptual and mathematical framework for purposes of clarity of presentation. Neither believed that these assumptions held for most biological systems. Clark stated that proportionality seemed too simplistic. Ariens explained that direct proportionality between [LR] complex and effect probably was the simplest case of a larger, more likely set of circumstances. To both scientists, proportionality seemed improbable when the observed effect was either complex in nature or an aggregate of other effects.

In addition, the hypothesis that chemical equilibrium applied to biological experiments seemed problematic to both theorists. Physically, what pharmacologists mean by equilibrium is not the same phenomenon that chemists characterize. Pharmacologists wait until they obtain a maximum effect at a given dose, and describe this as the "equilibrium effect." Because disposition is a kinetic process, the concentration of a ligand near the receptor is in a state of flux, whereas the observed effect may lag in time because of the kinetics of biological action, unrelated to the ebb and flow of ligand. This state of affairs is distinctly unlike that of a chemical reaction in which the reactants no longer change in concentration because of thermodynamic considerations. Even in the 1930s, pharmacologists understood physical chemistry and knew that unknown, complex kinetic processes occurred beneath the surfaces of cells.

Classical receptor theory made two major assumptions:

(1) The magnitude of effect is directly proportional to the concentration of ligand-receptor complex, and

(2) The maximum effect occurs when the ligand occupies all receptors.

The tenuous nature of the essential assumptions of classical receptor theory did not prevent it from having extraordinary experimental utility and explanatory power. For example, working with the assumptions of classical receptor theory, J. H. Gaddum formulated competitive drug antagonism equations that some pharmacologists still use today (Gaddum, 1937; Gaddum, 1943).

For purposes of risk assessment, the implications of receptor mediation for the analysis of dose-response relationships can be presented as either "classical" or "general" receptor theories. The older, classical theory is more intuitive and is a subset of the general theory. The general theory provides a more comprehensive explanation of the intricacies of receptor binding and resulting biological effects. The overall conclusions from either theory are the same with respect to the quantitative evaluation of the potency of receptor-mediated toxic effects.

According to the law of mass action, at equilibrium the rate of formation of a ligand-receptor complex will equal the rate of dissociation. (In this context, "ligand" means a low molecular weight substance that exhibits structural specificity in forming a complex with a receptor.) The rate of forward reaction is proportional to the concentrations of uncomplexed ligand and receptor. Using common chemical kinetic notation, where k_1 is the kinetic rate coefficient, the rate of the forward reaction is proportional to the concentrations of uncomplexed ligand and receptor ([L] and [R]),

$$V_{forward} = k_1[L][R]$$

The rate of reverse reaction is proportional to the concentration of complex ([LR]),

$$V_{backward} = k_2[LR]$$

At equilibrium, the forward and backward rates are equal, and thus:

$$k_1[L][R] = k_2[LR]$$

The equilibrium constant for dissociation of the complex, K, is thus equal to the ratio of the kinetic rate coefficients:

$$K_{equilibrium} = \frac{k_2}{k_1} = \frac{[L][R]}{[LR]}$$

The assumption in receptor kinetics is that change in external conditions will rapidly change the concentration of the ligand near the receptor. Thus we call the constant, K, an *apparent* dissociation constant. For a specific substance we may lack stoichiometric knowledge about the concentrations of receptors and ligands. Thus we simply use the applied concentration, delivered dose, or exposure. The assumption that a change in applied concentration or dose rapidly transmits a changed concentration in the vicinity of the receptor, even if nonlinearly, may not hold for substances that have poor solubility which store in fat, have an existing body burden, or induce storage proteins. Such factors introduce additional nonlinearity into the biological response because they buffer small incremental changes but transmit large incremental changes.

These equations of classical receptor theory are mathematically nonlinear and of a hyperbolic nature. So are almost all receptor-mediated dose-response relationships. When we plot the magnitude of effect or the prevalence of effect versus the logarithm of ligand concentration, these equations produce a sigmoidal curve. On examination, these sigmoidal curves have the characteristics of normal distributions. Thus, receptor-mediated toxicities have log-normal exposure-response relationships. Consequently, pharmacologists often plot their data as probit of response versus the logarithm of dose—that is, as cumulative normal distributions with exposure displayed as the logarithm of dose.

The major impetus for expanding classical receptor theory came when M. Nickerson and R. P. Stephenson independently showed that less than full receptor occupation could achieve a maximum biological effect (Nickerson, 1956; Stephenson, 1956). Each put forward the idea that excess receptor capacity was the general state.

Stephenson proposed modifying classical receptor theory by introducing two different assumptions:

(1) A ligand can produce E_{max} without occupying all of the available receptor molecules, and

(2) The magnitude of effect is not necessarily directly proportional to the extent of receptor occupation; rather, it is some monotonically increasing function of the concentra-

tion of ligand-receptor complex which provides a stimulus leading to the expression of an effect through a complex cascade of events.

Stephenson decomposed the function relating the magnitude of effect to the concentration of ligand-receptor complex into two functions, essentially by applying the chain rule. One of the two functions describes a linear, ligand-dependent mechanism; the second function describes a nonlinear, ligand-independent mechanism that usually involves a cascade of steps with complex interactions, feedback loops, and external physiological regulation. Thus, we need more sophisticated nonlinear equations such as those described in Chapter 2.

If a ligand causes two effects in an animal, and if an antagonist blocks both effects to the same extent, both effects probably result from binding to the same receptor. If the antagonist blocks these effects differentially, different kinds of receptors probably are involved. If K for an agonist is the same for both effects, the receptors are quantitatively identical. These aspects of receptor theory led to quantitative classifications of different kinds of receptors. Exposure-response relationships of effects in populations of experimental subjects, such as a group of mice, conform to cumulative log-normal distributions for nearly all receptor-mediated effects. Extensive evidence supports this statement. This function probably represents the variation of effect resulting from the progressive occupation of a population of receptors on different cells (or in different animals).

Many toxicological mechanisms generate log-normal distributions (Masuyama, 1984). For example, as we discussed in Chapter 4, metabolic turnover follows a first order relationship with time (t) that controls the concentration (C) of a toxic substance, according to the relationship,

$$C = C_0 e^{-kt}$$

where C_0 is the concentration at the initial time, t, and k is a constant. If k is either normally or log-normally distributed in the population, the exposure to elicit toxicity will distribute log-normally (Koch, 1966; Koch, 1969).

Similarly, any process with steps that involve independent, multiplicative sources of variation will present as a log-normal distribution. Thus, allometric relationships exhibit log-normal distributions (Mosimann, 1975). Cell growth complies with these conditions because the mass of tissue at one stage of growth actively contributes to growth at the next stage. The crucial information from this theory for the risk assessor is that receptor-mediated effects have log-normal exposure-response relationships. In addition, knowledge of the receptor involved gives the risk assessor substantial compression of information. Predictions about a substance that reacts with one receptor, such as the nicotinic class of cholinergic receptors, allow the risk assessor to draw on a more extensive body of information about similar substances to predict potencies and outcomes. Even when receptors are not involved, log-normal distributions still may describe exposure-response relationships accurately because the exposure-response relationship describes the distribution of sensitivity within a population from most sensitive to least sensitive, and because sensitivity results from the multiplicative interactions of a cascade of events.

It is worth recalling that toxicological effects are both exposure and time dependent. Within the limits imposed by population size and competing risks, the risk of any adverse effect increases, often exponentially, with age. This principle explains why the potencies for acute, subchronic, and chronic lethality correlate highly over many substances. The comparisons of receptor theory usually involve experiments of similar durations. Here we inquire into nonreceptor toxicities, where the effect of duration of exposure becomes more prominent. The classical theory in this area is Haber's rule. Haber thought that toxicity varied with exposure magnitude multiplied by duration of exposure. If exposure x time is a constant over one period, and if the duration increases, the exposure to elicit the toxicity should decrease proportionally. This relationship is the basis for units like ppm-years to describe exposure.

Theoretical work by Karl Rozman and his colleagues has recently led to a useful general theory. A Malthusian differential equation describes change under constraint; it has an exponential form. (Malthus was the scientist who long ago predicted that the world's population would expand exponentially if unchecked by natural forces until mass starvation set in.) Rozman noted that irreversible toxic outcomes such as lethality should yield a time dependence in which exposure x time becomes a constant. However, with decreasing exposure, nonlethal effects might not follow Haber's rule. Instead, the functional relationship involving exposure and time must account for reversibility. If so, the shape of the exposure-response curve will dominate the mathematical description of the effect of exposure and time on responses. In this sense the log-normal distribution described above is a statistical description of a population, whereas an exposure-time relationship is more of a statement about causality. Rozman could always normalize the magnitude or prevalence of effect (E) as percent of maximum, whatever the units and dimensions of the biological effect in question. So, he looked at the incremental change in effect with time,

$$\frac{dE}{dt} = bE\,(E\,max - E)$$

and integrated this expression with respect to the incremental change in effect with exposure,

$$\frac{dE}{dD} = kE\,(E\,max - E)$$

The net result of this mathematical operation is a Weibull distribution, exactly as described in Chapter 2,

$$P\,(d) = 1 - e^{-(e(b)) \cdot d\,(h \cdot 1)}$$

Rozman is now in the process of describing the empirical fit of many exposure-time relationships to this theory. This theory is therefore still a work in progress and has not reached the scientific acceptance accorded receptor theory. However, this theory is extremely useful to risk assessors in making predictions about the duration of exposure likely to elicit particular toxic effects.

The Toxicology of Carcinogens

Carcinogenic effects are the most frequent basis for federal regulatory action. The reasons for this are that many substances yield positive results in the tests that the regulatory agencies use to identify carcinogens, and the agencies apply linear, nonthreshold exposure-response relationships to these substances. The background data that support these policies are extensive. At FDA, the controversial Delaney clause bans the use of animal carcinogens as food additives. Congress has recently sought to remove the application of the Delaney clause to pesticides in the Food Quality Protection Act of 1996. The Federal Office of Science and Technology Policy developed a government-wide statement of principles in 1985. EPA published an extensive guideline in 1986 and has worked without closure on proposed revisions to this guideline for nearly half a decade.

The exact relationship between animal test data and human cancer risk (predictive value of an animal bioassay for cancer) is open to question, especially if data from different kinds of experimental studies are not combined in making predictions. In the regulatory arena, lifetime cancer bioassays begin with the estimation of a maximum tolerated dose (MTD), usually obtained from a subchronic bioassay of the same substance in the same species, sex, and strain. Body weight and survival are traditionally the most variables considered. The MTD is the highest dose of the test agent predicted not to alter the animals' normal longevity from effects other than carcinogenicity. The MTD concept has been widely criticized. High exposures may cause significant cell killing and compensatory mitogenesis, thus producing carcinogenic effects that are unique to the high exposures. Similarly, high exposures may overwhelm detoxification mechanisms that operate at lower levels of exposure. For example, after saturation of the normal metabolic pathway, overflow metabolism may produce through a different pathway a carcinogenic metabolite not seen at exposures that humans experience. However, because a small number of animals serve as surrogates for a much larger human population, we must use exposures that are much larger than those associated with human exposures to be able to statistically ascertain an effect.

A typical chronic bioassay design will include two species, two sexes of each species, three exposures, and a control group. Thus these studies usually involve sixteen groups of animals (2 x 2 x 4 = 16). With 50 animals in each group, the study will involve the maintenance of 800 animals, typically for a duration of approximately two years. At the end of this time, surviving animals are sacrificed and the number of animals with tumors are counted. Typically, approximately 100 kinds of tumors are enumerated. The typical statistical analysis involves the application of Fisher's exact test to each exposed group in comparison to the experimental controls, and pooled exposed groups to experimental controls. For very rare tumors, the practice often is to compare historical control groups from many experiments with the same strain and sex combination. In addition, most investigators apply some kind of trend test to the data for each kind of tumor. Peto's trend test has been used extensively. The question that then arises is how to apply weight of the evidence to the data, ranging from no increase in tumors in an exposed group to increased prevalence of many groups (both sexes, both species, or two or more tumor sites).

The National Toxicology Program (NTP) has a consistent policy regarding the explanation of levels of evidence of carcinogenic activity that is contained in all of their reports. NTP holds that negative results do not necessarily mean that a chemical is not a carcinogen. They use the concept of strength of evidence based on all available evidence. Their classification scheme involves five categories:

(1) *clear evidence*: dose-related effects of either: (i) an increase of malignant neoplasms, (ii) an increase of a combination of malignant and benign neoplasms, or (iii) a marked increase of benign neoplasms if there is indication of possible progression to malignancy;
(2) *some evidence*: increased neoplasms (malignant, benign, or combined) in which the strength of the response is less than that required for clear evidence;
(3) *equivocal evidence*: studies showing marginal increase of neoplasms that may be chemically related;
(4) *no evidence*: no chemical-related increases in malignant or benign neoplasms, and
(5) *inadequate study*: cannot determine presence or absence of carcinogenic activity.

The most difficult part of classification is the use of animal data to predict human outcomes. Regulatory agencies generally assume that if a substance produces a toxic effect in animals, it also will yield the same outcome in humans. Similarly, carcinogenic effects at any one organ site in animal studies do not necessarily imply carcinogenic effects at the same site in humans, just carcinogenic effects at some human organ site. These assumptions are controversial.

At present, approximately 50–60% of substances subjected to an NTP design of chronic bioassay yield carcinogenic effects based on positive bioassay results, and nearly 500 substances have gone through testing at the NTP alone. This result bothers many toxicologists. Intuitively, the common expectation is that few substances would be animal carcinogens. Current federal regulatory policies about carcinogens assume as much.

Suppose, hypothetically, that we subject many known noncarcinogens to a chronic bioassay based on the NTP design. For any one kind of tumor and for each of the sex-species combinations (such as male mice), we might impose some demanding statistical criteria for a positive result. We might demand that any increase in tumor prevalence must increase with dose among the three dosed groups and that the prevalence in one or more of the dosed groups must differ statistically from the control group with a probability of 0.05 or less. Because we have three dose groups, the probability of the trend test result is 1/6. (You can only arrange 1, 2, and 3 in the sequence 1-2-3 in one of six ways. The others, such as 2-3-1 do not yield an increasing trend with dose.) Our probability of 0.05 should result in a positive result in one out of 20 comparisons at random. However, with a trend among the three doses, if only the highest dose is positive, the expected random probability is 1/3 x 1/20 = 1/60. If the top two dosed groups are positive, the expected probability is $1/2 \times (1/20)^2 = 1/800$; and if all three dosed groups are positive, the expected probability is $1/1 \times (1/20)^3 = 1/8000$. Because we expect the highest dosed group or the two highest dosed groups or all three dosed groups positive to have a result consistent with our trend test, we want to calculate the sum of these probabilities, which is approximately 0.018. Because we expect only one out of six studies to have an increasing trend, the probability of meeting our demanding criteria at random is approximately 0.003. However, we have four sex-species combinations. The probability that all four sex-species tests will be negative is $(1-0.003)^4$, or approximately 0.988. (The probability that any one sex-species combination will yield a positive result for one kind of tumor is approximately 0.012.) If we look for 100 kinds of tumors in our bioassay, we expect negative results $(1-0.012)^{100}$, or approximately 0.30.

Approximately 70% of noncarcinogens should yield positive results. However, our hypothetical expectation is greater than the empirically observed result for the substances that NTP already subjected to testing. Something appears wrong with the current interpretation of bioassay data. The bioassays yield fewer substances with carcinogenic effects than we might expect at random, even if none of the substances is carcinogenic in reality.

John Ashby, his colleagues, and many other investigators have looked at the relationship between cancer induction and direct mutagenic effects of chemical substances, using high quality, consistent bioassay data from sources such as the NTP. A combination of potency in bacterial mutation assays (with or without metabolism of the substance) and the presence of certain chemical substructures is highly predictive of carcinogenic effects (Ashby and Tennant, 1988; Ashby and Paton, 1993). The substructures indicate potential electrophilic sites within the structure of the substance or its metabolites that will react with DNA. Substances positive in bacterial mutation assays have greater potency than other carcinogens (Rosenkranz and Ennever, 1990). However, many animal carcinogens are not positive in bacterial mutation assays and they lack the predictive substructures.

Overall, the data are consistent with two classes of carcinogens, DNA-reactive and DNA-nonreactive. Approximately half of the animal carcinogens among substances previously subjected to chronic animal bioassays are DNA-reactive, but the proportion of DNA-reactive and DNA-nonreactive substances in the chemical universe is not known. Predictivity of animal bioassay results differs between the two classes of substances. A higher proportion of DNA reactive substances induces tumors in test animals. Approximately 30% of DNA-reactive substances are not animal carcinogens. DNA-reactive substances also tend to induce tumors in both rats and mice and/or in two or more organ sites. In addition, most known human carcinogens are DNA-reactive substances. A smaller proportion of DNA-nonreactive substances are animal carcinogens, and DNA-nonreactive substances tend to induce tumors at only one site in one sex-species combination.

However, the current federal regulatory policy treats all carcinogens as intrinsically DNA-reactive substances. Thus, as explained in Chapter 2, the agencies assume nonthreshold exposure-response relationships for all animal carcinogens, and all except OSHA force low dose linearity onto the data by a straight line from the lowest exposure with a statistically significant increase in tumors, a single hit model, or some version of the linearized multistage (LMS) model. Of the three, the LMS has the greatest flexibility in fitting data. These exposure-response models are consistent with the idea of substances (or their metabolites) reacting with DNA to produce mutations, which in turn induce increased tumor prevalence. This mechanistic expectation reduces to the simple idea that mutational hits will occur in direct proportion to exposure and tumor prevalence will occur in direct proportion to mutational hits.

Neither the nonthreshold assumption nor the linear assumption is necessary for chemical substances (Wilson, 1996). The two assumptions fit better with data from radiation carcinogenesis; they almost certainly do not apply to carcinogenic effects of infectious diseases. The major reason for their popularity is that a linear, nonthreshold exposure-response model is unlikely to underestimate risk. The idea has grown that linear, nonthreshold models are somehow consistent with public health protection. We disagree. Instead, the rote use of the models distorts regulatory priorities, particularly in the area of environmental protection. Aspirin, for example, is not a carcinogenic substance. Almost everyone would agree that consuming one-thousandth of an aspirin has a negligible risk of poisoning anyone. Yet a linear nonthreshold exposure-response relationship assumes that if half a million people each consumed this amount of aspirin, one case of aspirin poisoning will occur.

Aspirin is, of course, not carcinogenic, and at least some carcinogenic substances do have linear, nonthreshold exposure-response relationships. However, setting priorities between noncarcinogens and carcinogens becomes much more difficult when highly different assumptions about exposure-response relationships are applied to the two classes.

The absence of data about the two assumptions may force regulatory agencies to adopt them, but the use of these models has become so automatic that the federal agencies seldom explore available information about the underlying assumptions when assessing the risk of a substance. Unfortunately, most of the data for these regulatory exposure-response relationships comes from NTP-design bioassays. This design has the intent of detecting qualitative results. In the opinion of many toxicologists, this design uses too few doses and produces data that are not suitable to explore quantitative exposure-response relationships. In the context of risk assessment, the problem is that these data yield high statistical uncertainty in the exposure-response relationship, so that it is not reasonable to expect tests of a threshold or linearity to yield much information about the two assumptions.

The consensus view is that carcinogenesis involves somatic mutations. The mutations are irreversible and heritable; they pass to progeny cells. Several different mutations are required to create circumstances of uncontrolled growth. Thus an organism may have many cells that have acquired a few of the necessary mutations, but a tumor will not arise. A cell that loses responsiveness to growth and location signals still has to overcome defense and defense mechanisms such as competition from neighboring cells, inflammation, and immunity. However, once the first tumor cell forms, with all of the necessary mutations, the carcinogenic process becomes inevitable, a progression from initiation to tumor to metastasis to death. Given the biological characteristics of the tissue of origin and the set of mutations, the progression becomes predictable. Recent cancer models therefore try to predict the arrival of the first tumor cell and allow for a standard latency period before the appearance of a pathologically detectable tumor. Cancer appears as an all-or-none effect.

New mutations can arise through several mechanisms, such as:

- reaction of a substance or its metabolites with DNA to form mutagenic DNA adducts,
- increased reaction of an endogenous substance with DNA through a change in metabolism,
- generation of mutated daughter cells through increased replication of cells with a background rate of mutation,
- induction of mutagenic errors by inhibiting DNA repair or recombination, and
- increased background mutagenic rates, for example, by diminution of the portion of the cell in the resting portion of its cycle (G_0) by forced cell replication (allowing less time for normal repair and recombination events)

Mutation is a complex process that has alternative pathways leading to the same endpoint, and involves cell survival and cell replication. The events leading up to mutation (DNA adduct formation) are toxic and can kill the cell. Any increase in mutation rate (versus the number of mutated cells) will increase the death rate of mutated cells. Cells that do not divide will not fix the mutation and are not on a pathway leading to uncontrolled cell growth. For purposes of more accurate risk assessments of carcinogens, it is important to decide which, if any, of the above mechanisms apply to a carcinogenic substance.

Safety Assessment

Federal agencies accomplish the regulation of substances added to foods through positive lists (see Chapter 3). The agencies (FDA and EPA) publish a list of allowed substances and the conditions that apply to their use. The list is enforced in several ways, but the most important one is the presence of a substance not on a list. The use of unlisted substances is not legal. Food containing an unlisted substance is adulterated by definition, and regulators will impound it. In addition, the federal agencies insure that listed substances are not misused, for example, through excessive application. The process of evaluating a substance for addition to a positive list is called safety assessment. To enforce the proper use of listed substances, the agencies also publish lists of tolerances, maximum amounts of approved substances allowable in different foods. For example, many pesticides have one tolerance on strawberries and a different tolerance on apples.

Tolerance setting is a complex activity, but at a minimum it always involves an estimate of the exposure of a substance that the average person in the population would acquire from normal consumption of, in this case, strawberries and apples. Regulators calculate tolerances by estimating the average consumption of each food, including indirect consumption through prepared foods like canned apples or strawberry ice cream. Knowledge of the food consumption habits of the U.S. population is therefore of great interest to FDA and EPA.

The allowable amount from all foods is based on the acceptable daily intake (ADI; EPA calls ADIs Reference Doses). The idea behind an ADI is that humans can ingest this level of a substance every day throughout their lives without experiencing much likelihood of an adverse effect, unless the effect relates to some unpredictable sensitivity, such as an allergy to the substance. Politicians usually write legislation that calls for safe food, whereas risk assessors know that no substance in foods is without risk. The ADI concept attempts to bridge the gap between reality and political language. The standard that supports an ADI was developed by FDA. It is called a "reasonable certainty of no harm." It is not worthwhile to read this phrase too closely. It actually is a bureaucratic code word for "following FDA's traditional practices." The toxicological practice of safety assessment is complex. We have simplified it into the following five steps:

(1) Submission of data to the regulatory agency. The petitioner or registrant must provide required studies to government. Regulatory agencies publish guidelines to facilitate the submission of adequate quality studies. However, the petitioner or registrant

is responsible for the safety of the substance. The regulatory agency is not bound by the guideline and may require additional studies. The required studies are mostly animal toxicology studies; a few relate to analytical chemistry and anticipated exposure. The studies are costly, lengthy, and financially risky for the sponsor. At present, the data to support a food additive petition usually cost approximately eight million dollars, and the data to support a pesticide registration application petition usually cost approximately twelve million dollars. It takes at least three years, typically closer to seven years, to complete all of the studies before submission to the regulatory agencies.

(2) Intensive review of the required studies. Government scientists carry out the review of submitted data, usually for quality first. If a required study is missing or fails to meet quality standards, the petition or registration application is rejected. In addition, an agency may exclude all substances with certain classes of toxic effects. FDA excludes carcinogens.

(3) Derivation of a no-observed-effect level (NOEL) or no-observed-adverse-effect level (NOAEL). The NOEL is the highest body-weight specific, external exposure (applied dose) that does not induce an adverse effect in any of the submitted studies. NOELs have units of mg/kg per day. Because the toxicology studies are conducted following guidelines put out by the regulatory agencies, the NOEL is not a theoretical threshold for the exposure-response relationship. Instead, it is a regulatory surrogate for a threshold. Because a NOEL is restricted to the doses used in the required studies, it may be lower than the "true" NOEL. The study duration for the NOEL must match the duration of the intended exposure, acute for acute, chronic for chronic. Despite much confusion and statements to the contrary, a NOEL is determined according to biological judgement, not statistical criteria.

(4) Reduction of the NOEL to an ADI by dividing the NOEL by a safety factor of one-hundred. (EPA calls it an uncertainty factor.) In essence, the NOEL/100 defines the ADI, meaning that regulators believe that a person can consume this dose each and every day over a lifetime without experiencing a toxic effect. The safety factor accomplishes several tasks—it adjusts for errors in extrapolating an animal NOEL to a human NOEL, and it makes some allowance for uncertainty in the safety assessment process. For example, if the extrapolation from animal NOEL to human NOEL does not follow body weight specific dose, part of the factor compensates for any systematic error that might overestimate the human ADI. In part, the safety factor accounts for experimental uncertainties in the data. On rare occasions regulators will increase the safety factor to account for an unusually severe adverse effect. The hundred-fold safety factor is not meant to protect against dose-independent, idiosyncratic outcomes such as allergies. In addition, the design of toxicology studies, not the safety factor, accounts for duration. Despite statements that a safety factor of ten is used with hu-

man data, in practice human data are used in combination with other data to estimate a human ADI directly. An ADI does not mean that a dose higher than the ADI will necessarily cause adverse human effects. The regulatory agency warranty runs in the opposite direction; doses lower than the ADI should be free of adverse human effects.

(5) Publication of a regulation with accompanying conditions. The regulation, as part of a positive list, may refer to allowable tolerances for foods, label directions, manufacturing restrictions, other conditions of use, and so forth. Despite these restrictions, listing in a regulation and compliance with a regulation is usually not a legal defense against the sponsor's liability for any adverse human toxic effects discovered later. The regulatory agency will compare the ADI to the exposure expected from the uses intended by the sponsor and will modify factors such as application rates if the anticipated exposures exceed the ADI. The alternative is for the sponsor to withdraw the application or petition.

The original development of FDA's safety assessment process was quite bold. All that we know at present is that the policy apparently has not failed in more than sixty years of experience, in the sense that no substance placed on the positive list later manifested adverse human effects when used as intended. Given that approximately 1000 substances have gone through the process, the overall five-step process has an approximate false-positive rate of less than 0.001.

The interpretation that two separate tenfold safety factors account for (1) interspecies differences in sensitivity, and (2) intraspecies variability of human response is largely a post hoc rationalization. What we really know is that the overall safety assessment process has a very low false positive rate when using a hundredfold safety factor. Using the chain rule, you might as well partition the hundredfold safety factor in other ways. For example, Renwick has suggested two independent safety factors for uncertainty about (1) disposition and (2) sensitivity. His approach is quite interesting, in part because toxicologists can bring data that is more easily obtained and independent to bear on these factors.

The statutory basis for ADIs, a "reasonable certainty of no harm," implies that the ADI will protect the range of normal human behavior. Regulatory agencies cannot expect to protect against all possible misuses or extreme behaviors. "Reasonable" does not contemplate zero risk. The concept of reasonableness comes from English common law, which uses it extensively. Risk-benefit balancing is inherent in the concept. The reasonable certainty of a no harm standard applies a common, not a variable, degree of protection from all substances on the positive list. Congress inherently recognized the benefits of food additives and pesticides, even in setting up these regulatory processes. Otherwise, why bother to have food additives and pesticides at all? In this context, the burden is the loss of products necessary to the food supply. The original authors of the safety assessment process, A. J. Lehman and O. G. Fitzhugh, noted that a safety factor of one hundred appeared large enough to prevent human health effects and simultaneously allow the use of substances necessary for food production. Subsequent experience has demonstrated that a hundredfold safety factor balances certainty about the lack of harm with the benefits of a better quality food supply.

Over the decades, FDA has developed a legal and operational history with the reasonable certainty of no harm standard. The case studies are complicated, and some are inconsistent with others. The standard is therefore difficult to describe more accurately than we have attempted without going into great detail. We have yet to see how EPA will interpret the reasonable certainty of no harm standard. The change to this standard for pesticides only occurred in 1996 with the passage of the Food Quality Protection Act. So far it appears that we will have two widely different standards. For example, FDA interprets reasonable use as approximately the upper 90th percentile of consumption within the U.S. population, whereas EPA has proposed the use of the upper 99.9th percentile of consumption

EPA programs other than the Office of Pesticide Programs also use safety factors to set standards for noncarcinogenic substances. However, these uses differ from safety assessment, as outlined above. Among other differences, the lists usually are negative in nature; that is, EPA publishes lists that indicate the maximum permitted exposures for forbidden substances. Substances not on the list are allowed. The routes of exposure often do not involve ingestion. The data supporting the list is highly variable in quality and extent. Generally, EPA does not require specific studies but uses the available scientific literature to support these regulatory decisions. In addition, EPA uses a more extensive list of uncertainty factors, including factors up to ten for:

- converting acute exposure to subchronic exposure,
- converting subchronic exposure to chronic exposure,
- compensating for missing data,
- using the lowest observed effect level, when a no-effect level is not available,
- producing a shallow dose-response curve, and/or
- providing additional sensitivity or exposure of children.

In practice, EPA reduces many of these factors from ten to three with partial data. With adequate and direct data, EPA sets the factor at 1. For example, with adequate quality chronic exposure data, the two conversion factors for acute and subchronic exposure are set to one. Further, if the human exposure of concern is chronic, EPA would not apply two uncertainty factors to convert acute exposure. The last of these factors, sometimes called the children's factor, has proven particularly controversial. Politicians, not scientists, developed and legally required the use of this factor for pesticides. The rationale to apply it more broadly is not clear. The use of a factor to account for uncertainty in exposure, not toxicology data, is a radical change in approach. Application of a tenfold safety factor to protect children mostly lacks a scientific justification.

In addition, EPA has worked on a procedure to interpolate data to estimate a no-effect dose that is independent of experimental doses, called the Benchmark Dose (BMD). The BMD resembles the LMS, but the model interpolates bioassay data to a uniform level of risk, such as 10% prevalence, and uses this point of departure (with or without statistical compensation for sample size) as a surrogate for a no-effect level. The BMD is an entirely statistical operation. It does not account for changes in severity as exposure changes, and it does not account for supplementary information, particularly nonquantifiable biological observations.

Safety assessment combines risk assessment and risk management. Therefore, EPA's widespread generalization of the safety assessment process is questionable. Safety assessment combines risk assessment and risk management into one process. Part of the rationale for the combination is administrative efficiency. The application of discretionary uncertainty factors eliminates this efficiency. Instead of preempting discussion, EPA's applications of uncertainty factors require detailed analyses and fragment the decision-making process. While legislators and regulators want procedural consistency, EPA's process sets policy and science in opposition. Default assumptions, including uncertainty factors, have their basis in policy. Additional scientific data and theory therefore cannot counteract a risk management policy. We can only wait until enough experience has accumulated with EPA's process to estimate its false positive and false negative rates.

Despite widespread assertions to the contrary, safety assessment is not a form of risk assessment, although the products of the two processes do overlap. The difference goes beyond the admixture of risk management into the safety assessment process. A safety assessment leads to a statement that a substance is not likely to have appreciable harm when used in certain ways. This statement does overlap with a statement about risk. If we can decide what "not likely" and "appreciable harm" mean quantitatively, we may be able to convert the data supporting a safety assessment into a statement about risk. The statement of risk likely would vary from substance to substance and from use restriction to use restriction. Safety assessment does not lead to an estimate of the probability of future harm. Instead, it specifies an excursion limit to dose. In particular, safety assessments do not inform circumstances when someone breaches the use restrictions. Do we ignore the violation, treat it as a subject for administrative fines and penalties, or send for ambulances?

Additional Information About Toxicology

This chapter introduced you to the field of toxicology, a vigorous and rapidly developing discipline. It is important for risk assessors because, as a practical matter, the data used in most risk assessments comes from toxicological experiments. Toxicology is more bound to hypothesis testing and is less of an observational science than epidemiology. Experimental data therefore often fit better with particular models of a process, leading to a better risk assessment model. However, for risk assessors the similarities are more prominent than the differences. Toxicological data are a source of information about potency, and the assessor often can extract potency estimates from experimental data without much deference to the investigator's rationale. This difference leads to a high level of frustration among risk assessors. Toxicological publications often omit the very data necessary to estimate probabilities and potencies because the investigators did not have risk assessment in mind when conducting the experiments or because the editor was not sympathetic to risk assessment uses of the data.

We saw that toxicologists rely heavily on expert judgement and therefore rely heavily on peer review. Risk assessors can extract useful potency information from many kinds of toxicological data, not just bioassays of intact animals. With no other prior information, risk assessors can apply log-normal or Weibull models to most bioassay data for noncarcinogens to achieve reasonable exposure-response or exposure-time-response models, respectively. Because toxicologists often pursue mechanism of action, information may lead to even better specific models for the potency of a substance. However, current regulatory practices apply linear, nonthreshold exposure-response models to carcinogens.

We suggested that two classes of chemical carcinogens apparently exist: DNA-reactive and DNA-nonreactive. The assumptions underlying linear, nonthreshold models are more consistent with DNA-reactive substances. However, some experts believe that almost all carcinogens have thresholds or have nonlinear exposure-response relationships. Linear, nonthreshold models project large numbers of cancer deaths at very low levels of exposure in large populations. A high proportion of total U.S. expenditures for environmental protection are intended to avoid these hypothetical cancer cases. These funds purchase little, if any, public health protection if most carcinogens have thresholds or nonlinear exposure-response relationships.

We saw that safety assessment differs from risk assessment in key characteristics. Safety assessment combines risk assessment and risk management. A safety assessment does produce a conditional statement about the probability of all future human health effects, but only for exposures below a region where the risks are negligible. Safety assessment involves many procedures that qualify the regulatory decisions that result, such as reasonable behavior and observance of label restrictions. These procedures are not necessary to the assessment of risk.

Of necessity, our introduction has been highly superficial. We have skimped on case studies and examples. We have not gone into methods in any depth, and we have not reviewed toxic effects in the different organs and tissues in detail. Extrapolation of human responses to a substance from animal outcomes likely will differ according to the organ involved. We have not gone into the variation of sensitivity to substances within animal and human populations. We only briefly covered mixtures and interactions between substances. Many assessments lead to point estimates of potency at each exposure. If a risk estimate incorporates a single exposure, a point value results instead of a distribution of risk. However, estimates of the uncertainty in these single values are difficult, if not impossible, to obtain. For example, the uncertainty in a linear, nonthreshold model is unidirectional and essentially infinite. Many sources of uncertainty in potency are not quantitative or readily reduced to coherent quantitative values. Because the joint probability of a potency distribution and an exposure distribution does not reduce to a single risk value that the assessor can cite with confidence. The net result of these difficulties is that risk assessment procedures in regulatory agencies incorporate many conventions and assumptions. The key idea to keep in mind as you learn risk analysis is that most regulatory estimates have little meaning outside of the conventions and assumptions.

Fortunately, one textbook has primacy in toxicology, *Casarette and Doull's Toxicology: The Basic Science of Poisons* (5th) (McGraw-Hill, New York (1996)). We recommend it unconditionally. It will provide the interested risk analyst with much more extensive information. Also, toxicology is a field with many professional societies. In the U.S. the most prominent ones are the Society for Environmental Toxicology and Chemistry, the Society of Toxicology (SOT), the American College of Toxicology (ACT), and the International Society for Regulatory Toxicology and Pharmacology (ISRTP). SOT has a division devoted to risk assessment, and the journals published by ISRTP and ACT often deal with topics relating to safety assessment. Other more specialized journals also exist. You can locate them at a local medical library or on the Internet.

Two good sources of information about the use of toxicology information in human health risk assessment are D. J. Paustenbach' Retrospective on U.S. Health Risk Assessment: How Others Can Benefit. *Risk* 6: 283–332 (1995) and Office of Technology Assessment's *Researching Health Risks* [OTABBS570] U.S. Government Printing Office, Washington, D.C. (1993).

Other important sources include:

Allen, B. C., K. S. Crump and A. M. Shipp. 1988. Correlation between Carcinogenic Potency of Chemicals in Animals andHumans. *Risk Analysis* 8(4): 531–557.

Ames, B. N. and L. S. Gold. 1990. Chemical Carcinogenesis: Too Many Rodent Carcinogens. *Proc. Natl. Acad. Sci.*87: 7772–7776.

Whittemore, A. and B. Altschuler. 1976. *Biometrics* 82: 805.

Wilson, J. D. 1989. In *Biologically-Based Methods for Cancer Risk Assessment*. C. C. Travis, Ed. New York: Plenum.

One email list specializes in toxicological information, sponsored by the Bionet Toxicology News Group, which has a web site at www.bio.net/hypermail/TOXICOLOGY (or contact Bionet by email at biosci-help@net.bio.net to subscribe). One think tank, Toxicology Excellence for Risk Assessment, posts extensive information on its Internet site, www.tera.org, and the Chemical Industry Institute of Toxicology has extensive information at www/ciit.com. Both sites are good places to start searching for information. Many academic institutions have toxicology departments, and they often support Internet sites.

The risk analyst should differentiate sources of information about toxicology in general from sources of information about the toxicology of specific substances and processes. A good place to start a substance-specific search on the Internet is the National Library of Medicine's search engine for Medline, a compendium of journal article citations, at www.ncbi.nlm.nih.gov/PubMed. Also take a look at Hardin Meta-Directory for Toxicology at www.arcade.uiowa.edu/hardin-www/md-tox.html.

Government agencies have extensive repository information about substance and processes, including the following:

U.S. National Institutes of Health, www.nih.gov

International Agency for Research on Cancer, www.iarc.fr

Agency for Toxic Substances and Disease Registry, atsdr.cdc.gov

Occupational Safety and Health Administration, www.osha.gov

U.S. Environmental Protection Agency, www.epa.gov

The National Center for Environmental Assessment and Risk Assessment Forum subsites of the main EPA site have extensive information about risk assessment guidelines.

References

Ariens, E. J. 1954. *Arch. Int. Pharmacodynamie*. 99:32.

Ariens, E. J. *1964*. *Molecular Pharmacology*. New York: Academic Press.

Ashby, J. and D. Paton. 1993. The influence of chemical structure on the extent and sites ofcarcinogenesis for 522 rodent carcinogens and 55 different humancarcinogen exposures. Mutat. Res. 286(1): 3–74.

Ashby, J. and R. W. Tennant. 1988. Chemical structure, Salmonella mutagenicity and extent of carcinogenicity as indicators of genotoxic carcinogenesis among 222 chemicals tested in rodents by the U.S. NCI/NTP. Mutat. Res. 204(1):17–115.

Clark, A. J. 1933. *The Mode of Action of Drugs on Cells.* Baltimore: Williams & Wilkins.

Gaddum, J. H. 1937. *J. Physiol.* (London) 89:7.

Gaddum, J. H. 1943. *Trans. Faraday Soc.* 39:323.

Koch, A. L. 1966. *J. Theoret. Biol.* 12:276.

Koch, A. L. *1969*. *J. Theoret. Biol.* 23:251.

Langley, J. N. 1905. *J. Physiol.* (London) 33:374.

Masuyama, M. 1984. *Biom. J.* 26:337.

Nickerson, M. 1956. *Nature* 178:697.

OSTP (Office of Science and Technology Policy). 1985. Chemical Carcinogens: Review of the Science and Its Associated Principles. *Federal Register* 50:10372–10442.

Rosenkranz, H. S. and F. K. Ennever. 1990. An association between mutagenicity and carcinogenic potency. Mutat. Res. 244(1): 61–65.

Stephenson, R. P. 1956. *Brit. J. Pharmacol.* 11:379

USEPA, Guideline for Carcinogen Risk Assessment. *Federal Register* 51:33, 992 (1986).

Wilson, J. D., *Thresholds for Carcinogens: A Review of the Relevant Science and Its Implications for Regulatory Policy.* Resources for the Future Discussion Paper, 96–21 (1996).

Chapter Eight

Risk Characterization

Bringing Potency and Exposure Estimates Together

Besides the probability that an individual faces a risk, it is important for us to estimate the total number of expected cases. Let us call the total number of estimated cases of disease or other deleterious effect the *population risk*. We can estimate this for a year or for a lifetime (generally the difference is a factor of 70 or the average expected lifetime). We can calculate population risk in different ways depending on the information available about the source, exposure rate, and other factors involved in the calculations. In every case, however, we will have to multiply and divide several factors, including:

- occurrence or concentration;
- ingestion, inhalation, or dermal absorption rate;
- exposure duration;
- exposure frequency;
- body weight;
- population exposed (including parameters such as age and sex);
- risk (from a dose-response model);
- amount that reaches the affected organ, and
- transport factors.

We learned earlier that each of these factors has an inherent uncertainty and distribution or range of values. The above quantities have different statistical characteristics. The purist statistician will argue that they cannot be combined or, at least, the uncertainties cannot be combined. What we offer in this chapter is a "practical" approach to combining uncertainties or errors. Our problem now is how best to combine these ranges in the overall estimate or calculation. You must realize that the overall distribution of error is just as important a quantity as the estimate of the risk itself. Without knowing how precise or accurate our estimate is, we are in the dark about how far off we could be.

We will look at three examples here that involve calculation of the overall uncertainty of population risk estimates. The examples will consider (1) drinking water exposure to trichloroethylene, where we consider the uncertainty to be predominantly from the range of exposure-response model estimates; (2) drinking water exposure to radon, where we use conventional error analysis, and (3) returning to the trichloroethylene in drinking water example, we will combine frequency distributions for all the other quantities involved and assume that the exposure-response modeling estimate has no uncertainty (the opposite of the first example). We note that the two examples involving trichloroethylene demonstrate differences in judgement about the relative importance of the different uncertainties.

Exposure-Response Modeling Errors—Trichloroethylene

In our analysis, the main factors involved in the determination of population risk for a specific contaminant are:

- P_c, the number of people exposed to the contaminant at concentration c;
- R_c, the individual risk rate for contaminant c (directly from epidemiology studies or derived for humans from animal studies); and
- O_c, other factors that are involved for some contaminants, including the percentage that gets to the organ affected, age, sex, and transport factors.

Then we see that the population risk:

$$\text{Population risk} = \Sigma\, P_c R_c O_c$$

where Σ reads as the summation of this product over all concentration ranges, c. Multiplying these factors is a simple task if we use the average value. Our problem here is to learn how to combine ranges of uncertainties.

Our estimate of population risk involves assumptions. For the particular example of trichloroethylene in drinking water these assumptions are:

(1) the average human ingests two liters of drinking water per day and daily inhales an average of twenty cubic meters of air,
(2) the health end point is the same for animals and humans,
(3) the exposure level for environmental contaminants in the entire U.S. can be approximated from knowledge from a few sampled locations,
(4) the toxicity at low levels of exposure can be interpolated from knowledge of toxicity at high levels of exposure,

5) benign and malignant animal tumors are both indicative of cancer, and their numbers can be combined,
6) exposure-response curves can be extrapolated into regions for which there are no experimental data,
7) the only difference between humans and test animals is one of scale, and
8) dermal exposure is insignificant compared to oral and inhalation exposure.

We see that the final estimate is dependent on several essential assumptions. Now let us consider the first example. In this example we base the estimate on the experimental exposure-response data resulting from exposure of mice to trichloroethylene in drinking water which are:

animal exposure mg/kg/day	animals affected/ total
0	1/20
1530	26/50
2700	31/48

Realize that these data were generated for the purpose of screening this contaminant to provide a crude quantitative measure of the relative risk. They were not generated for the purpose of modeling, but we will use (misuse?) them here for that purpose, because that is the general practice. They are typical of the data used in exposure-response models. Usually there are only three (sometimes four) values available—controls and two other exposures.

Our first task is to convert these data to a human equivalent exposure. First, since the experiment was conducted five days a week, we multiply the exposure by 5/7. The experiment only lasted 1 1/2 years and the average mouse lifetime is two years—thus we multiply by 1.5/2. Next we need to convert the weight from that of a mouse to that of a human.

There are a variety of ways for us to convert the exposure-response data from the size of a mouse to that of a human (see Chapter 5). Two well-known approaches are to use relative body surface area or relative body weight. To adjust for surface area, multiply by the cube root of the ratio of body weights, or $(0.033/70)^{1/3}$. Depending on the method used, a range of exponents can be used (see Federal Register, 1992, or Rhomberg and Wolff, 1998). The differences in the methods used here result in a contribution to the overall uncertainty that is far less than that from the other uncertainties, and we have arbitrarily chosen the surface area method. Finally, we assume a 70 kg adult consuming two L/day of water. The resulting values for the human equivalent dose are:

animal exposure mg/kg/day	human equivalent mg/kg/day	animals affected/total	mg/L in drinking water
0	0	1/20	0
1530	64	26/50	1.83×10^6
2700	113	31/48	3.23×10^6

These values are plotted in Figure 8.1 as Xs. We note that the control point cannot be plotted on a log-log scale.

We can see from Table 8.1 that the ingestion and inhalation routes of absorption dominate and are about equal. For our purposes here we will assume them to be equal.

Table 8.1 Comparative Model of Absorbed Dose from a Volatile Pollutant (100 mg/liter) in Drinking Water (from Cothern and Van Ryzin, 1985)

	Amount absorbed (mg/day)			
Route	Formula fed infant (4 kg)	Preteen (32 kg)	Adult female (60 kg)	Adult male (70 kg)
Ingestion	80	200	200	200
Inhalation				
Indoor air	10	20	70	50
Shower air	-	100	300	200
Dermal				
Bathing	0.06			3
Swimming	10–300			
Total Absorbed Dose (mg/kg/day)	20	10–20	10	10

Next we need to determine the estimated occurrence of trichloroethylene in drinking water. The data we use here were based on those from five federal surveys (see Cothern et al., 1984 for more details). In the second column of Table 8.2, we find the number of people being served at each mean drinking water concentration.

We estimate the population risk due to trichlorethylene in drinking water using the procedure described in the schematic of Figure 8.2. As we see in Table 8.2, the population risks are determined by a product of the number of people affected and their individual risks. In each case the concentration shown is the average in the range.

Table 8.2 Population Risk Estimates for Trichloroethylene in Drinking Water

Mean Drinking Water Concentration (micrograms/liter)	Number of People Being Served	Total Lifetime Individual Risk For the Mean Concentration: Low (Probit)	High (Weibull)	Lifetime Population Risk
<0.5	1.9×10^{8}	$<10^{-10}$	2.4×10^{-4}	<1 – 45,600
2.75	2.3×10^{7}	$<10^{-10}$	7.3×10^{-4}	<1 – 16,790
7.5	4.3×10^{5}	$<10^{-10}$	1.3×10^{-3}	<1 – 559
15	2.1×10^{5}	$<10^{-10}$	1.7×10^{-3}	<1 – 367
35	7.4×10^{5}	7×10^{-8}	2.3×10^{-3}	<1 – 1,702
45	2.6×10^{5}	3×10^{-7}	2.6×10^{-3}	<1 – 676
55	4.2×10^{4}	4×10^{-7}	2.8×10^{-3}	<1 – 117
75	1.3×10^{5}	6×10^{-7}	3.2×10^{-3}	<1 – 416
100	4.2×10^{4}	1.2×10^{-6}	3.7×10^{-3}	<1– 155
			Total[a]	<1 – 70,000

Analysis of effect of setting a standard and having no violations

Maximum Allowable Drinking Water Concentration (micrograms/liter)	Cumulative Cases Averted
100	<1 – 200
45	<1 – 1,000
7.5	<1 – 4,000
2.75	<1 – 20,000

[a] Rounded off to one significant figure

Now let us look more closely at Figure 8.1. The four models we use have a range of statistical characteristics. The Weibull, probit, and logit models are tolerance distribution models (assume a range of individual susceptibility), while the multistage model assumes the dose-response curve is the result of randomly occurring biological events. Al-

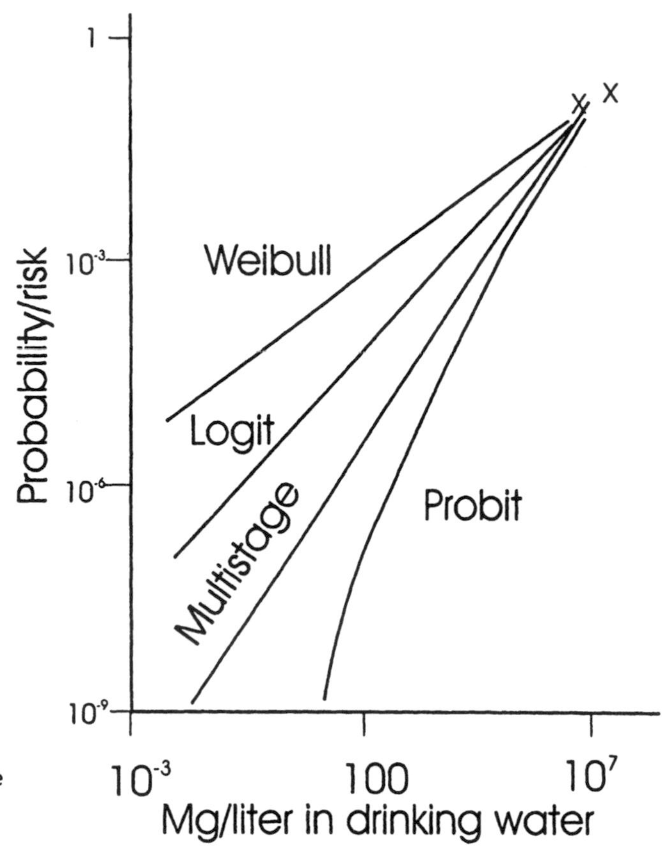

Figure 8.1 Exposure-response curves for exposure of mice to trichlorethylene

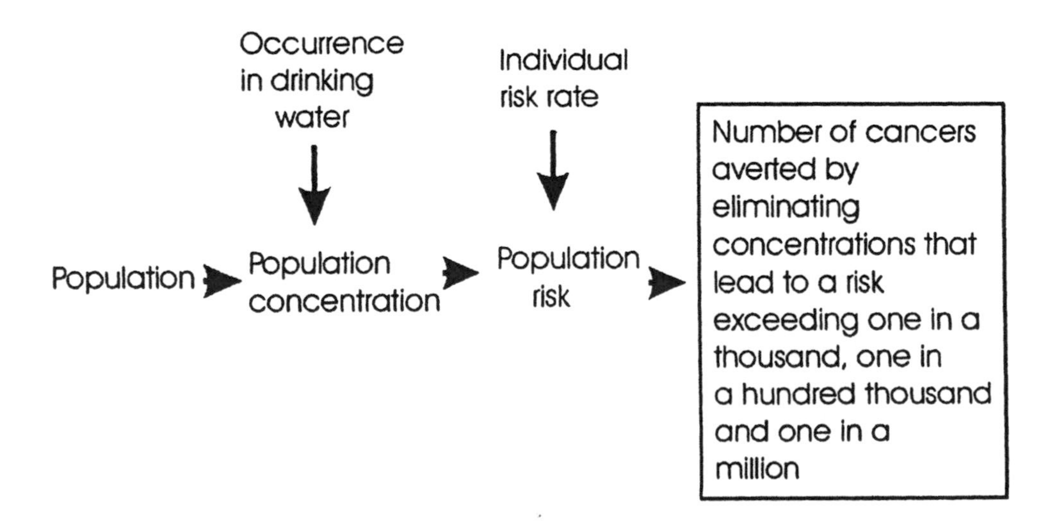

Figure 8.2 Conceptual model of the components of population risk

though not strictly comparable, they do show us plausible mathematical representations. The four model fits to the two data points spread out over much of the area shown. Assume that these four curves are representative of the range of models that can fit these

Table 8.3 Contributions to uncertainty from toxicological studies (Cothern, 1986a and b)

Many orders of magnitude

- Distribution of animals among dose levels and total number used
- Basing risk estimates on a curable cancer
- Animal variability
- Interspecies comparison

One to two orders of magnitude

- Purity of test compound
- Good laboratory procedures (GLPs)
- Synergisms and antagonisms

Little change up to one order of magnitude

- Selection of dose levels
- Examination of animals for cancer precursors
- Dietary considerations
- Time-to-tumor
- Adding benign tumors to malignant ones to improve the statistics
- Interference of diseases in animals
- No corresponding human tissue (e.g. zymbal gland)
- Body weight vs surface conversion

All or nothing

- Choice of endpoint
- Dose levels chosen
- Capabilities of personnel
- Choice of species, strain, age and sex of animals

same two data points. That being the case, the range of risk or probability for a given concentration can be determined by drawing a vertical line at that concentration. In the region of 10 to 100 mg/L we find that the range from highest (Weibull) to lowest (probit) is about six orders of magnitude (one millionfold).

We provide a list of all contributions to the uncertainty in the calculation of population risk in Tables 8.3 and 8.4. The six orders of magnitude we found for the contribution

Table 8.4 Sources of uncertainty for occurrence, population concentration, and exposure estimates (Cothern et al, 1984)

Factors	Occurrence (%)		Impact on estimate of* Population Concentration	Risk
Proportion of population sampled	5 (U)			
Representativeness of systems selected; geographic distribution, system size, and source of water 10(E)				
Sampling methods				
Site of sample collection	20(E)	→	50%	Factor of 2
Time of sample collection	20(E)			
Method of sample collection	10(U)			
Container type	10(U)			
Stability during storage	100(U)			
Sample analysis				
Recovery (%) from sample	10(U)			
Compound identification	10(E)			
Accuracy of quantitative determination	40(E)			
Oral exposure rates (intake of water, hot or cold water, absorption)			50%(O)	
Respiratory rates				Factor of 3

* U = leads to an underestimation of the risk; O = leads to an overestimation of the risk; E = could lead to an overestimation or an underestimation of the risk

from the exposure-response model estimates is clearly much larger than any other uncertainty listed in Tables 8.3 or 8.4. Thus, using the uncertainty in the model estimates alone dwarfs all other uncertainties.

The lowest concentration range shown in Table 8.2 (less than 0.5) represents the minimum reported concentration for the national surveys and does not necessarily represent the detection limit of the chemical in drinking water. We need to make an assumption about the possibility of concentrations below that value. We have several choices. We could assume a similar distribution to the one above the lowest level, that there are no concentrations below that level, or that all below that level are at one half that level.

For our calculations we assume they are all at 0.5 micrograms/liter. We assume that half the samples below the lowest reported value would be at half that value. In some cases this uncertainty can be quite large numerically.

In each case the curve giving the lowest individual risks and the curve giving the highest individual risks were used to determine the range of population risks (from Figure 8.1). As we noted earlier for the concentration values in this example, the uncertainties in the model estimates are much larger than any of the others involved, except for the assumption that trichlorethylene is a human carcinogen at all. When calculations yielded less than one person, the risk was listed that way rather than as a fraction of a person.

As we see in Table 8.2, the population risk estimates are listed as ranges. The range is indicative of the uncertainty in this estimate. For trichloroethylene in drinking water, we find that the lower estimate for each concentration range is so small as to be much less than one. The sum of the upper population risk estimates is 70,000 when rounded off to one significant figure. We note that the majority of the population risk is contributed by the lowest concentration category. This category represents calculations below the lowest reported value, and we must be careful because this may be an artifact of the mathematical analysis scheme used to estimate the population risk. The range of uncertainty we estimated is largely a reflection of the uncertainty in choice of mathematical model selected to estimate the individual risk.

Combining Estimated Error Ranges—Radon from Drinking Water

Radon is a colorless, tasteless, and odorless radioactive gas. We know of two health endpoints affected by radon in drinking water. It is associated with stomach cancer due to ingestion exposure and lung cancer from inhalation of particulates containing radon progeny. Radon is released into our indoor air by our common household activities, such as washing clothes and dishes or taking baths and showers. When radon decays, the resulting atom is ionized for a few microseconds and can attach to small dust particles in the air. The radon decay products are several different radioactive atoms which emit alpha, beta, and gamma rays that are believed to precipitate lung cancer. When we inhale very small particulates (less than ten microns in diameter), they will stay in the lung and can produce lung cancer. Larger particles are ejected from our bodies by the motion of the cilia (hair-like objects) in our respiratory system. For more details see Cothern (1987, 1989, and 1990).

To develop the population risk estimate for exposure to radon from drinking water, we will start with an analysis of the inhalation risk. A similar analysis for indoor air radon may be found in USEPA, 1992.

We volatilize radon from drinking water to indoor air when we take showers or baths, flush toilets, and wash clothes and dishes. The transfer factor from water to air has been measured; for every 10,000 picocuries in a liter of water, between 0.2 and 3.4 picocuries are released into a liter of air (USEPA, 1986). Using error notation we can write this as $(1.8 \pm 1.6) \times 10^{-4}$ pCi/L_a/pCi/L_w, where L_a is a liter of air and L_w is a liter of water.

Now look at the dose response curve for miners exposed to radon shown in Figure 8.3. The dose is measured in working level months (WLM). One working level is equivalent to 100 pCi/L_a (liters of air) if the decay progeny of radon are all in equilibrium.

As we might expect, breathing rates are different between miners and the general public. To compensate for this, we use an equivalent occupational working level/year of 12 to 24 months/year (or 18 ± 6 months/year). We see that miners breathe more heavily; thus, the general public would need to breathe more to be comparable. The extra time allows for a greater sensitivity of humans in homes. The only significant health endpoint for radon progeny inhalation is lung cancer and the annual lung cancer individual mortality risk from Figure 8.3 is in the range of:

$(1.5 \text{ to } 4.5) \times 10^{-4}$ cases/WLM (also see ICRP 50 or USEPA, 1992), or

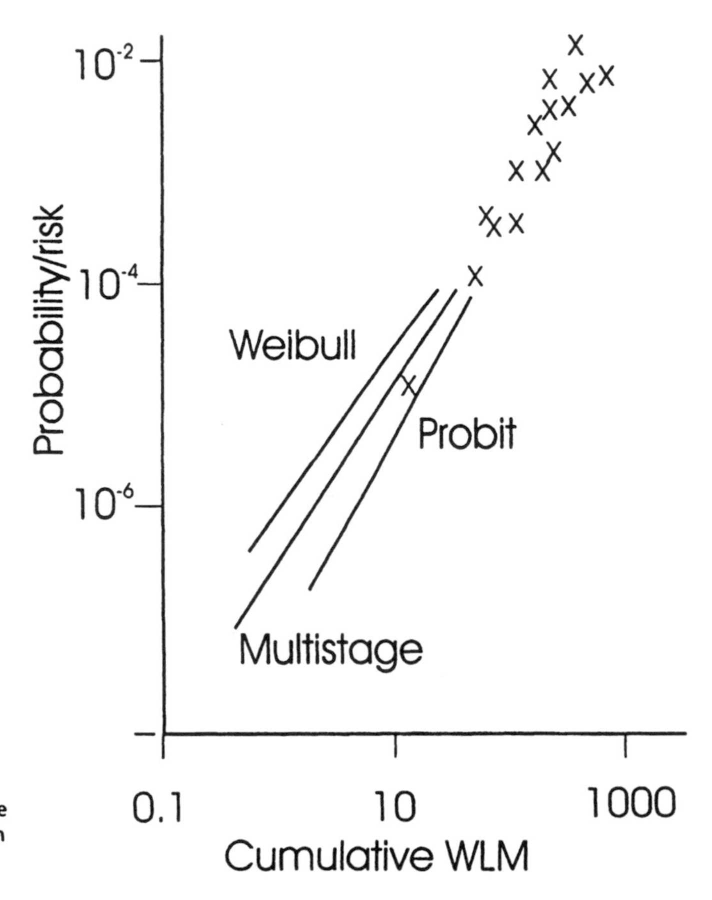

Figure 8.3 Dose-response curve for exposure of miners to radon gas

$(3 \pm 1.5) \times 10^{-4}$ cases/WLM (here we assume that the mean value is in the center of the range).

We can produce this estimate by assuming a linear fit of the data in the dose-response curve. There are two schools of thought here. One believes that the linear fit is correct and the other believes that a quadratic fit is more appropriate. We will use the linear fit for simplicity in this example.

When we combine all the above quantities for the individual risk per pCi/L_w, we have:

$$[(3 \pm 1.5) \times 10^{-4}/WLM]\ [1\ WL/\ 100\ pCi/L_a]\ [(1.8 \pm 1.6) \times 10^{-4}\ pCi/L_a/pCi/L_w]\ [(18 \pm 6)]$$

The units here cancel, leaving the risk per year per pCi/L_w per person. Because equilibrium between radon and its progeny is not generally achieved, we also multiply by a factor in the range of 0.3 to 0.7 or (0.5 ± 0.2).

Figure 8.4 shows the occurrence of radon in drinking water with an average in the range of 50 to 300 pCi/L_w (or $175 \pm 125\ pCi/L_w$). We can now combine all the factors to arrive at the estimate of the number of lung cancers for a population of 260 million (the approximate U.S. population), which is:

$$[175\ pCi/L_w][0.5][3 \times 10^{-4}\ cases/WLM][1\ WL/100\ pCi/L_a]\ X$$

$$[1.8 \times 10^{-4}\ pCi/L_a\ per\ pCi/L_w][18\ months/year][260 \times 10^6]$$

$$= 200\ lung\ cancer\ deaths/year$$

Note that this is an annual number. Population risk can be quoted in annual or lifetime quantities. The lifetime population risk would be 70 x 200 = 14,000/lifetime assuming that 70 years is an average lifetime.

Including all the error ranges, we now have:

$$[1/100][260 \times 10^6]\ X$$

$$[175 \pm 125][0.5 \pm 0.2][3 \pm 1.5][1.8 \pm 1.6][18 \pm 6] \times 10^{-8}$$

This calculation has the general form of two constants and five quantities that have error ranges or uncertainties.

One alternative for combining these errors is for us to multiply all the lower values for the final lower value and all the upper values for the final upper value. However, statistically this is much too extreme, as the maximum would not be true in every case.

Alternatively we can combine the uncertainties by using a standard error analysis method. If the distributions were normal, the errors would combine (the technical word is propagate) as the squares of the standard deviations which in this case are the ratios of the uncertainty to the mean value. Thus the standard deviation or σ for the first quantity is 125/175, for the second is 0.2/0.5, and so on. If we think of the five quantities as A, B, C, D, and E, then the square of the overall standard deviation is:

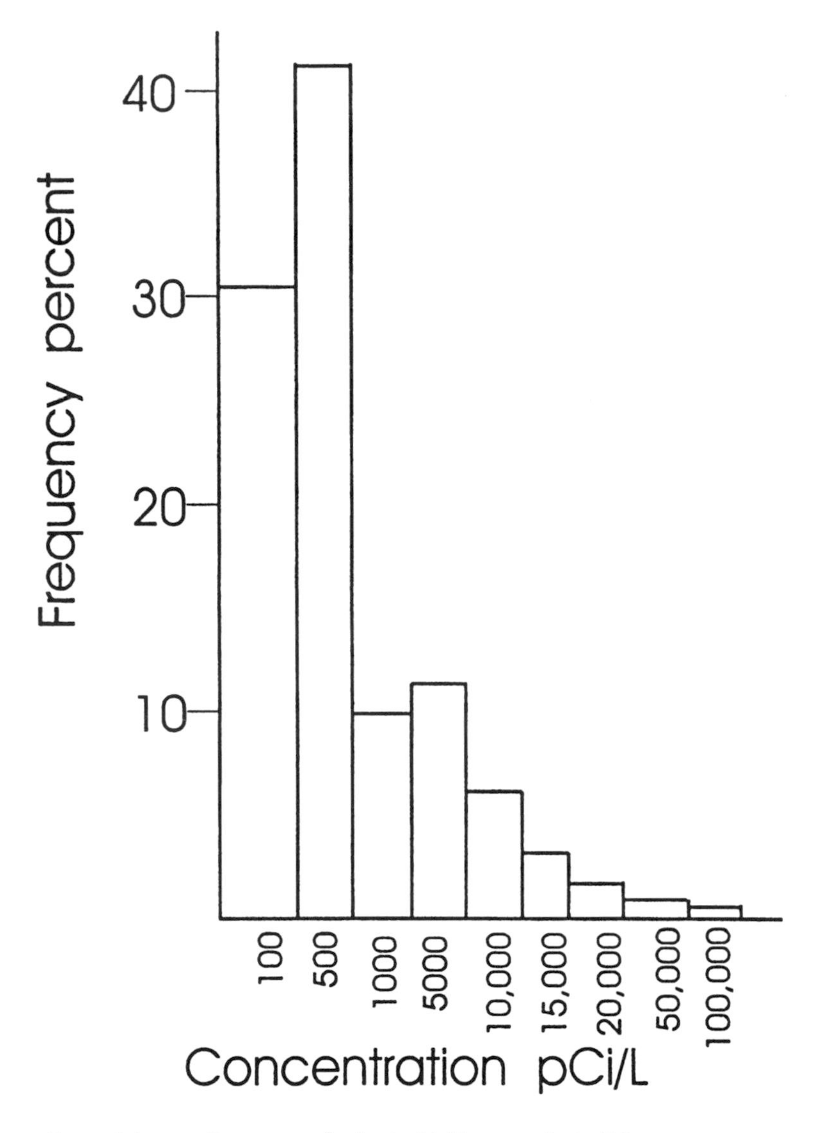

Figure 8.4 Occurrence of radon in drinking water in the U.S.

$$\sigma^2 = (\sigma_A)^2 + (\sigma_B)^2 + (\sigma_C)^2 + (\sigma_D)^2 + (\sigma_E)^2$$

or

$$\sigma = [(125/175)^2 + (0.2/0.5)^2 + (1.5/3)^2 + (1.6/1.8)^2 + (6/18)^2]^{1/2} = 1.3$$

This procedure is a standard one for propagating errors and works well when the variances or standard deviations are small. We will seldom find a case where the standard deviation is small, including the present example, and so we must do some adapting, as we often have to do for environmental estimates. We can use an upper limit of 200 + 200 x 1.3 = 460 or, rounded off to one significant figure, 500. For the lower limit we cannot use 200 – 1.3 x 200 as that is negative. Thus we have to invent something to reflect the uncertainty. The lower value could well be down an order of magnitude. Arbitrarily we could choose the value as 10. Thus, the estimated range for the number of lung cancer cases expected in the U.S. due to inhaled radon from the drinking water source is 10 to 500 per year.

One of the problems with the above analysis is that we not only assumed small standard deviations, but we also assumed that the quantities being multiplied have normal distributions. In general, environmental quantities are distributed lognormally. Thus, we could probably be closer to reality in our estimate by propagating the errors as though the quantities were distributed lognormally. Using lognormal distributions, the result is in the range of 3 to 85 excess lung cancer deaths per year.

In the dose-response curve used for radon from the miner data (Figure 8.3), the values span over two orders of magnitude in dose and are interpolated by only an order of magnitude or so into the unknown (this is still into the unknown and does not preclude the possible existence of a threshold). Note that for the previous trichloroethylene example, the range was six orders of magnitude.

Now, we combine the analysis of the stomach cancers expected from direct ingestion of drinking water containing radon with the above estimate for inhalation. To determine where and how fast the radon moves through the body, experiments were done using xenon as a surrogate. The bottom line is that the contribution to total population risk from ingestion of drinking water containing radon is about the same numerically as the inhalation risk (see Crawford-Brown, 1991). Thus the overall population risk due to radon in drinking water for the U.S. for one year is twice that for the inhalation risk, or in the range of 20 to 1000 cancers if the errors are distributed normally and 6 to 170 annually if the distribution is lognormal. This value can be compared to the estimate of 7,000 to 30,000 lung cancer fatalities per year from all indoor air radon (USEPA, 1994) and the total of about 130,000 fatalities from lung cancers per year overall.

Combining Error Ranges—Monte Carlo Analysis of Trichloroethylene

We can represent the population risk due to exposure to trichloroethylene (TCE) in drinking water by the expression:

$$q_1^* \frac{\text{TCE Concentration x Ingestion Rate x Exposure Duration x Exposure Frequency}}{\text{Body Weight x Averaging Time}}$$

where the averaging time is the total number of days per lifetime, or 70 x 365 = 25,500 days per lifetime. In Figures 8.5 through 8.9 we see the frequency distributions for all the quantities in the above equation, except the carcinogenic potential (q_1^*) which we assume in this example to have no uncertainty. Figure 8.5 shows the distribution of trichloroethylene concentrations. We need to realize that the curve is a mathematical representation of the individual Monte Carlo simulation of these curves. Figure 8.6 shows the frequency distribution of the tapwater ingestion rate; Figure 8.7, the frequency distribution for the exposure duration; Figure 8.8, the frequency distribution for the exposure frequency and, finally, Figure 8.9 shows the frequency distribution for the body weight. Our problem now is how to combine these frequency distributions.

One way we can combine the frequency distributions shown in Figures 8.5 through 8.9 is to use the Monte Carlo Method. In this method we choose a value at random from each of the five frequency distributions and the value from each is combined according to the risk equation shown at the beginning of this example. Then we choose another set of values at random and substitute those in the equation. With computers we can repeat this process hundreds, thousands, or even hundreds of thousands of times. The more times we repeat this process, the more accurate the resulting frequency distribution will be, as shown in Figure 8.10. We can convert this distribution to a probability distribution as shown in Figure 8.11 simply by summing the frequencies.

Using the mean lifetime risk of 1.2×10^{-7}, we multiply by 260 million people in the U.S., which yields an average of approximately 30 fatalities per lifetime. Using the 5 and 95 percentile values of 9×10^{-9} and 4×10^{-7}, the range of population risk would be 2 to 100 fatalities per lifetime. We can compare this range of uncertainty in the exposure-response model. Individual risk estimates were earlier shown to be six orders of magnitude with an overall population risk of <1 to 70,000 fatalities per lifetime. Again the pure mathematician will wince at this practical use of the quantities involved. For the purist, this estimate of uncertainty cannot be made. Our argument is that, although not purely correct, it is a reasonable representation.

The Monte Carlo method thus requires us to know the individual frequency distributions for all the quantities that are to be combined. They are seldom available; however, they can be estimated.

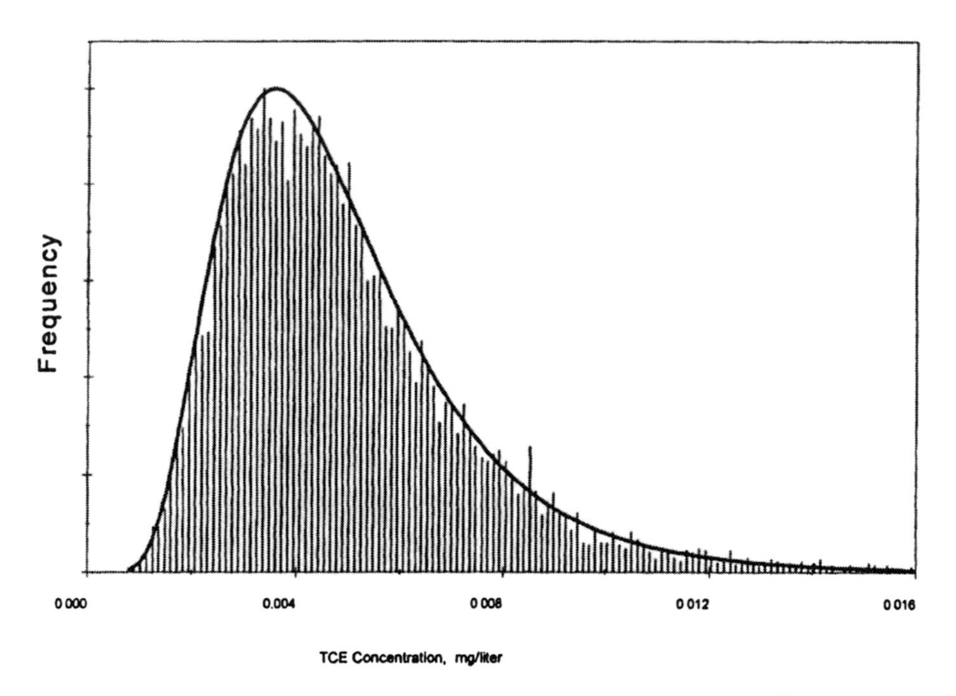

Figure 8.5 Frequency distribution of trichloroethylene concentration in drinking water

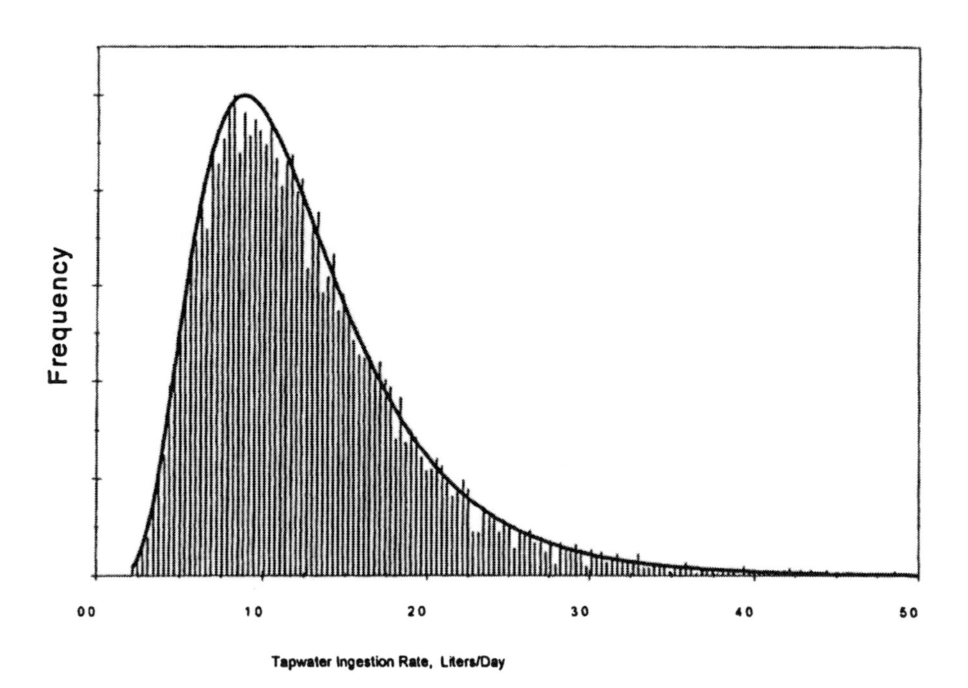

Figure 8.6 Frequency distribution of tapwater ingestion rate

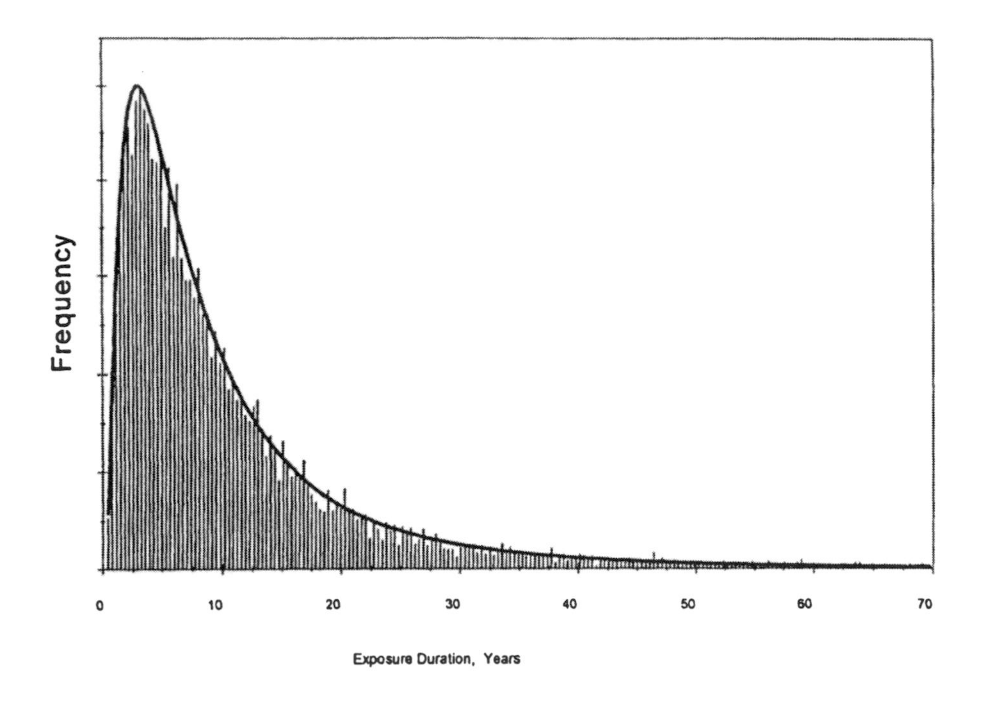

Figure 8.7 Frequency distribution of exposure duration

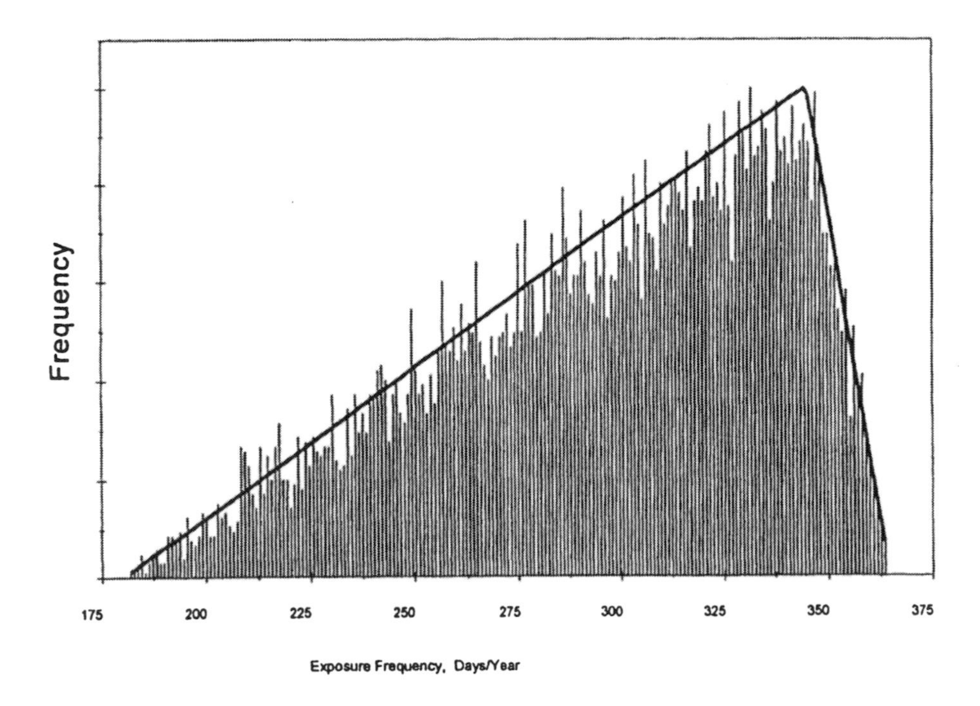

Figure 8.8 Frequency distribution of exposure frequency

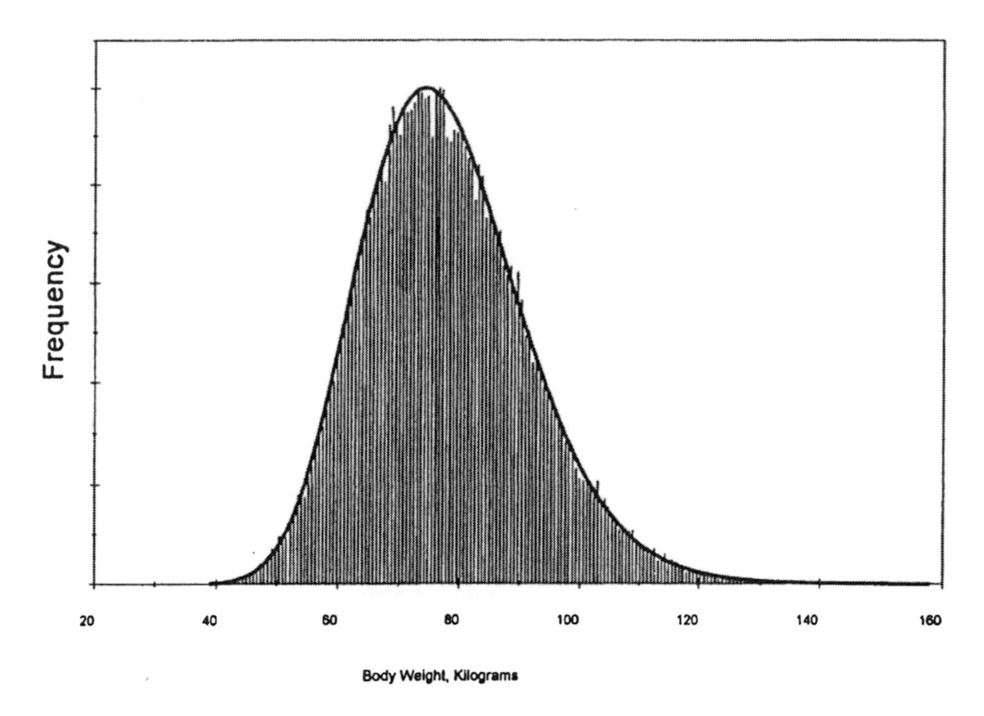

Figure 8.9 Frequency distribution of body weight

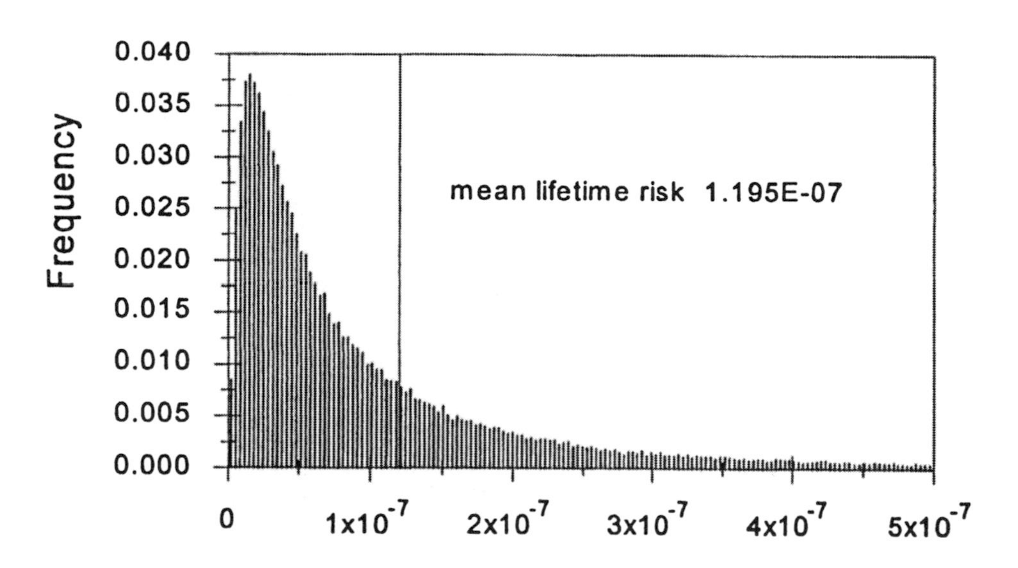

Figure 8.10 Frequency distribution of lifetime risk due to exposure to trichloroethylene in drinking water

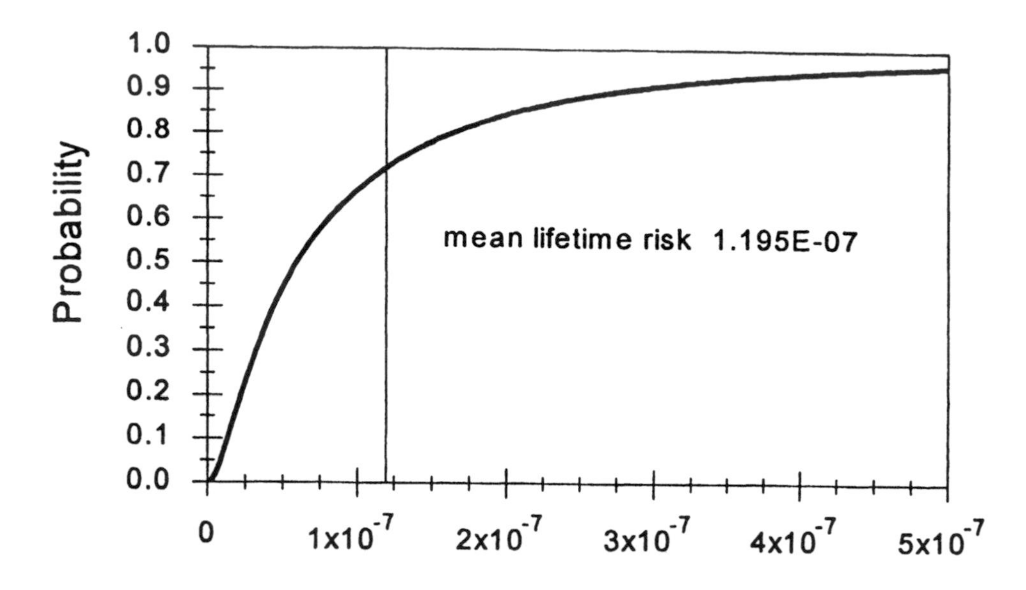

Figure 8.11 Probability distribution of lifetime individual risk for exposure to trichloroethylene in drinking water

Conclusions

For our purposes there is no completely satisfactory way of combining factors in population risk estimates to fully represent reality. Each method has its deficiencies. If no range or σ values are given, the single numerical estimate is open to the criticism of being an over- or underestimate. If the upper limit is used for all values, such as the 95% upper limit, the estimate is open to the criticism of being a gross overestimate. The Monte Carlo method is very attractive, but seldom are the distributions of the involved quantities known. In the TCE example discussed here, it was shown that the range of individual risk values from the fits of different models far exceeded any contribution to the overall uncertainty from the other quantities. However, we do not know if the models chosen are a valid range.

References

Cothern, C. R., W. A. Coniglio and W. L. Marcus. 1984. Techniques for the Assessment of the Carcinogenic Risk to the U.S. Population Due to Exposure from Selected Volatile Organic Compounds from Drinking Water Via the Ingestion, Inhalation and Dermal Routes, Background Document, NTIS PB 84-21394l, USEPA 570/9-85-001.

Cothern, C. R., and J. Van Ryzin. 1985. Risk Assessment For the Regulation of Carcinogens in the Environment, in *Hazard Assessment of Chemicals*, Vol. 4. J. Saxena, Ed., New York: Academic Press, 179–241.

Cothern, C. R. 1986a. Techniques for the Assessment of Carcinogenic Risk Due to Drinking Water Contaminants, *CRC Critical Reviews in Environmental Control*, 16: 357–400.

Cothern, C. R., W. A. Coniglio and W. L. Marcus. 1986b. Estimation of the Risk to Human Health Due to Volatile Organic Compounds in Drinking Water Using Trichloroethylene as an Example, *Environmental Science and Technology*, 20: 111.

Cothern, C. R. 1987. Radon in Drinking Water, *American Water Works Assoc. J.*, April.

Cothern, C. R. Uncertainties in Quantitative Risk Assessments—Two Examples—Trichloroethylene and Radon in Drinking Water. 1988. IN *Risk Assessment and Risk Management of Industrial and Environmental Chemicals, Vol. 15 of Advances in Modern Environmental Toxicology*, Eds. C. R. Cothern, W. L. Marcus, and M. A. Mehlman, Princeton Scientific Publishers.

Cothern, C. R. 1989. Health Effects of Inhaled Radon Progeny (a review), *J. of Environmental Science and Health, Environmental Carcinogenesis Reviews*, Part C7(1): 75–108.

Cothern, C. R. 1990. Indoor Air Radon (a review), Reviews of Environmental Contamination and Toxicology, *Springer-Verlag* 111: 1–60.

Crawford-Brown, Douglas J. 1991. Cancer Fatalities from Waterborne Radon (Rn-222), *Risk Analysis* 11:135–143.

Federal Register, Draft Report: A Cross-Species Scaling Factor for Carcinogen Risk Assessment Based on Equivalence of mg/kg$^{3/4}$/Day, 24152-24173, Notice (Friday, June 5, 1992).

International Commission on Radiological Protection. 1987. ICRP Publication 50, *Lung Cancer Risk from Indoor Exposure to Radon Daughters*, Oxford, U.K.: Pergamon Press.

Rhomberg, Lorenz R. and Scott K. Wolff. 1988. Empirical Scaling of Single Oral Lethal Doses across Mammalian Species Based on a Large Database, *Risk Analysis* 18:741–753.

USEPA. Federal Register, Water Pollution Control: National Primary Drinking Water Regulations; Radionuclides, Advanced Notice of Proposed Rulemaking, (Tuesday, September 30, 1986).

USEPA. 1992. Technical Support Document for the 1992 Citizen's Guide to Radon.

USEPA. 1994. Multimedia Risk and Cost Assessment of Radon in Drinking Water, EPA 811-R-94-001, March.

Chapter Nine

Comparative Risk Analysis

Introduction

In order to make policy, regulatory, and other risk decisions, we need to put risks in some kind of priority order. We can use the concept of *comparative environmental risk* analysis, which is a priority risk ranking, quantitative where possible, of the different risks to human health and the environment. Although this is not altogether a new idea, it is seldom attempted and rarely used. It provides a framework into which we can insert new and existing information concerning environmental risks. In Figure 9.1, you can schematically view the role of comparative environmental risk analysis. The bottom level is the most specialized, and as one moves up the levels shown, more information is included and the scope becomes more broad. Only when we approach the concept of a comparative environmental risk analysis can several or ultimately all of the contaminants and problem areas be compared on some equal basis.

When we look at the whole of human health and environmental risk analyses, we find very few comparative risk analyses. The major problem is lack of information (see Cothern, 1993).

In the field of environmental risk analysis and especially in the area of comparative risk analysis, someone inevitably points out that there are so many differences in causes and effects that we are comparing "apples and oranges". Although this is clearly a criticism that we need to address, this is not necessarily a fatal flaw. One view of this question is addressed in a recent *New Yorker* cartoon comparing apples and oranges and noting similar characteristics: edible, warm color, round shape, similar size, contain seeds, grow on trees, good for juice, name begins with a vowel, similar pesticide treatment, and unsuitable for most sports. We need to realize that comparisons are not perfect and are subject to misinterpretation. In the present work we assume that this is a problem and its consequences must be acknowledged in any analysis, but we do not consider it grounds for dismissal.

COMPARATIVE QUANTITATIVE RISK ANALYSIS
ultimate goal is to compare quantitative risk assessments for all contaminants to all living things on a common basis
QUANTITATIVE RISK ANALYSIS
involves risk/benefit analysis and economic, social and political elements
QUANTITATIVE RISK ASSESSMENT
estimate of risk to humans and other living things for individual contaminant or problem area based on extrapolations of animal studies and epidemiologic information combined with exposure level analysis
QUALITATIVE RISK ANALYSIS
examine the quality of life for humans and other living things including different endpoints
TOXICOLOGY
animal studies and epidemiology

Figure 9.1 Conceptual outline of the hierarchy of risk in analyzing environmental problems and supporting decision making. To get perspective of how and where each element enters, start at the bottom and read upwards which will determine the roles of gathering information and conceptualizing. Note that each level going up adds more information and complexity. The relative importance of different contaminants and environmental problems emerges as one goes up the levels. Only when the top level is reached can comparisons be made on some equal basis.

Comparative Environmental Risk Assessments

General Comparative Risk Assessments

In the spirit of going from the general to the specific, we will first consider comparative risks involved in all causes of death before focusing on those involving environmental threats. In Table 9.1, we see the proportions of cancer deaths attributable to specific factors. This comparative assessment of risk clearly shows us that cancer deaths attributable to environmental pollution are a small part of the overall picture, in the range of one to three percent.

Table 9.1 Proportion of Cancer Deaths Attributable to Different Factors (modified from Doll and Peto, 1981 and from Wilson, 1988)

Factor or class of factors	Best estimates	Percentage of all cancer deaths, range of acceptable estimates
Cigarette smoking	25	20 to 30
Alcohol	3	2 to 4
Diet	35	10 to 70
Food additives[1]	>1	-5 to 2
Reproductive and sexual behavior	7	1 to 13
Occupational	5	2 to 10
Pollution	1	>1 to 3
Industrial products	>1	>1 to 2
Medicines and medical procedures	2	1 to 3
Geophysical factors	3	2 to 3
Infection	3	1 to ?
Unknown	?	?

Note that the food additives category could be negative because of the possible protective effect of antioxidants

It is desirable to compare risks that have the same endpoint because one need not balance or weigh different health consequences. In only a few cases can comparisons be made meaningfully between different exposures for the same endpoint. Consider the following table:

Cause	Cancer deaths/yr.
active smoking	approximately 100,000 (lung)
passive smoking	approximately 3,000 (lung)
radon from soil	7,000 to 30,000 (lung)
radon from drinking water	30 to 600 (lung)
other drinking water contaminants	0.00001 to 1 (all cancers)

From this table we can conclude that the impact of cancer deaths due to radon in drinking water is minimal when viewed only from the perspective of lung cancer. However, compared to the other contaminants in drinking water involving other cancer endpoints, radon causes more deaths than all the other drinking water contaminants combined, and by far. (See for example Cothern and Van Ryzin, 1985.) The problem is that different cancer endpoints are involved in the latter comparison. In trying to get our perspective it is important to consider all available information.

Figure 9.2 and Table 9.2 show some selected commonplace and occupational risks of death. Such comparisons make clear to us the difficulty of comparing and interpreting these different endpoints.

Cohen (1991) assessed risks in terms of loss of life expectancy (LLE) since he thinks it is easier for the average person to understand a loss of 30 days of life rather than a mortality risk of 3×10^{-3}. Also he notes that the premature death of an elderly person is less regrettable than the death of a young person. (See Table 9.3.)

In the above analyses we attempt to compare a broad range of risks, but their intercomparison suffers accordingly in this attempt to broaden the perspective.

Generation of Electric Power

In 1978 a physicist at the Atomic Energy Control Board in Canada developed an assessment of the risk to the health and safety of workers and the general public involved in the production of electrical energy. The central idea was to compare all the different ways of generating electricity in terms of deaths per megawatt-hour or days lost per megawatt-hour. The goal of the report was to determine what might constitute an acceptable level of risk (Inhaber 1978 and 1979). This assessment was one of the first exercises comparing risks in a systematic and consistent fashion. The assessment involved a new methodology for processing and interpreting data, with the understanding that data relating to a wide variety of information can never be completely intercomparable. The methodology was called risk accounting and, according to the report, it can be employed to inform us of the risk inherent in competing energy systems. The methodology combined the two measures of deaths and man-days lost.

The report concludes that nuclear power and natural gas have the lowest overall risk of the 11 technologies considered (coal, hydroelectricity, methanol, natural gas, nuclear, ocean thermal, oil, photoelectric, solar, space heating, and wind). Ocean thermal energy ranked third.

A numerical table was not included but we can conclude that the range of estimates of total deaths per megawatt year times 1000 are roughly:

coal	5 - 150
hydroelectric	3 - 5
natural gas	0.2 - 2
nuclear	0.02 - 0.2
oil	2 - 120
solar	8 - 40
wind	20 - 70

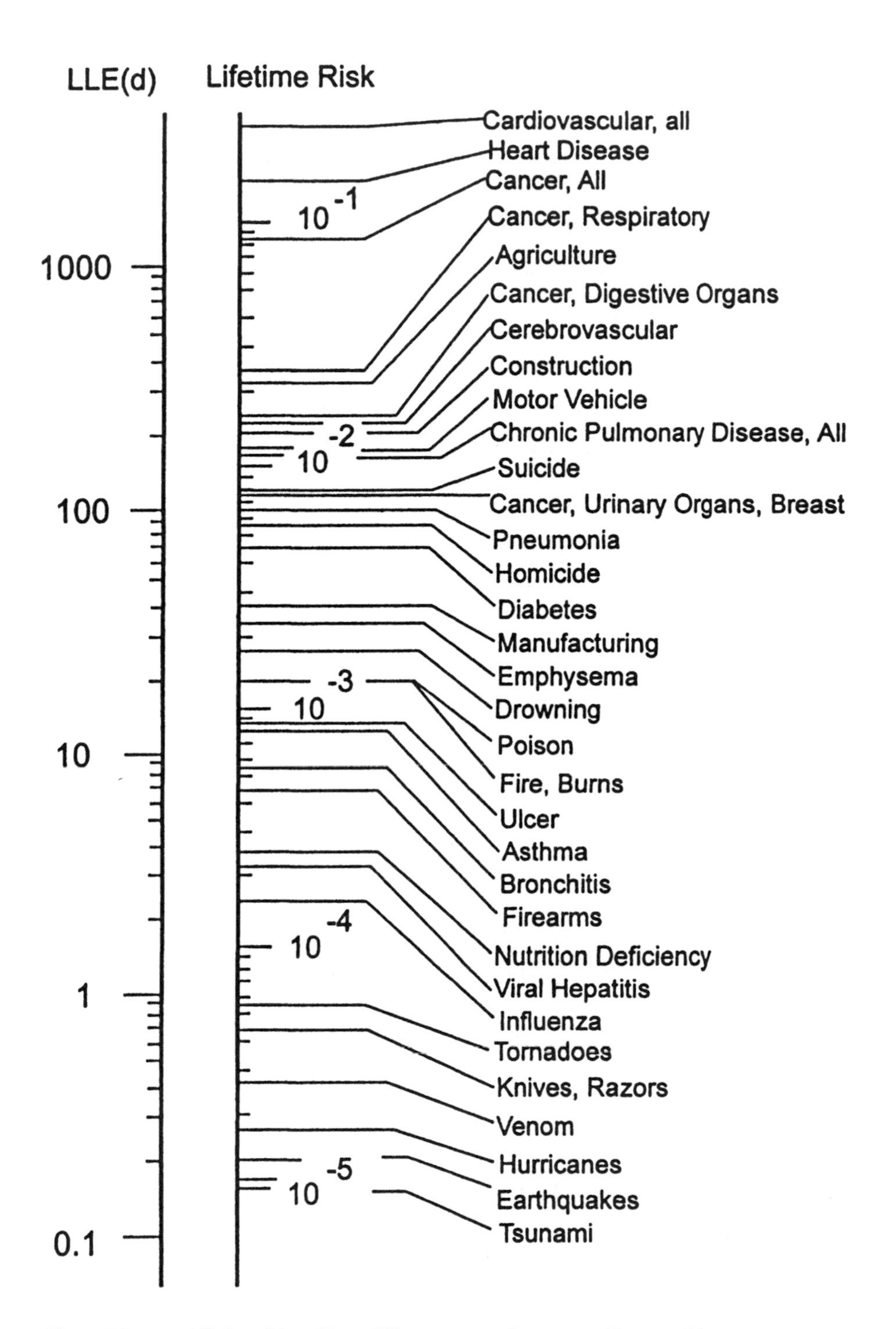

Figure 9.2 Lifetime risk and loss of life expectancy for a range of human risks

Table 9.2 Some selected occupational risks of death—annual per capita risk (from Crouch and Wilson, 1982)

Risk of death	Annual/person
manufacturing	8.2×10^{-3}
service and government	1.0×10^{-4}
agriculture	6.0×10^{-4}
construction	6.1×10^{-4}
football	4×10^{-4}
rodeo	3×10^{-5}
Cancer risks	
all cancers	2.8×10^{-3}
natural background ionizing radiation	2.0×10^{-5}
one 12 oz diet drink/day	$< 10^{-5}$
four tablespoons peanut butter/day	$< 8.0 \times 10^{-6}$

The total risk for each energy production method was calculated by considering six components: material acquisition (such as mining) and plant construction, emissions caused by material production, operation and maintenance, energy backup (required when the source is not continuous), energy storage, and transportation, including waste deposition.

Table 9.3 Loss of life expectancy (LLE) for selected causes (Cohen, 1991)

Cause	LLE days
Motor vehicle accidents	207
Home accidents	74
Drowning	24
Fire	20
Tornado	0.8
Venom, plant and animal	0.4

Some of the data used in the above analysis had unknown uncertainty, but the uncertainty in all contributions had to be estimated. We see that in some cases exposure and risk data from similar situations or scientific judgements were employed. The validity of the actual estimates are open to criticism, but we welcome the idea of including a measure of the overall uncertainty, and you should do this vigorously in all cases where it is desirable to compare environmental risk assessments.

In the above assessment, because of incomplete information it was necessary to make certain assumptions in order to obtain a numerical comparison. Assumptions made in this study were: (1) data from different countries were similar, (2) adequate resources were available (e.g. land, trees, steel, and other resources), (3) deleterious effects were considered for humans only, (4) risks incurred in producing, fabricating, and installing transmission lines were not included, (5) the need for back up systems in the future due to technological changes were not included, and (6) other potential changes in the future were not included. Also not included are effects of terrorist attacks or the contribution of social, psychological, and aesthetic risks, since we could not quantify them.

A Department of Energy report (House, 1981) published two years after the Inhaber report reviewed several comparative studies involving the impact of energy technologies. A common criticism of these assessments was that the data were incomplete, inadequate, and contained large uncertainties. The major criticisms of the Inhaber study were that the data used were developed for purposes other than this particular comparative assessment, tenuous assumptions were used, and the techniques used were static in time. We should not let these difficulties preclude attempts to compare risks, but we need to keep these limitations in clear view.

U.S. Environmental Protection Agency's Unfinished Business Report

We now turn our attention to the myriad risks encountered by the U.S. Environmental Protection Agency. We need to understand that EPA has limited resources and needs to develop ways to put these problems into some priority order. Problem areas encountered by EPA were comparatively assessed by a special task force in 1985 of about 75 agency career managers and analysts (USEPA, 1987). Their objective was to determine how to make the best use of finite resources to achieve EPA's mission and mandates on the basis of comparative risk. The overall task was limited in four ways: (1) they divided the environmental problems into 31 pieces that correspond roughly to existing programs or statutes; (2) they partitioned the types of risks into four areas (cancer, non-cancer health, ecological effects, and welfare effects such as visibility impairment, materials damage, etc.); (3) for practical reasons, they did not include economic controllability, technical feasibility, and qualitative aspects such as degree to which risks are voluntary, familiar, or equitable; and (4) they assessed risks as they exist today and did not include those that have already been abated.

The 31 areas considered were:

1. criteria air pollutants from mobile and stationary sources (includes acid precipitation);
2. hazardous/toxic air pollutants;
3. other air pollutants (sulfuric acid mist, fluorides, and emitting substances);

4. radon (indoor air only);
5. indoor air pollutants other than radon;
6. radiation from sources other than indoor radon;
7. substances suspected of depleting the stratospheric ozone layer (CFCs, etc.);
8. carbon dioxide and global warming;
9. direct, point source discharges (industrial, etc.) to surface waters;
10. indirect, point source discharges (POTWs) to surface waters;
11. nonpoint, source discharges to surface water;
12. contaminated sludge (includes municipal and scrubber sludge);
13. discharges to estuaries, coastal waters, and oceans from all sources;
14. discharges to wetlands from all sources;
15. drinking water as it arrives at the tap (includes chemicals, lead from pipes, biological contaminants, radiation, etc.);
16. hazardous waste sites—active (includes hazardous waste tanks) (groundwater and other media);
17. hazardous waste sites—inactive (Superfund) (groundwater and other media);
18. nonhazardous waste sites—municipal (groundwater and other media);
19. nonhazardous waste sites—industrial (includes utilities) (groundwater and other media);
20. mining waste (includes oil and gas extraction wastes);
21. accidental releases of toxic substances (includes all media);
22. accidental releases (oil spills);
23. releases from storage tanks (includes product and petroleum tanks—above, on, and under ground);
24. other groundwater contamination (includes septic systems, road salt, injection wells, etc.);
25. pesticide residues on foods eaten by humans and wildlife;
26. application of pesticides (risks to applicators, which includes workers who mix and load, as well as apply, and also consumers who apply pesticides);
27. other pesticide residues (includes leaching and runoff of pesticides and agricultural chemicals, air deposition from spraying, etc.);
28. new toxic substances;
29. biotechnology (environmental releases of genetically altered materials);
30. exposure to consumer products; and
31. worker exposure to chemicals.

The assessors arbitrarily ranked the above 31 categories as high, medium, or low using available quantitative data, qualitative information, and professional judgement. The major results of this study were: (1) no problems rank relatively high in all four types of risk, or relatively low in all four; (2) problems that rank relatively high in three of four risk types, or at least medium in all four, include criteria air pollutants, stratospheric ozone depletion, pesticide residues on food, and other pesticide risks (runoff and air deposition of pesticides); 3) problems that rank relatively high in cancer and non-cancer health risks but low in ecological and welfare risks include hazardous air pollutants, indoor radon, indoor air pollution other than radon, pesticide application, exposure to consumer products, and worker exposure to chemicals; (4) problems that rank relatively high in ecological and welfare risks, but rank low in both health risks include global warming, point and nonpoint sources of surface water pollution, physical alteration of aquatic habitats (including estuaries and wetlands), and mining waste; and (5) areas related to groundwater consistently rank medium or low.

The assessors observed the following concerning these results:

(1) The risks do not correspond well with EPA's current program priorities. Areas of relatively high risk but low EPA effort included indoor radon, indoor air pollution, stratospheric ozone depletion, global warming, non-point sources, discharges to estuaries, coastal waters and oceans, other pesticide risks, accidental releases of toxic substances, consumer products, and worker exposures. Areas of high EPA effort but relatively medium or low risk include RCRA sites, Superfund, underground storage tanks and municipal non-hazardous waste sites.

(2) Overall, EPA's priorities appear more closely aligned with public opinion than with the estimated risks. Recent national polling data rank areas of public concern about environmental issues as follows:
   - high—chemical waste disposal, water pollution, chemical plant accidents, and air pollution;
   - medium—oil spills, worker exposure, pesticides, and drinking water; and
   - low—indoor air pollution, consumer products, genetic effects, radiation (except nuclear power), and global warming.

These observations do help those in EPA determine priorities. However, they also show the conflicts between public opinion and risk levels.

Reducing Risk—Report of EPA's Science Advisory Board

Continuing our discussion of comparative risk at EPA, we find the next step involves EPA's Science Advisory Board. The Board had major goals of re-examining the earlier Unfinished Business report and making some recommendations. The resulting report was entitled Reducing Risk (USEPA, 1990).

The findings of the report include:

(1) Importance of Unfinished Business: They concluded that the study that produced the Unfinished Business report was a bold and needed step. Comparing relative residual risks posed by a range of different environmental problems was an important shift in national environmental policy.

(2) Problems in ranking risks: "As long as there are large gaps in key data sets, efforts to evaluate risk on a consistent, rigorous basis or to define optimum risk reduction strategies necessarily will be incomplete, and the results will be uncertain."

(3) The extraordinary value of natural ecosystems: Natural ecosystems provide resources that feed, clothe, and house the human race. They also act as sinks, to a certain extent, to absorb and neutralize the pollution generated by human activity.

(4) Time, space, and risk: Besides the number of people and other organisms affected, the likelihood of the problem actually occurring among those exposed, and the severity of the effects, we need to consider two other factors: temporal and spatial dimensions.

(5) Public perceptions of risk: Public opinion polls show that people are more worried about environmental problems now than they were 20 years ago, but those considered most serious by the public are different from those considered most serious by the technical professionals charged with reducing environmental risk.

(6) The strategy options for reducing environmental risk: Strategies include research and development; pollution prevention; provision of information; market incentives such as marketable permits, deposit-refund systems, removal of market barriers, and revising the legal standards of liability; conventional regulations; enforcement; and cooperation with other governmental agencies and nations.

The Reducing Risk report listed the following as high risk: ambient air pollutants, indoor air pollutants, pollutants in drinking water, habitat alteration and destruction, species extinction and overall loss of biological diversity, global climate change, and stratospheric ozone depletion.

This report demonstrates some of the difficulties involved in attempting a comparative quantitative risk analysis.

Disinfection of Drinking Water

As you are aware, drinking water has been disinfected with chlorine since the turn of the century to prevent the spread of infectious disease. In the 1970s, measurable levels of carcinogens, such as chloroform and other trihalomethanes, were found in the drinking water in the United States. These contaminants were shown to be present because of the chlorination process. Thus we need to determine the relative or comparative risk between the infectious diseases resulting from not chlorinating compared to the risk of cancer due to chloroform. This was done by Bull et al (1990). They observed that only for the disinfectant chlorine were the data available to make a quantitative estimate of the relative risk. While still not precisely estimated, the risk of infectious disease deaths from consuming unchlorinated water exceeds the risk of cancer deaths from consuming chlorinated water by many orders of magnitude, at least a thousandfold. Thus, the idea that communities should cease drinking chlorinated water is irrational. The EPA is currently in the process of determining if other disinfectants should be used in place of chlorine because of the production of chloroform. Better data is needed before a relative risk estimate can be made for all the possible alternatives (among which are chloramine, chlorine dioxide, and ozone). The current movement away from chlorination in the drinking water industry, without a comparative quantitative risk analysis of the alternatives, may lead to even more risk.

Hazard Ranking System

Consider the need for a priority ranking, or comparative risk analysis, as required by the Superfund Act (CERCLA or Comprehensive Environmental Response, Compensation, and Liability Act). Specifically, Congress required EPA to develop a National Priorities List (NPL) for the purpose of remedial action to protect public health, welfare, and the environment. EPA met this need by creating the Hazard Ranking System (U.S. EPA, 1990). The HRS estimates the qualitative potential for releases of uncontrolled hazardous substances that could damage human health or the environment. Thus, the HRS predicts the relative ranks of the risks of sites, not their absolute risks. EPA uses the HRS to put waste sites on the NPL.

For each waste site, the HRS calculation scores four separate exposure pathways: (1) groundwater migration, (2) surface water migration, (3) soil exposure, and (4) air migration. Each pathway receives a relative score between 0 and 100, based on the multiplication product of three factors: (1) likelihood of release, (2) waste characteristics, and (3) targets. Thus, the HRS uses a multiplicative model for each pathway. (See the section on models in Chapter 1.) The HRS applies a qualitative procedure to each pathway. In essence, a check listing process generates the factor scores. Release factors encompass both observed and potential releases. The HRS considers toxicity (to humans and the environment), mobility, persistence, bioaccumulation, and quantity in the estimate of waste characteristics. The target score includes the nearest individual, the population, sensitive environmental resources, and food chain. When a site lacks specific data, EPA provides assumptions to score the three factors. (See the section on qualitative approaches to risk in Chapter 2.)

Because EPA wanted to weight the overall HRS score such that a high risk from a single pathway would produce a high result, the calculation resembles a standard error. Thus, the final score is the root mean square of the four pathway scores. For example, a score of 100 for one pathway and zero for the other three pathways produces a final score of 50 [$50 = \sqrt{(10{,}000/4)}$]. (See the section on statistics for pedestrians in Chapter 2.) When the final score exceeds a threshold value set of 28.5, the site goes onto the NPL. EPA originally predicted that this threshold would create approximately 400 Superfund sites nationally.

Because the HRS normalizes the score for each of the four different pathways on a scale of 0 to 100, the scores do not necessarily reflect the absolute risks of the four pathways at a site. For example, we would not necessarily know from these scores whether a Superfund site generates more human health risk for the surrounding population from migration of groundwater, surface water, or air. However, overall HRS scores do crudely reflect risks of sites, and for one pathway, the scores at different sites should rank them identically to more detailed, quantitative assessments. To our knowledge, no one has carried out the necessary research to demonstrate this correlation.

To date, the HRS is the most detailed regulatory scheme for risk ranking. Methods similar to the HRS could form the basis for comparative risk assessments of a wide range of sources, not only environmental risks, particularly if the ranking process was confined to one pathway. In principal, you could widely apply its methods for purposes of comparative risk assessment.

Problems with Comparative Environmental Risk Assessments

We need to be aware of several characteristics that lead to problems in comparing different risks. In many cases, a single contaminant can cause multiple endpoints. For example, the following endpoints have been suggested for arsenic: blackfoot disease, cancers (kidney, liver, lung, prostate, skin), conjunctivitis, diabetes mellitus, dermatitis, EEG abnormality, hypertension, ischemic heart disease, lens opacity, melanosis, mental retardation, peripheral neuropathy, and Reynold's syndrome (see the case study for more details). Our difficulty is how to weigh the different endpoints so that the effects from that contaminant can be compared to others. Illness and death are two general endpoints that have different importance. No clear method exists for comparing morbidity and mortality. How many viral infections can be compared to a fatality? A crude question showing us this problem in the context of human health and ecotoxicity is Who cares if a fish gets diarrhea? The time for a health endpoint to manifest itself (the latent period) is different for different contaminants. It can take 10 to 20 years after exposure for a health effect to manifest itself. This makes our comparisons difficult unless a longer period, such as a lifetime, is considered.

We need to realize that there is an overemphasis on cancer to the detriment of other effects. Because of this imbalance, it is even more difficult to compare cancer and noncancer endpoints we find in other areas such as neurotoxicology, immunotoxicology, and developmental toxicology. Another problem facing us is the range of potencies that further complicates comparative risk assessments. For example, see Figure 9.3 where four different endpoints (A, B, C, and D) occur at different doses. We need very specific dose information to be able to make a comparative risk analysis.

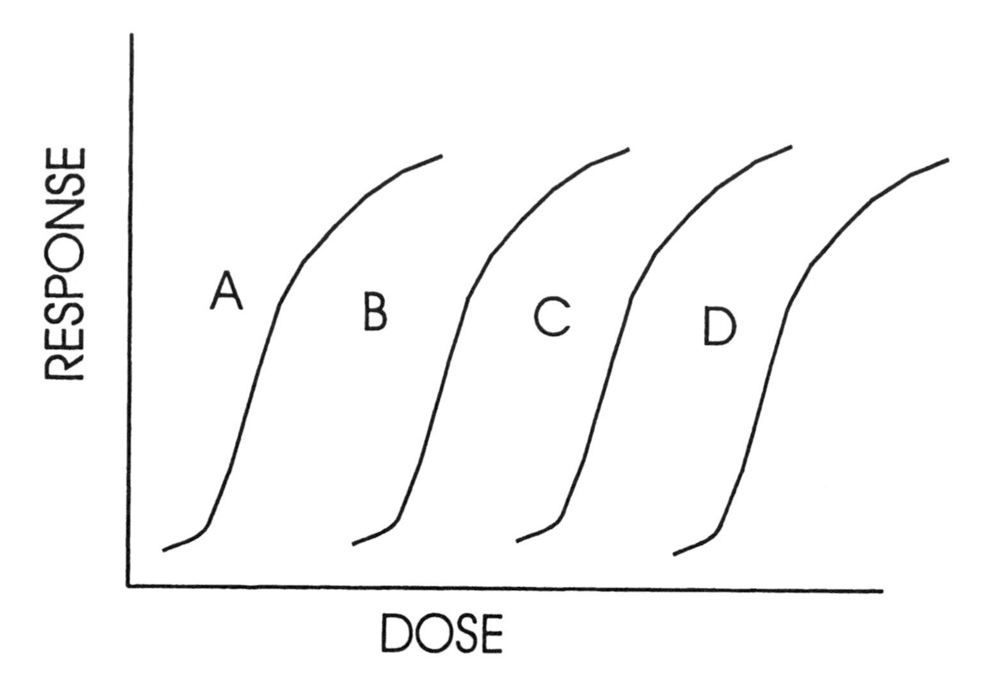

Figure 9.3 Four dose-response curves for four different endpoints for the same contaminant

The methodology we use to estimate risk in the area of ecological health has not been fully developed and its characteristics are different and much less quantitative (see chapter 10)(USEPA, 1990A). As an example, copper and zinc are harmless to humans but deadly to some forms of life in a river. Species specificity even further complicates our intercomparison between effects on human health and effects on ecological health.

Other complications are seasonal effects such as atmospheric concentrations, some diseases, and the growth of plants. We also find effects that are long term, such as desertification, genetic disorders, and evolutionary phenomena. Comparing different environmental problems in a temporal way generates complexities for us that are difficult to estimate. In a similar way, spatial distributions on local, regional, and global scales complicate comparative risk assessments. For example, we have a problem comparing risks from a local Superfund site to those for global climate change.

In any comparative risk assessment we need to consider synergisms and antagonisms that can cause large differences in estimates of environmental risk. This problem arises when we consider the real world where there are not individual contaminants but mixtures of contaminants. For very few mixtures are the potential synergisms and antagonisms known. Some of the synergisms we know the best occur with cigarette smoke. There are demonstrated synergisms between cigarette smoke and asbestos (Health Effects Institute, 1991) and radon (NAS, 1988a) that approach an order of magnitude. That is, the health effects of cigarette smoke alone and radon alone are less than the health effects when both are present, by a factor that could be as large as an order of magnitude. In the case of radon, we find the effect is large enough for some to argue that it makes more sense to ban cigarettes than to regulate the levels of radon in homes and offices. We also know of some synergisms and antagonisms in the area of foods (see Gold et al, 1993). But, outside of these examples, we know of few other synergisms, and we need to develop strategies to cope with this deficiency in risk estimation (NAS, 1988b). In general, we assume that the health effects from environmental contaminants combine linearly and, in some cases, we add an uncertainty factor to compensate for possible synergisms and antagonisms.

One of the most troubling potential problems in our comparative risk assessments are the surprises or effects that are completely unexpected. An example of this phenomenon is the side effect of thalidomide. Clearly we should expect other such happenings, and this possibly should be included in risk assessments. Another complication for us is the wide range of human sensitivities. Some humans are not bothered by a given level of a contaminant, while others show serious effects. Natural carcinogens exist, and we often overlook them in a risk assessment (Gold, 1993). These can provide us with enormous insight to the background of human exposure, especially in relation to foods. We too often consider risk analyses at a local or regional level. This makes comparisons to global risks, such as global climate change and stratospheric ozone depleters, difficult.

In the current environmental regulatory community, we often downplay, or even ignore, the health effects to living things other than human beings. This has slowed the development of ecological risk assessment and made our intercomparison between it and human health risk assessment even more difficult. Further, we find the quantification of ecological health more difficult to achieve. Two of the most important consequences of contamination to living things other than humans are loss of species and habitat destruction. But, some perspective can be gained if we realize that we would not consider the loss of 10% of animal outliers due to a contaminant in a system as serious as the loss of 10% of human lives.

There are a number of perceptions that impinge on quantitative environmental risk assessment and make comparisons more difficult. Some of these opposing perceptions are natural versus man made, voluntary versus involuntary, ordinary versus catastrophic, etc. (see Chapter 2).

Discussion

Change the Philosophical Approach?

Of the questions posed to us as risk analysts, few, if any, require a comparative analysis. Consider the following typical questions asked the risk analyst: the individual member of the public wants to know if this contaminant or situation is safe, the regulator or attorney wants to know what the numbers are, the lay person is concerned that these numbers are only estimates, the business person wants to know the cost, the politician wants to know how this will help or hurt his political position, the ecologist is interested in how other living things will be affected, and no one asks the relative importance of a specific contaminant or problem. Rather than answer such questions, we as analysts usually only calculate estimates in hope that they will satisfy some of these questions.

Instead of taking the data and information and constructing a comparative environmental risk analysis, would it be better for us to first ask the questions of what do we want to know and how does that fit into our understanding of what is happening in the environment? We can do this at the policy level with budgets and regulatory priorities (see Davies, 1996). What would be useful information to have? Is there a crisis, or crises? What are the goals and objectives? What is important to the different groups, including the public, media, government, industry, labor, environmental organizations, regulators, and so on. Some questions may not be answerable immediately and some may take much time and resources to answer; however, the framework of a comparative environmental risk analysis provides us a way to approach these and similar questions.

Some potential questions in such an approach could be:

- Why are cancer rates different in different geographical areas?
- If we cannot burn, bury, or otherwise dispose of waste, where should we put it (a kind of environmental gridlock)?
- Why are there changes in life expectancy?
- What are the causes of shortened life?
- What are the relative causes of suffering and disability?
- How can we maximize life span and health?
- How can we compare different endpoints?
- How many species are there? Are they changing in population? How does change affect the environment?
- How can the results of pollution prevention be measured?
- How can the population question be clearly understood?
- How can we measure the effect of genetic engineering?
- What are the important parameters in the description of indoor air (gases, pesticides, sick buildings, micro-organisms, dust, etc.)?
- What do the polls of public concern indicate as important? Concerning municipal waste, are we running out of dump sites? What is the best disposal alternative (incineration, burial, recycling, or other)?
- Is Superfund remediation saving lives? Is the asbestos remediation program saving lives? At what cost? What are the problem areas that will become more important in the future? How can we determine which areas are important from our current information concerning trends?

By asking ourselves comparative questions, the philosophic basis of risk analysis is shifted towards goals and objectives that will produce perspectives for us and better understanding of the effects and relative importance of those factors that affect public health and the environment.

Who Needs to See the "Big Picture"?

Who in this world, country, or locality, thinks in the larger sense about the "big picture", or where it all is going? Who thinks beyond today or, perhaps, the immediate tomorrow? D.H. Meadows et al. (1992) suggest taking a global view of our environment and insist that we will find that we have exceeded some of the limits.

One of the maxims of the environment we need to consider is that everything is connected to everything else. For example, in the early 1950s the Daydak people of Borneo were suffering from malaria. To combat this situation, the World Health Organization sprayed the area with DDT. The mosquitoes died and malaria declined. However, the DDT also killed parasitic wasps that controlled the thatch-eating caterpillars in the area. As a result, the roofs of the village's huts began to fall. The DDT also poisoned insects eaten by geckoes which were the food for the local cats. The cats started to die and rats flourished, resulting in sylvatic plague and typhus. To cope with this situation, the World Health Organization parachuted live cats into Borneo. We need to realize that there is a need for taking the larger view, which inevitably requires comparative risk analysis.

We all have emotional blocks as well as skill deficiencies that will need to be overcome if comparative quantitative risk assessments and analyses of the "big picture" are to be fully achieved. The fears and anxieties can be faced, and the lack of quantitative thinking can be overcome with education.

Few of us perceive a need to rank environmental problems. Some alternative reactions are that the "scientific" community understands that kind of thing, that it is too technical and, anyway, the government will protect us from such complex and unpleasant things. Many of us think that anything that can cause a deleterious problem should be eliminated, so why worry if one is worse than the other? Another component of this way of thinking is to argue that since there are no bodies on the floor (at least on obvious ones) as a consequence of environmental problems, then there is no need to be concerned.

Perhaps a more serious impediment (for all of us in environmental analysis) to being able to rank environmental problems quantitatively is that few of us think we have the training in quantitative and analytical thinking to achieve this broad perspective. Few have complete knowledge about probability, statistics, uncertainty, and logarithms in or out of school. Further, most of us do not realize that we use this knowledge in everyday life. Many would argue that it takes too much time to learn about these concepts and that they have more important things to do. It seems paradoxical that the very concepts that would help in getting perspective about life are rejected because we have too little perspective about the value of them.

What are the potential consequences of continuing the status quo and not trying to get the "big picture"? Some possible answers include:

- We may lose our bearings without knowing the bigger picture.
- The population could exceed the food supply.
- Public health could deteriorate.
- We could, through diminished species, upset the natural balance.
- We could exceed the limits of spaceship earth.
- We could change the global climate and reduce the amount of food available.

Observations

There have been comparative quantitative risk assessments of specific areas, such as energy, as some of the examples in this review show. However, there has been no comprehensive, broad effort to do this for all environmental problems (the Unfinished Business and Relative Risk reports discussed earlier were qualitative and limited in scope). There are good reasons for these limitations, which include lack of data, methods, and education, to mention only a few.

We can make several observations from the view presented by a comparative environmental risk analysis (since a complete comparative analysis has not been made, these observations are preliminary):

(1) We pretend that we are not part of nature and that we do not influence it, but in reality we have the opportunity to destroy the earth. Consider such examples as nuclear weapons, global climate, the burning of tropical forests, species diversity and extinction, stratospheric ozone, and acid rain.

(2) We ignore trends or seem oblivious to them. For example, our population continues to increase, yet the maximum number the earth can support could be of the order of ten billion. The rate of increase in the numbers of cars and trucks is greater than the rate of population increase. Carbon dioxide levels are increasing, and stratospheric ozone is also increasing.

(3) We have a limited time line even as our use and misuse of the environment is increasing. If we look only at the short term, we could miss recognizing some devastating occurrence or influence until it is too late.

(4) There is an overload of information and an increasing number of distractions to occupy us, and we have less leisure time, a limited ability to absorb new knowledge, and an increasing number of problems (too many problems to digest).

(5) The environment is so complex and diverse that we cannot understand all of the many interactions or relationships between its multiple parts. Even the smallest change, in the spirit of the present theories of complexity, may have large effects.

Thus, we must strive to achieve the best possible detailed knowledge of all systems, their components, and their interactions.

(6) It is all happening at an accelerating rate, a geometric progression, where we need logarithmic thinking to keep up.

To quote Al Gore (1992): "Too many of us, however, display a reverence only for information and analysis. The environmental crisis is a case in point: many refuse to take it seriously simply because they have supreme confidence in our ability to cope with any challenge by defining it, gathering reams of information about it, breaking it down into manageable parts, and finally solving it. The amount of information—and exformation—about the crisis is now so overwhelming that conventional approaches to problem-solving simply won't work. Furthermore, we have encouraged our best thinkers to concentrate their talents not on understanding the whole but on analyzing smaller and smaller parts."

The reluctance to compare risk assessments also may partially arise from political instincts to respond to public concerns but not take responsibility for technical matters. Referrals to agencies may collide with exaggerations of mission importance. Uncertainty about the conclusions that might emerge from comparisons can make both politicians and bureaucrats feel disinclined. In conducting comparative risk analyses we often do not consider the following factors: public values, the public's perception of risk, geography, socio-economic consensus, time, control, sociocultural conflicts, ethical concerns, burden of proof, uncertainty, distributive justice, and pollution standards.

Thus, we need to view the environmental situation broadly and use techniques like comparative risk analysis!

Summary

Putting all the risks encountered in our environment into perspective and comparing them using a common parameter is an attractive idea. One possible parameter for such a comparison is risk, as determined with the newly developed tools of risk analysis. There are very few comparisons of assessments in the literature. However, some attempts have been made, and the results have been published. The results of these efforts should be considered only after their validity is determined. It is the methodology that is considered in this book, and not which environmental problem is at the top of the list. These existing comparative risk assessments are reviewed and analyzed here. What emerges from our review is the question of why more comparative risk analyses have not been done. To answer this question, the problems encountered in such an assessment are discussed to get an idea of the magnitude of this task. In addition, the need for a philosophical basis for this newly emerging field of comparative risk assessment is examined, including a discussion of the need to see the "big picture." This perspective can help determine which are the most important problem areas for attention in terms of regulation, research, and analysis. Our conclusion is that this approach is useful. It needs further development and refining, and a firm philosophical basis. The myriad of analytical problems need not be fatal to the development and use of comparative environmental risk analysis.

Conclusions

Is comparative risk analysis possible? Although no completely perfect one has been achieved, it is clearly possible to produce, within limitations, useful analyses. Will it be understandable? Until we begin to think strategically and change our way of thinking to include the "big picture," comparative risk analyses will likely be misunderstood. An increasing complexity of problems may force us to make this change.

References

Bull, R. J., C. Gerba and R. R. Trussell. 1990. Evaluation of the Health Risks Associated With Disinfection, *Chemical Reviews in Environmental Control* 20:77–113.

Cohen, B. L. 1991. Catalog of Risks Extended and Updated, *Health Physics* 61:317–335.

Cothern, C. R. and J. Van Ryzin. 1985. Risk Assessment for the Regulation of Carcinogens in the Environment, *Hazard Assessment of Chemicals*, Vol. 4, Academic Press.

Cothern, C.R., Ed. 1993. Comparative Environmental Risk Assessment. Boca Raton, FL: Lewis Publishers.

Crouch, E. and R. Wilson. 1982. *Risk/Benefit Analysis.* Cambridge, MA: Ballanger Publishing.

Davies, J. Clarence. 1996. *Comparing Environmental Risks: Tools For Setting Government Priorities, Resources For the Future.* Washington, DC.

Doll, A. and R. Peto. 1981. *The Causes of Cancer*, Oxford: Oxford Press.

Federal Register, USEPA, 40 CFR Part 300, Hazard Ranking System: Final Rule, Federal Register, 51532–51667 (Friday, December 14, 1990).

Gold, L. S., T. H. Stone, B. R. Stern, N. B. Manley, and B. N. Ames. 1993. Possible Carcinogenic Hazards from Natural and Synthetic Chemicals: Setting Priorities, IN *Comparative Environmental Risk Assessment.* Ed. C. R. Cothern. Chelsea, MI: Lewis Publishers.

Health Effects Institute. 1991. *Asbestos in Public and Commercial Buildings: A Literature Review and Synthesis of Current Knowledge*, Cambridge, MA: Health Effects Institute-Asbestos Research.

House, P. W., J. A. Coleman, R. D. Shull, R. W. Matheny, and J. C. Hock. 1981. *Comparing Energy Technology Alternatives From An Environmental Perspective.* Washington, DC: U.S. Department of Energy, Office of Environmental Assessments, DOE/EV-0109.

Inhaber, H., *Risk of Energy Production.* 1978. AECB-1119/REV-1, Ottawa. Canada: Atomic Energy Control Board.

Inhaber, H. 1979. Risk with Energy from Conventional and Nonconventional Sources. *Science* 203:718–723.

Meadows, D. H., D.L. Meadows, and J. Randers. 1992. *Beyond the Limits.* Post Mills, VT.: Chelsea Green Publishing.

National Academy of Sciences. 1988a. *Health Risks of Radon and Other Internally Deposited Alpha-Emitters (BEIR IV).* Washington, DC: National Academy Press.

National Academy of Sciences. 1988b. *Complex Mixtures: Methods for In Vivo Toxicity Testing.* Washington, DC: National Academy Press.

National Academy of Sciences. *Biologic Markers in Pulmonary Toxicology.* Washington, DC: National Academy Press.

USEPA. 1987. *Unfinished Business: A Comparative Assessment of Environmental Problems; Overview Report.* Washington, DC: Office of Policy Analysis, Office of Policy, Planning and Evaluation.

USEPA. 1990. *Reducing Risk: Setting Priorities and Strategies for Environmental Protection,* Science Advisory Board (A-101), SAB-EC-90-021.

Chapter Ten

Ecological Risk Analysis

Introduction

Many hold the anthropocentric view that humans are the center of life and all else is here to serve us. Others, the authors included, think that we are part of nature and the sooner we accept that fact, the better our lives will be. As far back as 1990, the Science Advisory Board of the U.S. Environmental Protection Agency recommended that "EPA should attach as much importance to the reducing of ecological risk as it does to reducing human risk" (U.S. EPA, 1990). Another reason that we should be concerned with ecological issues is that they are essential parts of several laws, including the National Environmental Policy Act (NEPA); Clean Water Act (CWA); Marine Protection, Research and Sanctuaries Act (MPRSA); Endangered Species Act (ESA); Comprehensive, Environmental Response, Compensation and Liability Act (CERCLA); Resource Conservation and Recovery Act (RCRA); Federal Insecticide, Fungicide and Rodenticide Act (FIFRA); Toxic Substances Control Act (TSCA); and Clean Air Act (CAA).

Too often we stumble across our language. What does the term environment mean to you? To many it is all the living things on this planet and where they live—everything in the biosphere, atmosphere, oceans, lakes, soil, troposphere, and stratosphere. The principle goal of EPA is to protect the environment. Many of us separate humans from other living things and refer to "them" as the ecology. The focus of this chapter is on all other living things besides humans. We will use the term ecological for this purpose. Be aware that for most of us this understanding is not universal.

Laws Requiring Ecological Risk Assessment

What are the requirements of the laws that involve ecological risk assessments? The National Environmental Policy Act (NEPA) of 1969 challenges us to be more than just conservationists of wilderness areas and to become the active protectors of earth, land, air, and water. The objective of this act is to encourage productive and enjoyable harmony between humans and the environment, and to prevent or eliminate damage to the ecology and biosphere. Besides setting the policy concerning the ecology, this act requires an environmental impact statement for any federal action which significantly affects it, such as the construction of dams, or the use of pesticides on public lands to prevent or control outbreaks of insects.

The objective of the Clean Water Act (CWA) is to restore and maintain the chemical, physical, and biological integrity of the Nation's waters. This includes provisions for the protection and propagation of fish, shellfish, and wildlife in our rivers, lakes, and streams.

The Marine Protection, Research and Sanctuaries Act (MPRSA) of 1972 allows EPA to permit ocean dumping only where such dumping will not unreasonably degrade or endanger human health, welfare, or the marine environment. Included in the coverage of the act we find reference to potential changes in marine ecosystem diversity, productivity, and stability as well as species and community population dynamics.

An act that we find directly speaks to ecological risk assessment is the Endangered Species Act (ESA). Under this act, EPA must consult with the Departments of the Interior and Commerce on any action which may jeopardize the continued existence of any endangered or threatened species, or result in the destruction or adverse modification of critical habitats.

Two acts that EPA uses to regulate waste speak directly to ecological risk. The Comprehensive Environmental Response, Compensation and Liability Act (CERCLA), also known as the Superfund Act, has the objective of protecting public health and the environment. Here environment is defined as navigable waters, waters of the contiguous zone, ocean waters, and any other surface water; groundwater; drinking water; land surface or subsurface strata; and ambient air. Part of the hazard ranking system used to list Superfund sites includes ecological factors and sensitive environments. The other major waste act, the Resource Conservation and Recovery Act (RCRA), has the objective of protecting and preventing adverse effects on health and the environment. The 1984 amendments to RCRA significantly expanded the scope to include the consideration of ecological impacts and incorporate ecological endpoints.

The Federal Insecticide, Fungicide and Rodenticide Act (FIFRA) includes a pesticide registration procedure that requires attention to unreasonable adverse effects on the environment. Here environment is water, air, land, and all plants, humans, and other animals living therein, and the interrelationships which exist among these.

The Toxic Substances Control Act (TSCA) calls on EPA to regulate industrial chemical substances and mixtures which present an unreasonable risk of injury to health or the environment. Here environment is water, air, and land, and the interrelationships which exist among them and all living things.

Finally, the Clean Air Act (CAA) regulates hazardous air pollutants that may cause adverse environmental effects. The Clean Air Act is intended to preserve, protect, and enhance air quality in national parks, national wilderness areas, national monuments, national seashores, and other areas of special national or regional natural, recreational, scenic, or historic value. EPA protects the public welfare through the Secondary National Ambient Air Quality Standards where public welfare is defined to include effects on soils, water, crops, vegetation, man-made materials, animals, wildlife, weather, visibility, and climate.

What Is Ecological Risk Assessment?

Ecological risk assessment is defined by EPA's Risk Assessment Forum as "A process that evaluates the likelihood that adverse ecological effects may occur or are occurring as a result of exposure to one or more stressors" (U.S. EPA, 1998). We can define environment as the circumstances, objects, or conditions surrounding an organism, and use the term ecology to refer to the relationships between organisms and their environments.

In assessing ecological risk we will consider chemical, physical, and biological stressors. We take chemical stressors to include natural and man-made toxic chemicals. These include industrial chemicals, pesticides, fertilizers, pharmaceuticals, munitions, metals, smog, auto exhaust, and radionuclides. We also include phenomena such as acidic deposition or precipitation that can decrease the pH of streams and ponds, making them unsuitable for aquatic life (see the case study on this topic in Chapter 13). Other chemical stressors include eutrophication, where increased nitrogen and phosphorous from agricultural drainage lead to biological overproduction (e.g., algae). These nonpoint sources emanate from broad areas such as farms. For physical stressors we include, among others, highway construction and activities that remove or alter habitats, such as logging, dredging, and filling wetlands. We consider biological stressors to include the introduction of exotic plants and animals which, lacking the natural checks and balances, out compete and replace native species. Also included are genetically engineered microorganisms.

The word ecology derives from *oikos,* meaning house, and *logos,* meaning governing rules. The primary levels of ecological organization include species, populations, communities, and ecosystems. A *species* refers to a group of actual or potential interbreeding organisms that are reproductively isolated from other organisms. *Populations* are groups of organisms of the same species occupying a particular space over a given interval of time. We define *communities* as the populations of different species that live and interact with one another in complex associations. A *niche* is the role or function of a species in a community. An *ecosystem* is a community, its environment, and the relationship between the two considered as a single unit.

Next we will discuss ecological characteristics and stressors as a way of organizing our thoughts. We will then move into a discussion of screening and testing for risk purposes. Finally in this chapter, we will discuss EPA's guidelines that help us illustrate the principles of ecological risk analysis thought.

Ecological Characteristics

We will conduct ecological risk assessment using characteristics of ecological systems that include habitat, communities, niche evolution, and ecosystems. The habitat provides the local factors necessary for survival. These include hiding places, nests, birthing sites, shelters from ambient weather, covers for the growth and survival of shade-tolerant species of vegetation, structural features needed for song perches, and food sources.

One aspect of our ecology is constant change. Change is an integral part of our surroundings; it is one of the few constants. *Succession* refers to the gradual replacement of one community by another as an environmental condition changes.

Evolution is also an integral part of our environment. Three principles of evolution are important to keep in mind when analyzing ecological risk. First, no two individuals are exactly alike. Second, there is competition for survival. Third, some individuals have traits better suited for survival in a particular environment.

Three factors that are directly involved in the functioning of the environment are abiotic factors, photosynthesis, and nutrients. Important abiotic factors are soils, sediment, water, solar radiation, and minerals. Photosynthesis is the process by which plants convert light into chemical energy, which is stored as glucose, and is thus involved in the storage and flow of ecological energy. Nutrients are essential to plants and animals, and are recycled and used over and over again. There is constant recycling between organisms and their environment. Nature is organized so that the waste from one organism is the food for another through a series of ecological cycles.

The two major types of ecosystems are terrestrial and aquatic. Examples of terrestrial ecosystems are grasslands, deserts, coniferous forests, deciduous forests, alpine, tundra, and rain forests. Aquatic ecosystems include lakes and ponds, streams and rivers, wetlands, estuaries, and open seas.

Ecological Stressors

As mentioned earlier, we are considering three stressor types: chemical, physical, and biological. If the chemical stressors are bioavailable, they are active and can cause an effect; conversely, if they are not bioavailable, they cause no effect. Chemical stressors can also *bioconcentrate* as we move up the food chain, *bioaccumulate* in a given species, or be *biomagnified* through metabolism or other changes. That is, they can build up in concentration, just accumulate, or be magnified by exposure to other organisms.

Chemical stressors can cause effects at the organism, population, and community levels. At the organism level, the effects include mortality, behavioral changes, and physiological impairment on processes such as growth and reproduction. At the population level, effects include decreased birth rates, increased mortality rates, increased dispersion, and local extinction. At the community level, effects include structural changes, such as population loss, functional changes such as niche loss, and habitat destruction.

In order to be able to analyze physical stressors we need to know their severity. The important factors in describing physical stressors are severity of the impact, size of the affected area, frequency of the disturbance, and intensity of the physical force. Physical stressors can cause erosion (removal by water or wind), siltation (removal of soil by erosion moving into streams and rivers), and increased light intensity (removal of vegetation in higher soil increases the temperature reducing soil moisture).

Biological stressors can cause predation, parasitism, pathogenesis, and competition for resources such as food and space. Unlike chemical and physical stressors, biological stressors can reproduce, adapt, and spread, adding new dimensions to the ecological risk assessment process.

SCREENING AND TESTING OF CHEMICALS

We will next develop procedures to cope with the range of ecological effects due to man-made chemicals. For example, using the methods described here, EPA screened and assessed over 30,000 new industrial chemicals (Zeeman, 1997). The approach described here is from EPA, and is analogous to the one developed by the National Academy of Sciences.

The general approach we use here involves six specific areas: (1) appropriate ecological endpoints, (2) a tier-testing scheme, (3) testing guidelines, (4) structure-activity relationships, (5) hazard-assessment factors, and (6) ecological risk assessment methodologies (Zeeman, 1995).

The ecological endpoints we will consider are mortality, growth, development, and reproduction. These seem to be the critical ones in evaluating the potential impacts at the population level. We will use them here to assess the potential to cause adverse environmental effects that may be of regulatory significance.

Let us now consider the tier-testing scheme shown in Figure 10.1 which is designed to help us determine the kind and amount of testing needed to develop data adequate to measure the potential hazard of a chemical. This scheme provides for sequencing (tiering) of the testing so that quick and inexpensive screening tests are performed first. The criteria or triggers for additional testing and the logic for moving from one tier to another are described in Figure 10.2.

Tier 1: toxicity tests
- aquatic vertebrate acute toxicity
- aquatic invertebrate acute toxicity
- aquatic plant acute toxicity
- terrestrial vertebrate acute toxicity
- terrestrial plant toxicity

Tier 2: toxicity tests
- additional aquatic vertebrate acute toxicity
- additional aquatic invertebrate acute toxicity
- additional aquatic plant acute toxicity
- additional terrestrial vertebrate acute toxicity
- additional terrestrial plant toxicity

Tier 3: toxicity tests
- aquatic vertebrate chronic toxicity
- aquatic invertebrate chronic toxicity
- aquatic bioconcentration reproduction
- terrestrial vertebrate
- terrestrial plant uptake

Tier 4: field tests

Figure 10.1 The EPA Ecological Testing Scheme (from Zeeman, 1995)

Table 10.1 EPA Assessment Factors Used in Setting Concern Levels of Risk (from Zeeman, 1995)

Available data on chemical or analogue	Assessment factor
Limited (e.g. only one acute LC_{50} via SAR or QSAR)	1000
Base set acute toxicity (e.g. fish and	
daphnid LC_{50}'s, and algal EC_{50})	100
Chronic toxicity data	10
Field test data for chemical	1

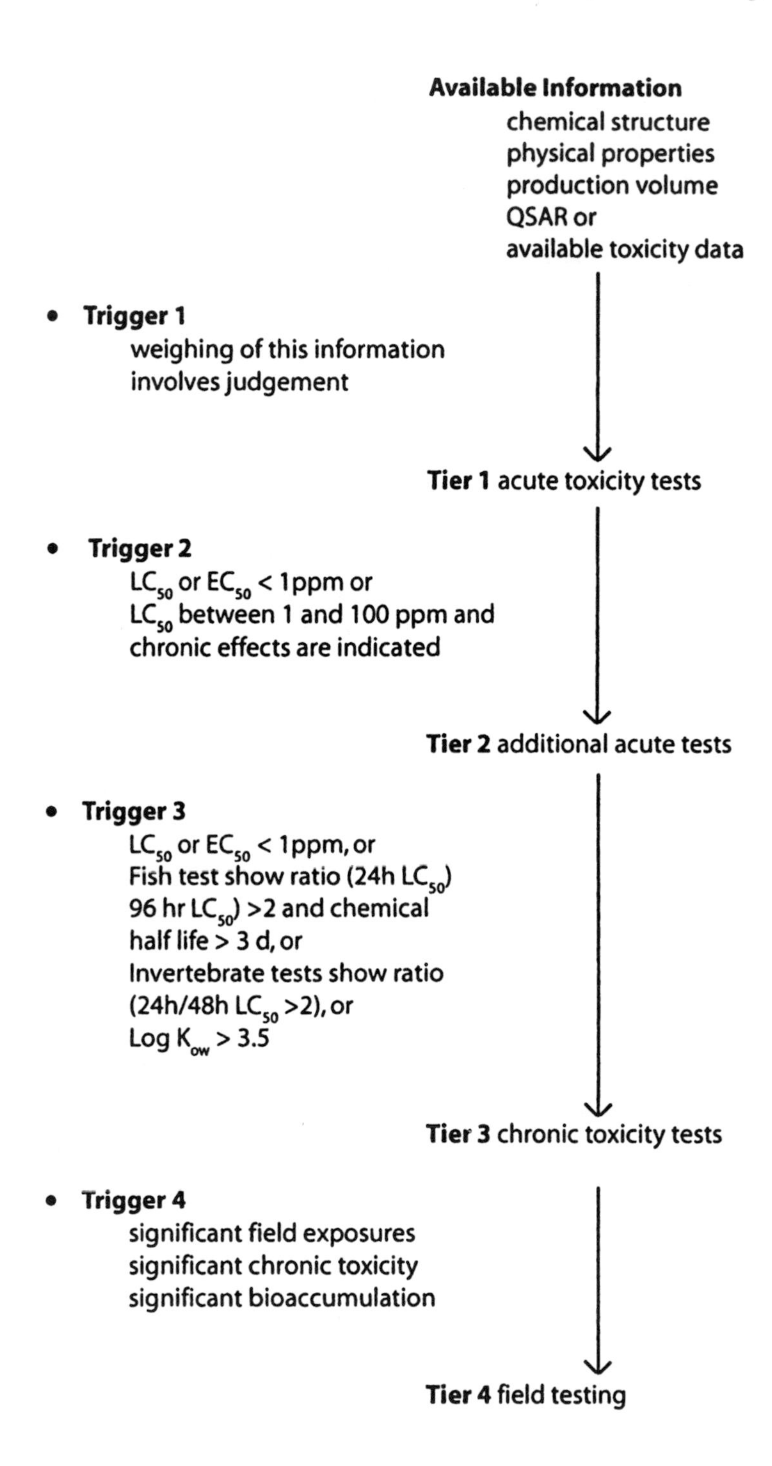

Figure 10.2 Triggers for aquatic Tier Testing System (from Zeeman 1995)

We can use surrogate test species for laboratory testing. For example, fish may suitably represent most aquatic vertebrates found in the water column (alternatives are whales, amphibians, etc.).

Test guidelines provide procedures for conducting acute and chronic toxicity tests using vertebrates, fish and birds and other biota (U.S. EPA, 1985). These include bioconcentration tests using fish and oysters, bioassays using freshwater and marine algae, and plant toxicity tests. Endpoints include mortality, impairment (i.e., LC_{50}, LD_{50} or EC_{50}), and effects on growth, development, or reproduction (see Suter, 1993).

In many cases we do not have the data and information concerning the toxicity of a chemical, but such data may exist for chemicals with similar structures. We can then infer the toxicological consequences of exposure to a chemical from that of a chemical with a similar structure. This inference is called a structure activity relationship (SAR). In some cases the dose-response relationship for a chemical with similar structure is known, or quantitative data are available. In that case a statistical inference called a quantitative structure activity relationship (QSAR) can be developed and used.

In ecological risk assessment, quantitative measures similar to those we developed for human health risk assessment are seldom known or available except for pesticides. However, it is possible to rank chemicals according to how hazardous they are to the ecology. The factors shown in Table 10.1 can be used to divide risk levels to adjust for this uncertainty.

Finally we need to develop an overall scheme or ecological risk assessment into which the above information and data are to be inserted. We will discuss this scheme next.

The Ecological Risk Analysis Process

We need to develop a process or plan for analyzing ecological risk. We will use one developed by EPA. For the guidelines promulgated by EPA, the two major elements are characterization of effects and characterization of exposure (U.S. EPA, 1998). The process is divided into three phases: problem formulation, analysis, and risk characterization.

To move through this process, it is helpful to have a checklist of questions to ask. This helps us to look for aspects that we may have missed. Many of the questions found in the sidebars on the following pages are general and apply to ecological as well as human health risk assessments.

First consider the overall ecological risk assessment process as shown in Figure 10.3. Here we see a systematic way to approach any ecological risk assessment. We consider the characterization elements as part of the central analysis phase. Next consider the details of each phase as portrayed in Figure 10.4. Here we find the three phases of risk assessment: problem formulation, analysis, and risk characterization. The information in Figure 10.4 found outside of these three areas shows us the critical interactive activities that influence why and how a risk assessment is conducted and how it will be used.

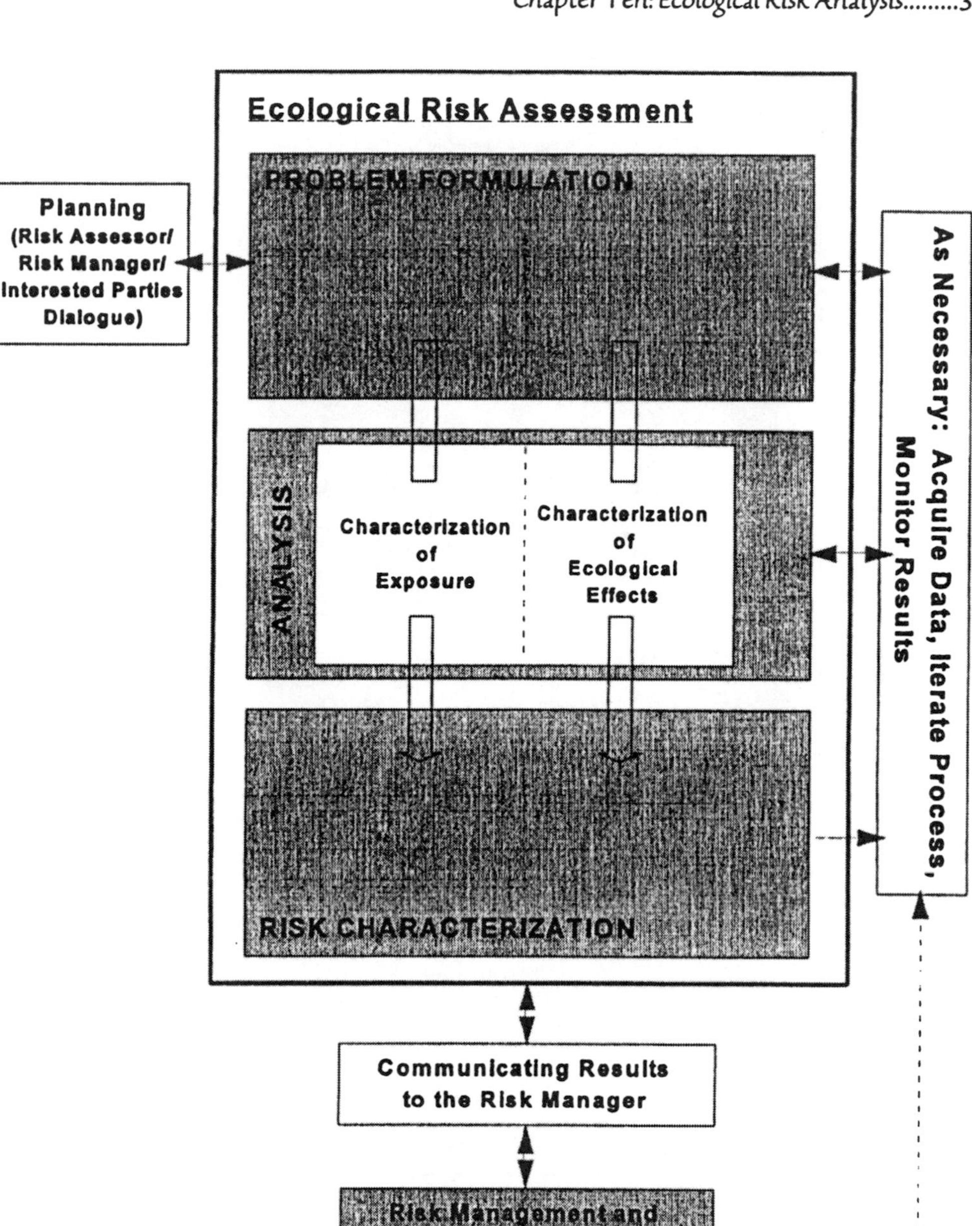

Figure 10.3 The framework for ecological risk assessment

Consider problem formulation, the first phase, found at the top of the figure. The steps we need to follow in problem formation are threefold, as the figure demonstrates. First, we need to determine the purpose for the assessment; second, we must define the problem; and third, we must determine a plan for characterizing risk. We determine the purpose by integrating available information on sources, stressors, effects, and ecosystem and receptor characteristics. We use this information to generate two products: risk as-

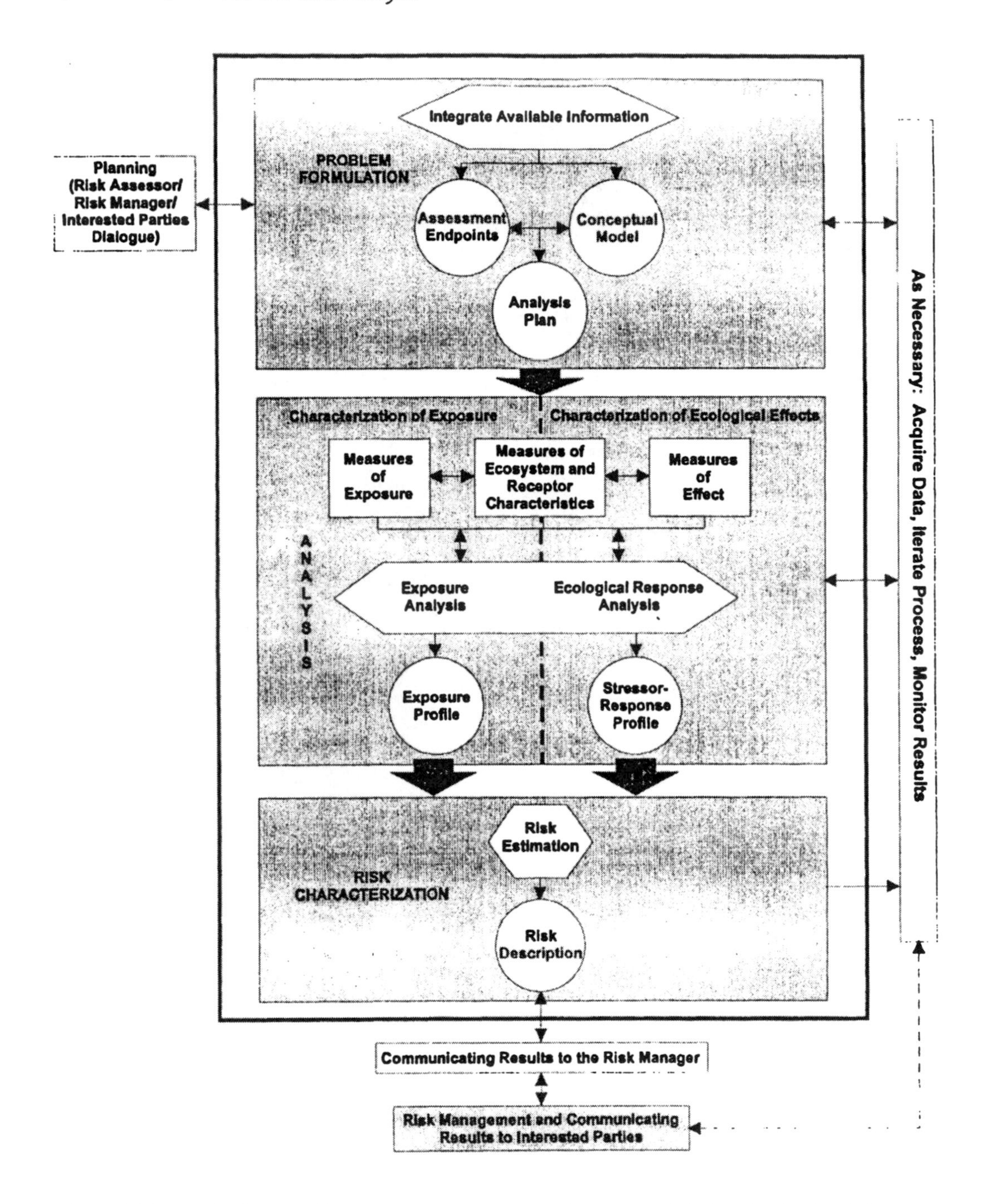

Figure 10.4 The framework for ecological risk assessment with an expanded view of each phase (Within each phase, rectangles designate inputs, hexagons actions, and circles outputs.)

sessment endpoints or effects, and possible conceptual models. We can then incorporate these details in a plan for analysis of the risk.

In the analysis phase, depicted in the middle of Figure 10.4, we develop the details of the overall ecological risk assessment. This phase uses the products or output of the problem formulation phase. During the analysis phase, we evaluate data to determine how exposure to stressors could likely occur (characterization of exposure) and, given this exposure, the potential and type of ecological effects that can be expected (characterization of ecological effects). We first need to determine how good the data are in describing exposure, effects, and what the ecosystem and receptor characteristics could be. Next we determine the place of all the information in the conceptual model we are developing. For convenience, we separate the products from these analyses into two profiles; one for exposure and the other for stressor response. We will eventually use the products as the basis for risk characterization.

We finally pull together all we have developed so far in the third and final stage, risk characterization. We include in risk characterization a summary of assumptions, scientific uncertainties, and strengths and limitations of the analyses. The final product is a report that communicates to all relevant and interested parties the full analysis of a particular problem.

An important part of the ecological risk analysis is communication. We can see the interactions among risk assessors, risk managers, and other interested parties in two places in the schematic representation of Figure 10.4. The left side box represents planning, where interested parties come to agreement about the management goals, the purpose for the risk assessment, and the resources available to conduct the work. Following risk characterization, the results of the risk assessment are formally communicated by risk assessors to risk managers and other interested parties (see Chapter 12). We show these activities outside the ecological risk assessment process diagram to emphasize that

Questions for risk managers

- What is the nature of the problem and the best scale for the assessment?
- What are the management goals and decisions needed, and how will risk assessment help?
- What are the ecological values (e.g., entities and ecosystem characteristics) of concern?
- What are the policy considerations (law, corporate stewardship, societal concerns, environmental justice, or intergenerational equity)?
- What precedents have been set by similar risk assessments and previous decisions?
- What is the context of the assessment (e.g., industrial site, national park)?
- What resources (e.g., personnel, time, money) are available?
- What level of uncertainty is acceptable?

Questions for risk assessors

- What is the scale of the risk assessment?
- What are the critical ecological endpoints and ecosystem and receptor characteristics?
- How likely is recovery, and how long will it take?
- What is the nature of the problem (past, present, and future)?
- What is our current knowledge of the problem?
- What data and data analyses are available and appropriate?
- What are the potential constraints (e.g., limits on expertise or time, availability of methods or data)?

risk assessment, risk management, and risk communication are distinct activities. Risk assessment involves the evaluation of the likelihood of adverse effects, while risk management involves the selection of a course of action in response to an identified risk that is based on many factors (e.g., social, legal, political, or economic) in addition to the risk assessment results. Risk communication is a two-way process which involves all the interested parties. We will see in Chapter 11 that risk assessment and risk management are not as separate as it appears at this stage. Risk communication is discussed more fully in Chapter 12.

In the bar along the right side of Figure 10.4 we have highlighted data acquisition, iteration, and monitoring. You will use monitoring data in all phases of a ecological risk assessment and identify changes in ecological conditions to evaluate a risk assessment's predictions. Also, you will use monitoring data to determine whether mitigation efforts were effective, verify whether source reduction was effective, or determine the extent and nature of ecological recovery.

Problem Formulation

As we learned earlier, problem formulation involves three products: (1) ecological risk assessment endpoints that adequately reflect management goals and the ecosystem they represent, (2) conceptual models that describe key relationships between a stressor and assessment endpoint, and (3) an analysis plan. We need to realize that problem formulation is frequently interactive and iterative rather than linear. Thus you may find a re-evaluation step during any part of problem formulation.

One important facet of problem formation is to decide which of the characteristics are most critical to ecosystem function and find agreement among professionals or the public on which are most valuable. You will note that all ecosystems are diverse, with many levels of ecological organization (e.g., individuals, populations, communities, ecosystems, landscapes, etc.) and multiple ecosystem processes. Your choices or plans are critical, however, because they become the basis for defining assessment endpoints and will determine how you transition between broad management goals and the specific measures used in ecological risk assessment.

You can use three principal criteria to select ecological values: (1) ecological relevance, (2) susceptibility to known or potential stressors, and (3) relevance to management goals. You will need a conceptual model or models to organize the information, make it more understandable, and help identify unknown elements. The complexity of the conceptual model depends on the complexity of the problem: the number of stressors, number of assessment endpoints, nature of effects, and characteristics of the ecosystem.

We think of conceptual models as consisting of two principal components: a set of risk hypotheses that describe predicted relationships between stressor, exposure, and assessment endpoint, along with the rationale for their selection, and a diagram that illustrates the relationships presented in the risk hypotheses.

The analysis plan is the final stage of our problem formulation. In developing an analysis plan, we evaluate risk hypotheses to determine how they will be assessed using available and new data. Our plan should include a delineation of the assessment design, data needs, measures, and methods for conducting the analysis phase of the risk assessment. Our analysis plan also should include pathways and relationships identified during problem formulation that will be pursued during the analysis phase. We should emphasize those hypotheses considered more likely to contribute to risk.

Analysis

In the analysis phase, we examine the two primary components of risk, exposure and effects, and their relationships between each other and ecosystem characteristics. We can use many types of data for ecological risk assessment. The data may come from laboratory or field studies or may be produced as output from a conceptual model.

Most of the data will be reported as measurements for single variables such as a chemical concentration or the number of dead organisms. In some cases, however, there will be variables that are combined and reported as indices. You can use indices such as the Index of Biotic Integrity (IBI) to access trends. These indices have several advantages, including the ability to (1) provide an overall indication of biological condition by incorporating many attributes of system structure and function, from individual to ecosystem

Source and Stressor Characteristics

- What is the source? Is it anthropogenic, natural, point source, or diffuse nonpoint?
- What type of stressor is it: chemical, physical, or biological?
- What is the intensity of the stressor (e.g., the dose or concentration of a chemical, the magnitude or extent of physical disruption, the density or population size of a biological stressor)?
- What is the mode of action? How does the stressor act on organisms or ecosystem functions?

Exposure Characteristics

- With what frequency does a stressor event occur (e.g., is it isolated, episodic, or continuous; is it subject to natural daily, seasonal, or annual periodicity)?
- What is its duration? How long does it persist in the environment (e.g., for chemical, what is its half-life, does it bioaccumulate; for physical, is habitat alteration sufficient to prevent recovery; for biological, will it reproduce and proliferate)?
- What is the timing of exposure? When does it occur in relation to critical organism life cycles or ecosystem events (e.g., reproduction, lake overturn)?
- What is the spatial scale of exposure? Is the extent or influence of the stressor local, regional, global, habitat-specific, or ecosystem-wide?
- What is the distribution? How does the stressor move through the environment (e.g., for chemical, fate and transport; for physical, movement of physical structures; for biological, life-history dispersal characteristics)?

levels; (2) evaluate responses from a broad range of anthropogenic stressors; and (3) minimize the limitations of individual metrics for detecting specific types of responses.

However, you will find that indices also have several drawbacks, many of which are associated with combining heterogeneous variables. Realize that the final value may depend strongly on the method used to combine variables. Some indices (e.g., the IBI) combine only measures of effects. For others we find unequal severity and a range of stressors and effects. For us to determine causality, indices may need to be separated into their components, or analyzed using multivariate methods (Suter, 1993). Another problem is double counting due to multiple causes for one effect and multiple effects due to one cause. For these reasons, we need to use professional judgment.

We observe that stressors can be transported via many pathways. Thus we need to take care to ensure that measurements are taken in the appropriate media and locations, and that models include the most important processes, based on our judgement

For a chemical stressor, our evaluation usually begins by determining which media it can go to. We need to consider the physicochemical properties such as water solubility and vapor pressure. For example, we find chemicals with low solubility in environmental compartments with higher proportions of organic carbon such as soils, sediments, and biota. From there we may examine the transport in the contaminated medium. Because chemical mixture constituents may have different properties, our analysis should consider how the composition of a mixture may change over time or as it moves through the environment. We next need to evaluate the fate and transport of chemicals

Ecosystems Potentially at Risk

- What are the geographic boundaries? How do they relate to functional characteristics of the ecosystem?
- What are the key abiotic factors influencing the ecosystem (e.g., climatic factors, geology, hydrology, soil type, water quality)?
- Where and how are functional characteristics driving the ecosystem (e.g., energy source and processing, nutrient cycling)?
- What are the structural characteristics of the ecosystem (e.g., species number and abundance, trophic relationships)?
- What habitat types are present?
- How do these characteristics influence the susceptibility (sensitivity and likelihood of exposure) of the ecosystem to the stressor(s)?
- Are there unique features that are particularly valued (e.g., the last representative of an ecosystem type)?
- What is the landscape context within which the ecosystem occurs?

Ecological Effects

- What are the type and extent of available ecological effects information (e.g., field surveys, laboratory tests, structure-activity relationships)?
- Given the nature of the stressor (if known), which effects are expected to be elicited by the stressor?
- Under what circumstances will effects occur?

(including bioaccumulation). For additional information concerning fate and transport, refer to the exposure assessment guidelines (USEPA, 1992); also see Chapter 4. We know that bioaccumulation and biomagnification are important characteristics in that area.

By considering the attributes of physical stressors, we can deduce where they will go. For example, suspended particles deposit in a stream according to size. We may require no modeling for pathways involving physical stressors that eliminate ecosystems or portions of them (e.g., fishing activities or the construction of dams). The activities are either yes or no.

Biological stressors can be dispersed in two ways; either by diffusion or jump-dispersal. When a stressor is spread in a gradual way by reproductive rates or motility, we describe it as diffusion. Jump-dispersal involves erratic spreads over periods of time, usually by means of a vector like wind, tides, eruptions, and other natural occurrences. Other examples include the gypsy moth diffusing via egg masses on vehicles and the zebra mussel via boat ballast water. Since some biological stressors may move both by diffusion and jump-dispersal, we may find dispersal rates very difficult to predict. In the analysis we should consider factors such as vector availability, attributes that enhance dispersal (e.g., ability to fly, adhere to objects, disperse reproductive units), and habitat or host needs.

As assessors we should consider the additional factors of survival and reproduction for biological stressors. In order to survive adverse conditions, organisms use a wide range of strategies; for example, fungi form resting stages such as sclerotia and chlamydospores, and some amphibians become dormant during drought. We can measure survival rates of some organisms under laboratory conditions. However, it may be impossible for us to determine how long resting stages (e.g., spores) can survive under adverse conditions where many can remain viable for years. Also, reproductive rates may vary substantially depending on specific environmental conditions. Therefore, life-history data such as temperature and substrate preferences, important predators, competitors or diseases, habitat needs, and reproductive rates are of great value to us. Thus, we should interpret these factors with caution, and the uncertainty involved should be addressed by using several different scenarios.

Now let us consider the ecosystem characteristics that can influence the transport of all types of stressors. Our challenge is to determine the particular aspects of the ecosystem that are most important. In some cases, we know the ecosystem characteristics that influence distribution. For example, fine sediments tend to accumulate in streams in areas of low energy, such as pools and backwaters. Other cases need more of our professional judgment. We need to know whether the ecosystem is generally similar to or different from the one where the biological stressor originated. Our professional judgment determines which characteristics of the current and original ecosystems should be compared.

Secondary stressors can greatly alter our conclusions about risk, and they may be of greater or lesser concern than the primary stressor. For chemicals, the evaluation usually focuses on metabolites, biodegradation products, or chemicals formed through abiotic processes. As an example, microbial action increases the bioaccumulation of mercury by transforming inorganic forms into a bioavailable organic species such as methylmercury, thus making it more toxic. Many azo dyes are not toxic because of their large molecular size, but they become toxic in an anaerobic environment where the dye is hydrolyzed into more toxic water-soluble units.

Secondary stressors can also be formed through ecosystem processes. Nutrient inputs into an estuary can decrease dissolved oxygen concentrations because they increase primary production, algal blooms, and subsequent decomposition. Although transformation can be investigated in the laboratory, we usually find that rates in the field differ substantially. Some processes may be more difficult or impossible for us to replicate in a laboratory. When evaluating field information, it may be difficult for us to distinguish between transformation processes (e.g., oil degradation by microorganisms) and transport processes (e.g., volatilization). Although they may be difficult to distinguish, we should be aware that these two different processes will largely determine if secondary stressors are likely to be formed. A combination of these factors will also determine how much secondary stressor(s) may be bioavailable to receptors. These considerations reinforce our need to have a chemical risk assessment team experienced in physical, chemical, and biological processes.

Physical disturbances can also generate secondary stressors, and identifying the specific consequences that will affect the assessment endpoint can be a difficult task. The removal of riparian vegetation, for example, can generate many secondary stressors to adult salmon, including increased nutrients, stream temperature, and sedimentation as well as altered stream flow. However, it may be the temperature change that is most responsible for adult salmon mortality in a particular stream.

We have previously mentioned the importance of knowing cause and effect. If we do not have a sound basis for linking cause and effect, uncertainty in the conclusions of an ecological risk assessment is likely to be high. A relationship generates strong evidence of causality when it demonstrates consistency of association through the reproduction of results.

Risk Estimation

For us, risk estimation is the process of integrating exposure and effects data, and evaluating any associated uncertainties. The process uses exposure and stressor-response profiles developed according to the analysis plan. We can deduce risk estimates using one or more of the following techniques: (1) field observational studies, (2) categorical rankings, (3) comparisons of single-point exposure and effects estimates, (4) comparisons incorporating the entire stressor-response relationship, (5) incorporation of variability in exposure and/or effects estimates, and (6) process models that rely partially or entirely on theoretical approximations of exposure and effects.

References

Suter, Glen. 1993. Editor and Principal Author, *Ecological Risk Assessment*. Boca Raton, FL: Lewis Publishers.

USEPA. 1985. Federal Register, Toxic Substances Control Act Guidelines, Part 797, Final Rule, 50:39252–39516.

USEPA. 1990. *Reducing Risk: Setting Priorities and Strategies for Environmental Protection*; SAB-EC-90-021, Washington, DC: Science Advisory Board.

USEPA. 1998. *Guidelines for Ecological Risk Assessment*, EPA/630/R-95-002F. Washington, DC: Risk Assessment Forum.

Zeeman, Maurice. 1995. EPA's Framework for Ecological Effects Assessment, Chapter 8 IN *Screening and Testing Chemicals in Commerce*, OTA-BP-ENV-166. Washington, DC: Office of Technology Assessment, 69–78.

Chapter Eleven ....

Risk Management: Values and Decisions

Introduction

What Do Values and Ethics Have to Do With Risk Management?

We should include values and ethics in the regulatory decision-making processes for three reasons: (1) they are already a major component, although unacknowledged, (2) ignoring them causes almost insurmountable difficulties in risk communication, and (3) it is the right thing to do (see Cothern, 1996).

Webster's dictionary defines *ethics* as questioning what is good or bad, or right or wrong, in terms of a moral system of principles. Ethics should be distinguished from social sciences, such as sociology and psychology, which attempt to determine why individuals or groups make statements about what is good, right, or obligatory. Some ethical systems that put forth their own definitions of "good" include: Aristotelian ethics, utilitarianism, Kantian ethics, natural rights theory, and Rawlsian contract theory.

Different people looking at the same set of environmental data and information can come to different conclusions because of different value systems. *Values* are primarily based on cultural, ethnic, and religious principles. Values and value judgements enter at every stage of regulatory decision making; thus, they affect outcomes in a real, continuous, and profound way. Even the selection of which problems to study involves a value judgement. Our choice of assumptions or default values involves a value judgement. Because worlds, nations, states, localities, and any two professionals can have different value systems, the place of value judgements in regulatory decision making is central. There are no value-free inquiries.

Although we seldom acknowledge them, values and value judgements pervade the processes of risk assessment, risk management, and risk communication as major factors in regulatory decision making. When viewed from the perspective of values, the artificial lines between risk assessment, risk management, and risk communication disappear. Almost every step of an assessment involves values and value judgements. However, we seldom acknowledge that they even play a role. When weighing different risks we use value judgements to determine which are more important and by how much. We cannot, and should not, exclude values and value judgements from the environmental decision-making process, as they are fundamental to understanding the political nature of regulation and decisions that involve environmental health for humans and all living things.

One of our major problems in risk communication is the failure of different groups to listen to each other. For example, many animal rights groups object on ethical and moral grounds to the use of animals in toxicological testing. The American Medical Association and other scientific groups have mounted a response that argues that many human lives have been saved (or, rather, these lives have been lengthened) by information gained from animal testing. Both sides have a point, but neither is listening to the other's values. These opposing views represent two different value judgements, and these values are the driving force in the arguments between the different groups.

It is essential to understand the place of ethics and values in risk assessment and acknowledge them in any analysis. A risk analysis must take into consideration values such as safety, equity, fairness, and justice, as well as feelings such as fear, anger, and helplessness. Our values and feelings are often the major factor in effectively communicating a risk-based problem.

Following a discussion of values, we will describe some conceptual models of regulatory decision making, factors in making decisions, costs, and our conclusions about risk management.

Values

Introduction

In general, values operate throughout decision-making processes, and often permeate these processes. For example, in situations where data and information are missing or nonexistent, we may err on the safe side by assuming the worst possible case. The values we act upon are often not representative of carefully worked out value systems. This is one reason why there may be differences between personal values and those of the community. Most of us find it difficult to say in detail what our own values are because there are few, if any, unified, generally accepted moral or religious concepts in the United

States and many other western countries. In addition, we seldom acknowledge that we use ethical principles in decision making. Our western, democratic tradition emphasizes justice, fairness, equality, democracy, autonomy, and responsibility. We believe that these are good values, and that societies (including our own) should be evaluated according to the extent that they promote such values. These underlying principles are part of our regulatory decision making process, but we seldom acknowledge this fact. Other values we consider important include health, quality of life, responsibility, truth, equity, stewardship, honesty, sanctity of the individual life, and spiritual and emotional balance.

In our regulatory processes there is a requirement that facts and values be separated, although this may not be always possible. We can distort important questions about ethics and values through the use of "shoptalk," language of technical "experts," which further complicates this separation. In our translation of environmental problems into technical and scientific language, we can lose or distort the values.

We are all biased, and our biases have important implications for environmental risk decision making. We may be biased because of our educational backgrounds and bring different values to the activity of environmental risk decision making. The scientist focuses on experimental data; the psychologist on feelings; the theologian or philosopher on the meaning or quality of life; the journalist on reporting daily events; the economist on allocation of scarce resources; an individual on "not in my back yard" (NIMBY) problems; the attorney on procedure and winning; and so forth. It is not for us to be judgmental, but to acknowledge reality.

Scientists are taught to value scientific truths above all other truths because scientific truths are, ideally, never accepted until they have been publicly tested. However, the "truth" of our ethical positions cannot be empirically verified in the same way as the "objective" scientific truth. For this reason, many scientifically trained people express open hostility to ethical discourse and value judgments. They often describe ethical questions as "soft" or "fuzzy" while we think of scientific questions as "hard." Scientists openly value rationality, tolerance of diversity (disagreements among scientists), freedom of inquiry, cooperation, and open communication. They tend to consider that the expert formulations of scientists are more rational and valid than the more intuitive and subjective judgments of the lay public (Fiorino, 1989). Remember that causality is a very demanding requirement and is often not understood by the public.

Values in Toxicology

Let us consider the development of animal bioassays. The first question is which contaminants should we test? Should we test the contaminants of interest to the public, political leaders, scientists, or media? The choice is likely to be based on our values. At what dose should we begin testing? The general procedure is to start chronic exposures at a dose half of the maximum tolerated dose. Why choose that dose? Toxicologists and others know this has to do with the amount of money available for testing. We desire to spend less money on any one contaminant so we can test for more contaminants later. This is obviously a policy decision with moral dimensions. Which animal species do we test and which strain should we test for? In answering this question, we value consistency as well as saving funds and other resources. How many different dose levels should we test? How long should the study go—for the lifetime of the animal, 90 days, 30 days, or other? Again, our valuations of consistency and cost help answer these questions. Which endpoints should we look for? We value the concern of the public for cancer; thus we often neglect the endpoints of neurotoxicology, immunotoxicology, and developmental toxicology, arguing that we have limited funds and that political pressure pushes us to focus on cancer.

In determining the kind of exposure, our values are also involved. Should we test oral, inhalation, dermal, or lavage routes? In many cases, this is determined by the values of the person or people paying for the test. If the test is required in the development of a drinking water standard, for example, then inhalation and dermal exposure routes may not be considered. Other characteristics that require our value judgments in risk assessment include such factors as repair mechanisms. In some cases we find thresholds that produce a sharp drop in the dose-response curve. In other cases, we find sharp changes in the dose-response curve, usually occurring for mechanistic reasons. For example, the curve may rise slowly until the dose is too large and overwhelms a detoxifying mechanism. Then the curve will rise more steeply. An example is found in the exposure to arsenic, where a hockey stick shaped exposure-response curve significantly changes the estimate (see the case study of arsenic in Chapter 13). U-shaped and biphasic curves present a unique problem (Calabrese, 1992 and 1994). Too little or too much may both be bad. For example, we can die of thirst without water, but drown with too much. All too often, we do not know if a threshold exists. We value safety by assuming the worst case (i.e., that a threshold does not exist). To improve our statistics, we combine the data for benign tumors with those for malignant ones. This need for more statistical validity is also a matter of judgment. These choices are made by scientists.

We are less willing to experiment on humans, but are willing to experiment on other animals. This is a value judgement that places other living things on a lower level. In setting standards for environmental contaminants, we have a choice of using the average person or the most susceptible—this choice is a value-laden decision.

Values in Environmental Risk Decision Making

When we consider living things, a possible organizing idea for us is the value of integrity. This proposal is systematically examined by Laura Westra (1994) who derives her argument from a quote by Aldo Leopold in *A Sand County Almanac*: "A thing is right when it tends to preserve the integrity, stability, and beauty of the biotic community. It is wrong when it tends otherwise." This practical philosophical proposal is nonanthropogenic in its eventual direction, and involves cultural, ethical, philosophical, scientific, and legal aspects. The values involved in the idea of integrity include freedom, health, harmony, biodiversity, sustainability, life, morality, and scientific reality.

The importance of human life also is a value-laden concept. When we make decisions, do we view one human life as sacred as many human lives? Or, do we balance risks and save as many people as possible? Values associated with life and death are important in environmental risk decisions. With our societal denial of death, we credit standards with "saving" lives when these lives are only extended. We seem much more concerned with contaminants or health effects that shorten life as opposed to those that cause sickness. It is a value judgement when we think contaminants that cause cancer are of more concern than those that cause neurotoxic, immunotoxic, or developmental effects. Is it ethically sound to allow exposures to rise to the level given by a standard?

It is a value judgement and perception of the public that estimates based on risk assessments are not believable because the public does not trust the scientists that generate these risk assessments. Thus, our personal judgements and biases can distort our decision making by rejecting estimates that might be quite valid.

Value judgements enter into other components of our quantitative risk assessment protocol, including decisions concerning acceptable levels of uncertainty, causal links versus correlations, synergism or antagonism, latent periods, morbidity versus mortality, hormesis, thresholds, and comparisons between different health endpoints (or deciding which are the more important) (Cothern and Schnare, 1984). In each of these cases, we must make value-laden assumptions to complete the analysis.

One clearly value-laden decision for us is determining acceptable risk. What is a safe level? Each of us has a different level of risk that we would find acceptable. There is no universal acceptable level. Our individual decisions are affected by values when we attempt to determine if a situation is voluntary or involuntary, old or new, catastrophic or ordinary, and known or unknown (Latai, 1980).

In subjective areas, we all have thoughts and opinions that influence our risk decisions. Our values in relation to the following concepts will affect our decisions in life:

- Freedom: Many promote autonomy to such an extent that they conclude, "I don't care what the risk is, I am free to not use my seat belt, smoke cigarettes, abuse alcohol, etc."
- Equity: Those who do not value equitable practices may favor putting incinerators, chemical factories, or waste sites near the poor and politically weak.
- Trust: Many people do not trust scientists or the government.
- Quality of Life: Many of us face the decision of whether or not to live in polluted urban areas.
- Safety: Should we err on the safe side by using the value-laden, linear, non-threshold, dose-response relationship assumption?

- Stewardship: Does our land ethic influence conservation of wetlands and protection of all living things? Is species extinction important to us?
- Natural is Good: But what about radon and natural carcinogens? Is sustainability important to us?

  indoor air pollution—"my home is my castle" sentiments will influence value judgements.

  upstream or downstream—values differ with respect to position.

  anthropocentric or biocentric—values dramatically change when the focal points.

Too often we must use default assumptions, such as non-thresholds or the idea that whatever happens to animals will happen to humans, and these assumptions are accepted as "science policy" or "expert judgment". Without careful scrutiny of these assumptions, we can create politically controversial results which are challenged as arbitrary rules that have no basis in either science or public policy (Jasanoff, 1993). By examining the value dimensions of this process we can get a better and more useful perspective concerning the environmental risk decision process.

The value judgements we make in risk assessments and risk decisions have a strong effect on their nature, character, and outcome. The value-laden approach is widely used in making environmental risk decisions without much acknowledgment.

Some Conceptual Decision Models

Some conceptual models of risk analysis do not contain any explicit mention of values, value judgements, ethics, or perceptions. However, these are often the main bases used in making decisions. For example,

- Alar was banned to protect children, but the ban did not include adults;
- the linear, non-threshold dose-response curve and the use of combined upper 95% confidence limits are based on safety not science;
- the Superfund program started with the idea that "if I can sense it, it must be bad," while indoor radon has met with widespread apathy because it cannot be sensed;
- the idea of zero discharge is based on the sanctity of the individual and a mistaken concept of detection limits;
- forests and wetlands are preserved because of stewardship, and
- nuclear power is avoided because fear of catastrophe is so excessive that some call it a dreaded fear.

However, we have in the regulatory decision process many opportunities for value judgements: "Perhaps fifty opportunities exist in the normal risk assessment procedures for scientists to make discretionary judgements. Although scientists are presumed to bring to this task an expertise untainted by social values to bias their judgement, they are not immune to social prejudice, especially when their expertise is embroiled in a public controversy" (Rosenbaum, 1991).

We will now examine the function of values, value judgements, ethics, and perceptions in decision models. These characteristics are directly involved in current risk decisions, but existing models usually do not mention them. We attempt to disguise these characteristics of values and ethics in some decisions with scientific or technical labels. Are current and future environmental problems and decisions more complex and of a different character that those of the past? If so, then a new decision paradigm will be needed. Current environmental problems are characterized by levels of complexity and uncertainty never before experienced by any society. Values and ethics seem like perfectly good ways to balance and choose in the decision making process, and since they are widely used, why not acknowledge this fact and formally include them in the models?

We know of several models that impact the process of regulatory decision making. These models are not necessarily all that exist, and we present them here as examples of the many kinds that include thinking in the area of values, value judgements, ethics, and perceptions.

(1) Ideal Model

The ideal situation is when all possible information is known about a risk, including the scientific and technical aspects; health consequences of possible actions and their alternatives; exposure routes of all possible causes; present and future costs; social, political, and psychological consequences of all decisions; and all other possible relevant information. Since this situation seldom occurs, in general, only fragments of the necessary data are available. It is folly to think that the ideal situation can ever be achieved. Decisions will have to be made with imperfect information and incomplete data. To keep perspective, it is well to use the "perfect" or "ideal" as a goal and develop methodologies that can help you move closer to it.

(2) The National Academy of Sciences' "Red Book" Model

In 1983, the National Academy of Sciences created a regulatory decision model (the "Red Book") that starts by moving from hazard identification to dose-response assessment and combining the resulting potency estimate with exposure to yield a risk characterization, as illustrated in Figure 11.1.

The regulatory decisions that emerge from this analysis use inputs from this risk characterization with possible control options and non-risk analyses such as economical, political, statutory, legal, and social considerations.

Besides giving the field of risk assessment some coherent terminology, the Red Book clearly implied a separation of risk assessment from risk management, unless you believe that the estimation of a risk somehow automatically will lead to a decision. The values in a risk assessment most often take the form of *science policies*, assumptions based on general scientific knowledge and used in the absence of data. Usually, when an assump-

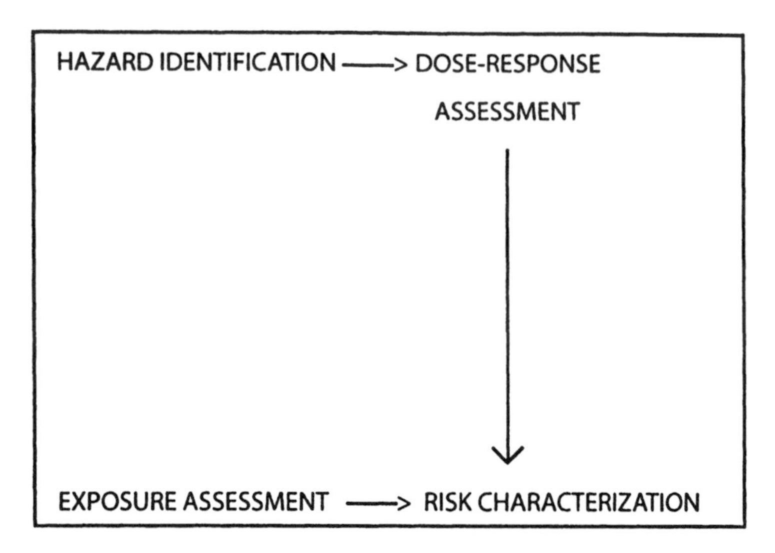

Figure 11.1 National Academy of Sciences regulatory decision model

tion could have several values, the value that provides the greatest public protection is preferred. Where the science policy states a generally preferred value, most assessors call this value the *default assumption* or *default option*. This approach makes risk characterization the crucial step in risk assessment, as it contains the information passed on to risk managers.

The Red Book model also inherently prefers assessment procedures that support the evaluation of a risk under at least two circumstances—the current, status quo situation and an alternative. Otherwise, risk managers cannot use the risk characterizations.

The discerning reader also will note that the Red Book model calls for a hazard identification step. We have avoided discussing *hazard* in this book, since it is not necessary for risk assessment. Mathematically, the concept of a hazard is defined in contradictory ways in the literature. It is a difficult, if not intractable, subject; it implies an exposure-response relationship at some unknown exposure. The concept of a hazard assumes that some substances have biological effects that are intrinsic characteristics similar in nature to chemical properties, like molecular weights. This assumption seems dubious. Instead, it seems more likely that substances have biological effects under specific conditions or circumstances.

(3) Cost-Benefit Analysis and Risk-Risk Balancing

The idea of this model is to compare the benefits of a decision (such as preventing death or disease, reducing property damage, or preserving a resource) to the costs. This approach can be used to determine the best solution among several options in relation to costs and benefits or net risks. Any situation involves limited resources and knowing how the costs and benefits compare is helpful. However, many of the benefits or risks are difficult, if not impossible, to quantify. For example, it is hard to determine the benefit of preserving a species, the aesthetic value of a forest, or the value of swimming or boating in a river or lake. Also many comparisons are difficult, such as the relative benefit of averting sickness or death, cancer, or a birth or developmental defect. Almost all of the problem areas in cost-benefit analysis or risk-risk balancing involve value judgements.

In cost-benefit analysis, risk managers monetize both the costs and the benefits of the current situation and at least one alternative (Kopp *et al.*, 1997). Then, the portrayal of each option becomes easy. It involves comparing commensurate values, dollars to dollars, either as net values or as ratios. The preferred option is the least costly. Similarly, in risk-risk balancing, risk managers evaluate commensurate values, the risks removed or decreased, as compared to the risk added or increased (Graham and Wiener, 1995).

Risk managers can interconvert risks and monetary values in many ways, but all are subject to technical controversies. For example, economists can convert risks to money, because loss of income shortens life spans or because people are willing to pay to avoid risks. Some people object to the use of cost considerations at all. However, sometimes their objection is more about the options posed. Perhaps an option they prefer is not under consideration. They divert their protest into an objection about monetisation because it is an easier target. Similarly, the idea that cost-benefit analysis requires that benefits outweigh costs is bogus. Instead, it simply implies that an option with costs that outweigh benefits by a factor of two is preferable to another option with costs that outweigh benefits by a factor of ten.

At present, Executive Order 12866 requires that federal agencies prepare cost-benefit analyses in support of new, major regulations, but not necessarily that the agencies factor cost-benefit considerations into their decision making. Except for FDA, federal agencies seldom engage in risk-risk balancing, and even there, comparison of risks under different scenarios, such as a health risk with a new drug compared to the health risk without the new drug, usually is done informally.

(4) A Framework Model

Risk assessment is an example of what some call a regulatory or mandated science. In regulatory sciences we try to fill the gap between theoretical or laboratory science, and reliable and defensible regulatory or management decisions. We realize that pure science is not value free; regulatory or mandated science certainly is not.

We can divide classical risk assessment into two stages: factual judgements, which are free of values, and evaluations, which are value-laden. As we learned earlier, the Red Book model separates risk estimation from risk management. The Red Book does not fully acknowledge the role of value-based judgement. Values can provide feedback between risk assessment and risk management without us realizing.

The alachlor controversy is an example of the breakdown of the Red Book's risk assessment model for decision making (Brunk, 1991). It was not a conflict between those who accepted the verdict of the risk assessment and those who did not. It was also not a conflict between those who understand the objective risks and those who were guided by subjective perceptions. It was a political debate among different value frameworks, different ways of thinking about moral values, different concepts of society, different attitudes towards technology, and different ideas about risk taking.

Authors who analyzed the alachlor controversy concluded, "A more realistic model of risk assessment, one that is sensitive to the role of values in the estimation of risk, is urgently needed" (Brunk, 1991). They recommended a framework that includes the acknowledgment of the connections between the scientific and social policy elements.

The components of their framework model include attitudes towards technology (positive or negative); uncertainty; statistical lack of knowledge or incomplete knowledge; methods to use; risk; causality, including confidence; burden of proof (deciding who has it and what the criteria are); and voluntariness (John Stuart Mill's liberalism) or social order.

The principle lesson learned from this analysis and proposal is not that we need to start a global debate on the meaning of rationality, the merits of technology, or the importance of voluntary risks—these issues are too broad. However, these are among the value issues that need to be addressed by risk assessors. "Sensitivity to the biases that are introduced by broad attitudes concerning rationality, technology, and the liberal state should bring recognition by risk analysts that their activity is not, as they imagine, neutral and value-free" (Brunk, 1991).

(5) A Channel Model

There are several elements or "channels" that can be used to move from an environmental risk decision problem to possible solutions. We show several of these in Figure 11.2 as horizontal elements in moving from the problem to the solution.

The model is arbitrarily separated into two areas: the so-called "objective value" elements such as risk, cost, and feasibility, and the "subjective value" elements such as social, political, psychological, and safety elements. Although seldom explicitly mentioned, we suggest that all of the elements involved in environmental risk decisions involve values, perceptions, and ethics.

We too often make a decision concerning an individual environmental problem using only a few, or even only one, of the many elements shown in Figure 11.2. Many of the values found in the horizontal channel of Figure 11.2 are ignored or overlaid with what the decision maker knowingly or unknowingly thinks are more important values. There appears to us to be no element that does not involve a value or ethical dimension. For example, the relative value of cancer, neurological, developmental, or immunological

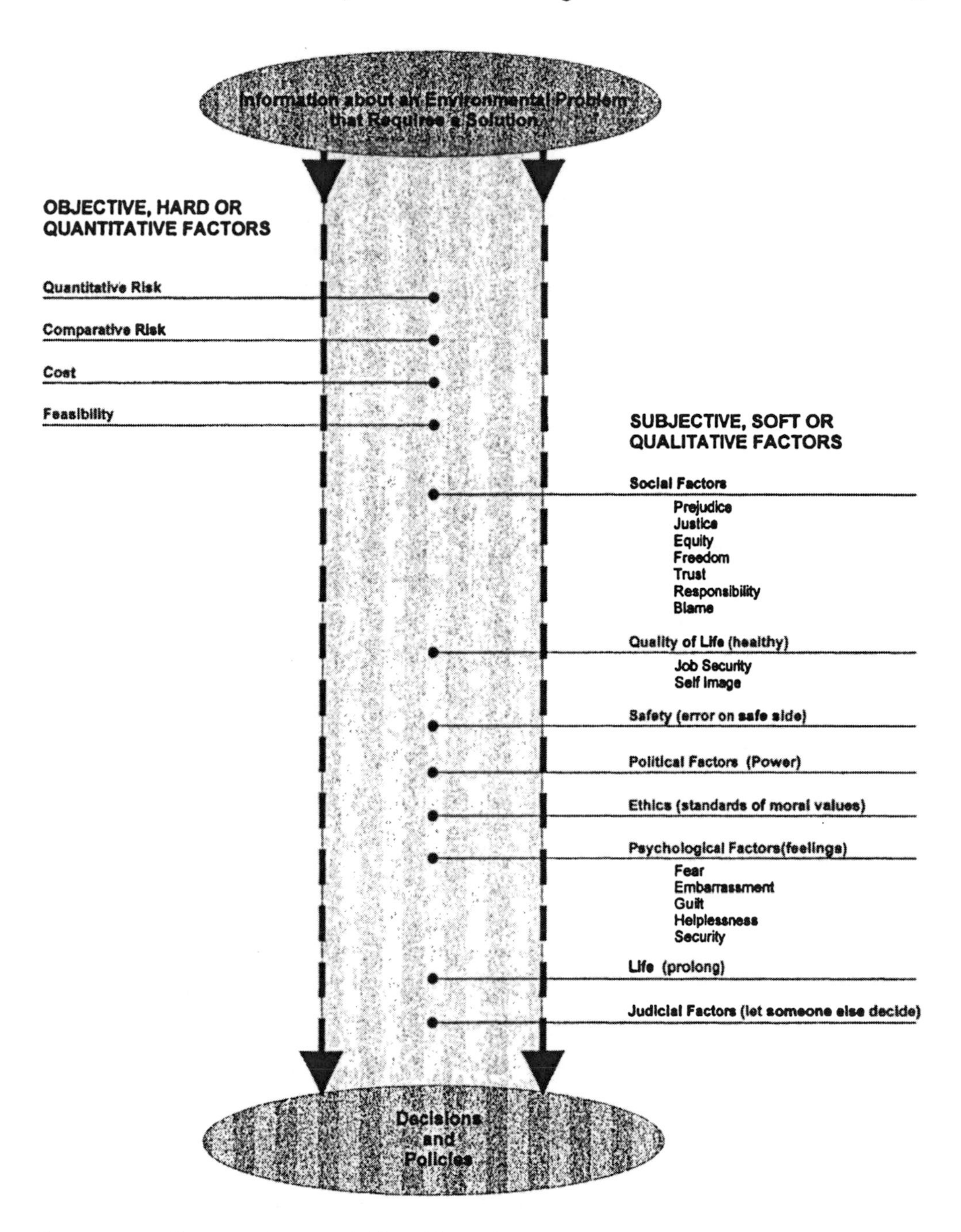

Figure 11.2 Channel Environmental Risk Decision-Making Model

endpoints as well as the relative importance of mortality and morbidity are value decisions made in the risk assessment process. The choice of model to describe the flow of a contaminant through the environment (and default assumptions), and the interpolation of a dose-response curve into the unknown, both involve values and perceptions. To discount something is to make a value-laden decision concerning the cost estimate.

Many feel it is not practical, especially for long periods of time like 100 or 200 years, to include discounting of costs. This choice is made through a value judgement.

(6) An Overlay Model

We next consider a variation of the channel model where each horizontal element is value-laden. Values, ethics, and value judgements are added as an overlay to the analysis. By adding the values at the end, we can easily lose sight of the critical features of a problem and focus almost completely on the value or ethic. An example of this approach is the use of the value of zero risk. Perhaps it is ethically sound, but it is not physically possible.

To the uninitiated or uninformed (or those who do not appreciate or understand the complexities), the most desirable decision concerning environmental risk is the one that would result in no risk, zero risk, or zero discharge. The laws in Massachusetts and Oregon that use the idea of toxics use reduction (TUR) rest on a simple argument: the use of every toxic chemical should be reduced or eliminated. Other attempts to effect zero risk include the Delaney Amendment for food additives, effluent guidelines for discharges into water, resistance to de minimus regulations (or any other minimization approach), and resistance to fluoridation of water, to name only a few.

To overlay information concerning an environmental problem with a value such as zero risk prevents us from getting perspective. This simple-minded approach prevents any understanding of the risks actually averted or the cost of doing so.

(7) Continuous Model

Now let us consider the continuous model where values, perceptions, and ethics enter the process in several places and do so continuously. These elements are inserted by many different individuals, in the form of assumptions or defaults, at different places in the overall process. These individuals include scientists (physical, biological, and social), economists, attorneys, politicians, regulators, engineers, managers, and many other professionals. Few of these individuals are professionally trained in the use of values and ethics.

What we need is a model (such as that shown in Figure 11.3) that includes inputs at each stage from those trained in the use of values, value judgements, ethics, and perceptions. We emphasize that this model is one view, or quick "snapshot", of a continuously and rapidly changing process.

The first step is to gather the known information and put it onto a "common table" so that it can be compared, weighed, and balanced. All too often we assemble only some of the existing information. However, in the sense of an ideal model, let us assume that much of the existing information is assembled, including inputs from all sectors (federal, state, and local government; regulators; industry; academia; labor; public; journal-

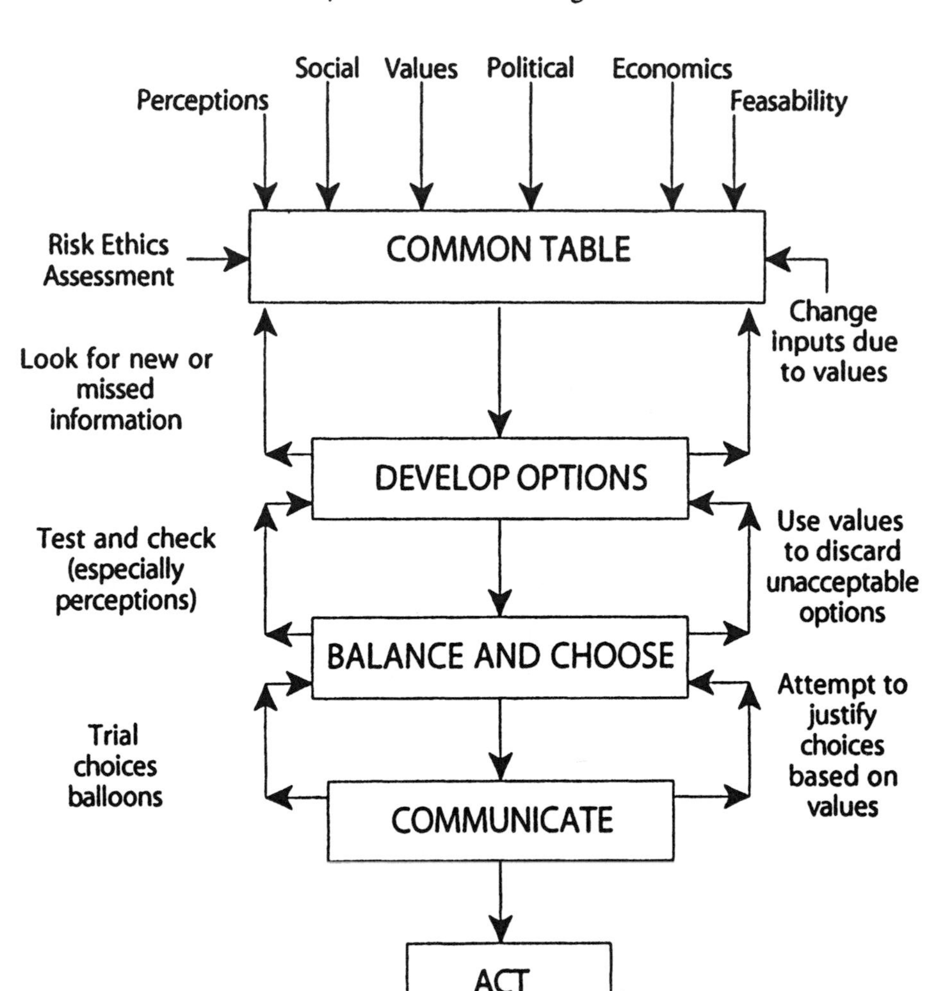

Figure 11.3 Snapshot in Time of the Continuously Moving Process of Environmental Risk Decision Making

ists; environmentalists; and any other stakeholders in potential decisions). The risk assessment input includes scientific and technical data and information concerning exposure and health effects to humans and other living things, qualitative and quantitative risk assessments, as well as comparative risk assessments.

We generate options in the next phase by considering the information and data gathered. These options are affected by the values of those generating them. We scrutinize the options to determine what information is missing, and research is instituted to develop it. At the same time, the values impinging on this decision are considered in a process that could be described as "YES, BUT ...", and new and different inputs are selected and added while existing ones are removed due to value judgements created in the process of developing options. The first tension in the process is generated by the conflict between the values and value judgements of the decision makers involved.

The third step is to choose among the options. In this process, values and value judgements affect our choices, and some options are discarded as unacceptable. An important step here is to test and check the options.

The option we choose is then floated as a trial balloon to test its acceptability, and an attempt is usually made to justify the choice based on values and value judgements. Some of the possible value judgements used include: catastrophic versus ordinary effect, interdependence of all living things, cancer versus non-cancer endpoints, equity, emotional and spiritual balance, exceeding natural resource limits or the sustainability of the earth, fairness, health, honesty, justice, protection, quality of life, relative value of humans and other living things, natural versus man-made, privacy, responsibility, safety, sanctity of individual lives, severity of the effect, stewardship, and voluntary versus involuntary.

(8) The National Academy of Sciences "Orange Book" Model

The Red Book omitted much consideration of risk characterization. Because risk characterization often is the crucial element in providing useful information for risk management purposes, in 1996, a second NAS committee prepared *Understanding Risk: Informing Decisions in a Democratic Society*, usually referred to as the Orange Book because of the color of its cover. To determine what kind of risk characterization a regulatory agency should prepare, the Orange Book had to describe risk management. Figure 11.4 depicts the process of risk management in the Orange Book.

(9) Other National Academy of Sciences Reports about Risk Management.

Several other NAS Committees have made important contributions to regulatory decision making. Two of them, the Committee on Principles of Decision Making for Regulating Chemicals in the Environment in 1975 and the Committee on Risk and Decision Making in 1982, thoroughly discussed the subject. The models in both books resemble the one displayed in Figure 11.4.

(10) The Presidential Risk Assessment and Risk Management Commission Model

The Clean Air Act Amendments of 1990 called for a Presidential Risk Assessment and Risk Management Commission to study the two processes and report its findings to Congress. In 1997, the Commission published a detailed report, summarizing its recommendations by means of a circular model with six steps for (a) evaluation of previous actions, (b) definition of a problem and its context, (c) assessment of risks, (d) examination of options, (e) making decisions, and (f) taking actions. The model proceeded from step (f) back to step (a). The model had stakeholder input as an activity crucial to all six

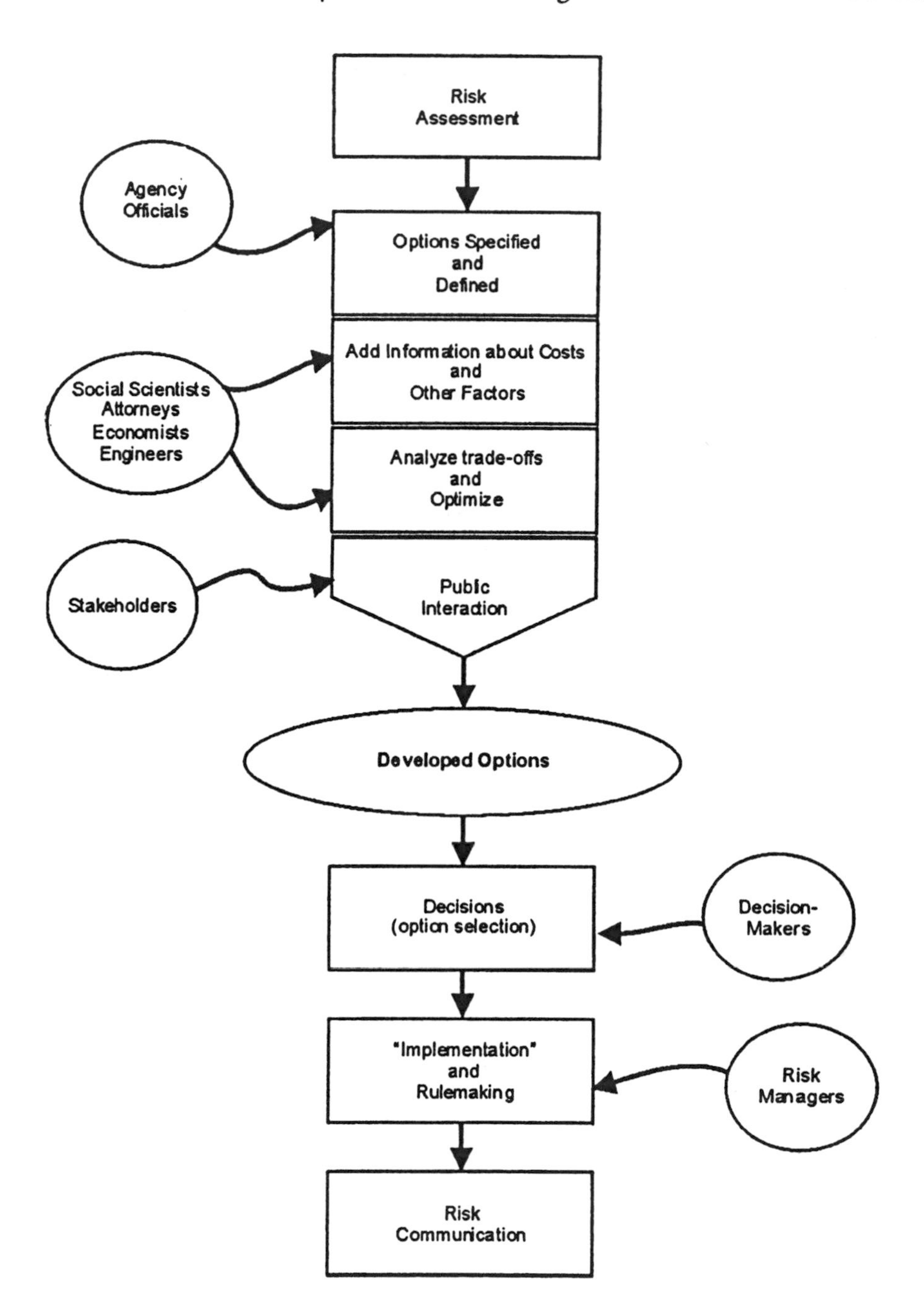

Figure 11.4 An Anatomy of Risk Management

steps. To us, stakeholder involvement is close to, but at one remove from, acceptable risk. Instead of directly accessing the views of the public about a risk, the stakeholders stand to lose or gain directly from a regulation. They transmit their views, and thus amplify some public concerns and muffle others.

We described acceptable risk briefly in Chapter 1. This concept had an important historical influence on the development of risk analysis and remains a powerful idea today.

In our view, however, the use of acceptable risk as a guiding principle in regulatory decision making is flawed. It implies that risk managers should defer their choices to the results of polls. If this is done, the role of the risk manager becomes unimportant. While public opinion is important, it changes in volatile ways, particularly about emotional subjects like risk, and is not a good basis for technical decisions. The public lacks skill in science-based decision making. In our view, the Commission's model overemphasized stakeholder involvement.

(11) Unreasonable Risk

Unreasonable risk has a long history in tort law, the part of our legal system that governs the duties of citizens to each other in a civil society (see Chapter 3). It is based on the idea that a risk is unreasonable in comparison to some other risk that is reasonable (Gough, 1981). Unreasonable risk intrinsically involves risk-benefit balancing (Davies, et al., 1979). For example, to see whether the failure to exercise precautions created a legal liability when a storm struck and a barge broke loose from its mooring, causing damage, Judge Learned Hand compared the probability of the "natural disaster" times the monetary damages to the cost of a reasonable precaution, securing the boat better (Hand, 1947). The product of probability times damages under the first scenario (a boat adrift) was unreasonable in comparison to a second scenario (a secured boat), which was reasonable.

Imagine that you, a risk manager, need to decide between two options. You write brief descriptions of the two scenarios, each on a separate sheet of paper. Underneath you write brief summaries of what you know about the risks and benefits under each scenario. Now you compare the two sheets and select one as more reasonable. Notice that you do not need quantitative information. Also, risks are not commensurate with benefits, unlike the case for risk-risk or cost-benefit balancing. Unreasonable risk lets you easily integrate qualitative factors such as values and ethics into your decision.

(12) Risk Magnitude Models

Because risk management seems so complex and difficult, much effort has been devoted to models that create bright lines, somewhat like the one in safety assessment (see Chapter 7). For example, significant risk suggests that an agency should not initiate the regulation of a substance unless it exceeds some level, as determined by a decision maker (Byrd and Lave, 1988). *De minimus* risk suggests that an agency should not regulate a substance or process that produces a trivial level of risk, a level essentially the equivalent of no risk from a detection limit perspective. Discernible risk suggests that agencies apply different models to risks that might be detected than to risks that cannot be detected (Cross et al., 1991). All three models are essentially screening devices. Their major problem is removal of values from the process of deciding what to regulate.

Despite the advice of experts, politicians often attempt to insert specific risk values, such a one-in-million (10^{-6}) lifetime risk of death, into legislation as regulatory goals. These criteria prevent the application of risk management to regulation. Instead, numerical goals force value judgements into the risk assessment process.

Making Regulatory Decisions

Schools of business management and of engineering have intensely studied the process of making decisions. This work led to a branch within the field of operations research called decision analysis (Raiffa, 1970). Much of decision analysis is based on Bayes Theorem, an application of the probability calculus that we have chosen not to cover in this introductory text. Even without getting into the details of decision analysis, however, you can understand the basic requirements of good decision-making. A mathematical proof is not required.

In brief, you would (a) attempt to formulate all of the exclusive, available options to reduce risk, (b) obtain data about these options and evaluate them, and (c) prefer the option with the best net value (for you, your agency, the federal government, Congress, or the public, depending on which perspective is most appropriate). If you fail to notice an important option, evaluate your options poorly, or prefer an option that is not the best one, you will make a deficient, less than optimal decision.

The preceding section about decision models emphasized that the creation or specification of options is a key step in risk management. Even models that rely on risk magnitude alone essentially are devices to sort options or confer a preference on one option. So, if you realize that risk management involves options, you are one step ahead. Each option is a scenario that portrays the imposition of a distinctive, different regulatory control. One of the options will be life as we know it today, the status quo.

Some regulatory approaches do not directly involve risk analysis. Like safety assessment and public health actions, which we covered in Chapters 6 and 7, these other approaches either involve completely different concepts, or they truncate some steps in risk analysis, skipping over them. For example, the Material Safety Data Sheet (MSDS) plays a role in occupational safety and community right-to-know processes (Marsick and Byrd, 1990). MSDSs skip from information about a hazard directly to risk communication. The underlying assumption seems to be that, given hazard information, everyone will spontaneously know what to do. The Hazard Analysis and Critical Control Points (HACCP) process plays a critical role in the assurance of meat quality. HAACP jumps from a hazard directly into risk management, where it imposes a check listing approach to minimize the possibility of contamination (Coleman and Marks, 1999). The Toxics Release Inventory (TRI), another part of community right-to-know, skips from information about releases, a component of exposure assessment, directly into risk communication.

The specification of options is a key step in risk management, but it also is an art. Failing to specify an option is the most common flaw in risk management. Useful options describe distinctive, separate scenarios; regulatory options seldom, if ever, have a continuous nature. Even setting standards—for example, going from 10 ppm to 5 ppm of permissible exposure to a substance—seldom involves dialing new numbers into equations. Instead, it usually involves different control technologies, feasibility, agency policies, statutory constraints, costs, public perceptions, political sensitivities, self compliance possibilities, deadlines, resources, and even legal considerations (whether the agency will get sued or not).

For example, under the Clean Air Act, a risk manager may have three possibilities for the control of emissions at each of 1,000 plants. They might be: (a) do nothing, (b) install scrubbers, or (c) install better valves and pipes to prevent fugitive emissions. The cost ramifications of these three options will vary dramatically from plant to plant. For a few plants, option (a) may even cost the most. If the risk manager has to choose among the three options, the same option will get applied to all 1,000 plants. However, if the reductions of risk for each plant and option were displayed as a probability distribution function, risk reduction will appear continuous. This appearance is deceiving. At a hypothetical plant in Virginia, the best option may achieve a large risk reduction, but at another in Texas, an entirely different option may achieve virtually nothing. Further, if costs get consideration, the best public protection might involve the imposition of both (b) and (c) at a few plants, but no controls at many others.

Clearly, forcing a choice of one kind of control option and applying it universally to all plants is inefficient. However, a risk management decision to allow plant mangers the latitude to achieve the most efficient solution by trading emissions also will prove difficult. The value and ethics of some public interest stakeholders will cause them to vigorously oppose trading schemes.

The specification of options is not as difficult as it might look here. Certain obvious options always exist. For example, the risk manager can delay selecting an option (do nothing), which is always an option itself. Similarly, the risk manager can confirm the status quo, conduct more research to remove some uncertainty, delegate or transfer responsibility (even wait until a legal deadline passes, forcing a judge to decide), better enforce existing regulations (increase fines and penalties), publicize information, require warnings or labels, negotiate a standard with stakeholders (instead of imposing a control process), or force cessation, by seizing, destroying, or banning. This last, most draconian option often is associated with the regulatory process in the minds of the lay public, but it is seldom employed except in local or restricted enforcement proceedings.

When you start to accumulate information relevant to the options, which is sometimes called parsing the options or developing the options, you will find many different kinds of information involved beyond risks, including control technologies, feasibility, internal organizational policies, statutory constraints, administrative barriers, costs, public perceptions of risk, political sensitivities, self compliance, and sometimes even legal considerations (whether your agency will be sued or not). Any or all of these complex data may assist in selecting the best regulatory option. So far, these factors do not explicitly involve any ethical or value systems. So, while risk assessment is the first step in risk analysis, and while the objective of risk management is risk reduction, risk management usually involves much additional information, unless some legal, regulatory, or policy restriction prevents its consideration.

Eventually, you need to choose the "best" option. In risk management, choosing seldom is a matter of optimization. If it was, computers would make risk management decisions, not people. Only two of the regulatory decision models in the previous section lead to optima (cost-benefit or risk-risk balancing and unreasonable risk). The rest suggest an endless conveyor belt of regulation.

Decision analysis usually involves monetary utilities to sort through options. A utility is the probability of the value of something times its value, an application of the same kind of logic that Judge Hand used in deciding that the failure to tie a barge down was unreasonable. He multiplied the probability of the barge breaking loose during a storm, times the expected damage if the barge did break loose. He compared that utility to the cost of tying the barge down securely times the probability of that scenario, in this case 1.0, because it was certain. The utility of an unsecured barge exceeded the utility of a secured barge.

Because utilities usually involve money, they can be either positive or negative. The application of the principles of decision analysis to risk analysis is based in part on the analogy between the two disciplines. A utility is like a risk, as defined in this book, but with the outcome already multiplied by the probability. (Part of the reason that we prefer not to multiply outcome times probability, but have the outcome accompany the probability, is that we can multiply later, treat the product like a negative utility, and engage in loss minimization.) If money and risk were the only criteria to apply to a decision, the analogy would hold up better. However, risk management usually involves many kinds of data that can often be incommensurate and may not yield to quantitative evaluation. Even so, the decision making under decision analysis and risk analysis resemble each other (Weinstein, 1979).

Decision analysis is a mature, advanced discipline that has much to offer to regulators (Lave, 1981). For example, decision making under uncertainty is a major topic: how to make the best decision when the data about the options are vague. Also, recall that more research is always is an option. Under cost of information approaches, a decision analyst will estimate the cost of making a decision with poor data. The decision analyst will compare the cost of a delay plus the cost of research that would yield a better decision.

Imagine that you have a mandate to regulate some process and that you elicited twelve fairly exclusive options to achieve this requirement. Initially you acquired information about the increased and decreased risks under each scenario. This approach led you to turn the data over to some risk assessors and ask them to calculate the risk-risk balance for each option. Initially you believed that you wanted to select the option with the most favorable ratio of risks for the public. However, just to check on the assessors, you also turned the same information over to some economists and asked them to calculate the cost-benefit balance for each option, to confirm the best option. Unfortunately, when both groups return with their calculations, the "best" option from a risk perspective turns out not to be the "best" option from an economics perspective. If you think you have narrowed your options to the best two and that you can discard the other ten, you will have made another error. The "best" option over both domains turns out to be a third scenario, which ranked second best from both risk and economic perspectives. Now, things might get ugly. Suppose that your engineers tell you that the controls required under all three "best" options have poor feasibility. The regulated industry will oppose them strenuously.

This hypothetical risk management story is typical. It occurs because risk management decision making involves different criteria. Multicriteria decision making (MCDM) also is a specialized topic within the field of decision analysis. Suffice it to note that rigorous optimization may prove difficult, if not impossible. Different criteria often will lead to the selection of different "best" options. Even when looking only at risk, you will discover that you often cannot interconvert outcomes, such as days of hospitalization for a curable disease versus the number of deaths, when the disease is not treated.

Again for the sake of our hypothetical discussion, assume that you have finally selected an option that involves closing a few plants. This option seems severe for a few people—the owners, employees, and people who depend on the plants for their local economies. While their risk increases, the net risk reduction for the entire population far outweighs the adverse effects in this small, local population. You would simply multiply individual risk at present, without regulation, times the total population, and compare it to individual risk after the plant closings times the total population. The decreases in population risk after closing the plants far outweigh the population risk increases for a much smaller population.

Even if you dislike thinking about the sacrifice of this small group, preferring the best net risk reduction seems obvious. If you agree with this statement, you are a confirmed utilitarian. Someone who believes in libertarian ethics would strongly disagree with your choice. Government, libertarians insist, has no ethical right to impose a risk on anyone, particularly if they have not made this choice personally. A libertarian will prefer a regulatory option with some risk reduction for every single person involved.

A similar controversy erupted recently about air bags in cars. It turns out that air bags increase the risk of injury or death after an accident for certain kinds of passengers—

pregnant women, children, and small adults—although net population risk decreases, particularly for passengers simultaneously wearing seat belts. For experienced risk managers, the idea that some risks will increase with a mandated control and others will decrease seems self-evident and obvious. The very concept of pursuing net risk decreases for the public, a utilitarian concept, seems intrinsic to a regulatory mission.

Since the federal government mandated air bags in cars after strong pressure from public interest groups, the discovery that air bags are not always good has proven difficult for many to countenance. In addition, Libertarians held that for ethical reasons, government should not have imposed a sweeping risk reduction measure like air bags on the general public. Instead, the ethical role of government was to make accurate data about the benefits of air bags available to consumers and to insure that air bag manufacturers did not sell defective products. In other words, government's legitimate role was to insure a fair and efficient marketplace, not to override individual choice. The trend between libertarian and utilitarian views changes over time in the U.S. We emphasize individual liberty, yet we exhibit preferences for regulations that involve utilitarian risk reductions, essentially trying to have it both ways.

Several institutional factors in risk management merit attention here:

(1) Some agencies have separated risk assessment from risk management, whereas others have integrated them. For some agencies, integration is so thorough that every employee participates in both activities. Asking them to separate risk assessment from risk management is like asking the right side of their brain not to communicate with the left. Clearly, separation has both advantages and disadvantages. One advantage is documentation. If risk assessment and risk management are separate, information has to pass back and forth in documents. However, separation can distort regulatory decisions. Risk assessors, worrying that managers will do "the wrong thing," may attempt to fix the process by exaggerating risks or concealing uncertainties. In turn, no risk manager wants to be held responsible for deaths in the future. So, the risk management process inserts even more precautions, just to be on the safe side. Society eventually gets strange, overly risk averse, decisions. Eventually the public regards the agencies as uncaring and blames risk assessment methods for unpopular decisions.

(2) Some agencies have a single person as the appointed risk manager (sometimes a small committee serves the same function), whereas others have organization-wide involvement in risk management. Clearly, dispersion has both advantages and disadvantages. An agency will take responsibility as an organization, instead of scapegoating departing political appointees, and participation in decisions improves morale. However, dispersed processes have difficulties in dealing with uncertain information, conflicted values, or ethical considerations.

(3) Some agencies make a single kind of regulatory decision routinely and repetitively, whereas others make a few, unique but momentous decisions (Wilson and Holbrow,

1997). Obviously, making many similar decisions will allow agency managers the luxury of creating policies and delegating responsibility for the decisions with a fair confidence in obtaining desired results. However, the same routine makes it easy to miss exceptions.

(4) Congressional requirements or court decisions may impose restrictions on the information allowed in parsing options. For example, Clean Air Act decisions may not involve information about costs. Such prohibitions can place a thoughtful manager in a quandary. For example, many analysts believe that aggressive implementation of the Clean Air Act, as written, would shut down industry. Yet, a shut down of industry clearly would increase population risks. Similarly, other statutes impose risk management criteria that distort decision making.

(5) Some statutes have stopping points, whereas others lack them. For example, the Toxic Substance Control Act requires EPA to balance impairment of innovation against risk reduction. The net effect is to create a stopping point to risk reduction, even with fragmentary and uncertain data. Without a stopping point, an agency will simply choose the option with the lowest risk, until someone devises an option with lower risk. Then, the agency will migrate to the new option. The net result is that the public never gets a stable standard or policy. The uncertainty about regulation makes commerce very difficult. In science, new information may support a new standard, but the new information will remove uncertainty and can either decrease or increase the standard. Of the regulatory decision models in the previous section, only cost-benefit or risk-risk balancing and unreasonable risk reach stopping points.

For example, the Endangered Species Act (ESA) calls on the Department of the Interior to declare some species endangered and seize private property to protect them. While protection of a species at risk of extinction seems laudable, ESA is a nightmare for property owners who "lose the lottery." The only stopping point to increased confiscation of personal and family assets is public insurrection (DeLong, 1997).

Overall, risk management is a simple process with only a few steps, even if it is complex in application.

(1) Specify the options.

(2) Evaluate the options.

(3) Choose the best option.

(4) Manage the process of adopting the chosen option.

The second and third of these steps admit values and ethics into risk management. For some reason, the public associates a lack of passion with fair or balanced decisions. A few unfortunate persons acquire an inability to perceive or have emotions as a result of damage to a specific brain region, the amygdala. Physicians sometimes diagnose these lesions, because the patients exhibit impaired decision making. Thus, too little emotion is just as damaging as too much emotion for decision making.

Costs

Regulations cost money, and costs are a resource constraint as well as a value. The costs we reference are not agency budgets. Instead, we mean the expenditures that compliance with regulations necessitate. We have no opinion about whether the current level of spending in the U.S. is a reasonable expense. However, questions about the appropriate level of expenditure are distinctly different from questions about whether current expenditures are wasted. Can we obtain more protection for the same expenditures? Can we obtain the same protection for less expenditure? These questions are matters of cost effectiveness.

Summaries of the cost per life saved for different regulations are available (Tsengs et al., 1995). The range is dramatic, from a few dollars to many millions of dollars per life. The total costs are staggering, even if only for health risks, which include consumer products, nuclear power, pharmaceuticals, foods, transportation, and work places.

We should not leave the subject of decision making without an example of the costs of cleaning up the environment. The cost of environmental regulation has been estimated to be in the range of an annual 150 to 200 million U.S. dollars. However, this estimate does not reflect the difference between the market prices and services, and the externalities from the full environmental and social costs. Table 11.1 shows several indirect, external, and hidden costs that we all pay for environmental pollution and degradation. As you can see, these costs amount to hundreds of billions of U.S. dollars annually (Corson, 1998).

Conclusions

William W. Lowrance (1976) gave the following guidelines for scientists concerning the impact of the intertwining of facts and values in developing public policy:

> "Recognizing that they are making value judgements for the public, scientists can take several measures toward converting an arrogation of wisdom into a stewardship of wisdom. First, they can leaven their discussions by including critical, articulate laymen in their group. Second, they can place on the record their sources of bias and potential conflicts of interest, perhaps even stating their previous public positions on the issue. Third, they can identify the components of their decisions being either scientific facts or matters of value judgement. Fourth, they can disclose in detail the specific bias on which their assessments and appraisals are made. Fifth, they can reveal the degree of certainty with which the various parts of the decision are known. Sixth, they can express their findings in clear, jargon-free terms, in supplementary non-technical presentations if not in the main report itself."

Table 11.1 Costs Related to Environmental Degradation

Category	Billions of Annual U.S. Dollars
Noise pollution (Cobb et al., 1995)	12
Cleanup of contaminated groundwater resources (NRC, 1997)	16-33
Cleanup of military and civilian hazardous waste sites (Russell et al., 1992)	10-33
Cleanup of existing hazardous waste sites (Estes, 1996)	20
Damage to health from water pollution (Estes, 1996)	1
Loss to commercial fisheries from water pollution (Estes, 1996)	2
Aesthetic cost of water pollution (Estes, 1996)	2
Impairment of recreational activities from water pollution (Estes, 1996)	11
Water pollution (Cobb et al., 1995)	35
Damage to crops and vegetation from air pollution (Estes, 1996)	9
Agricultural damage from air pollution (Estes, 1996)	13
Household soiling from air pollution (Estes, 1996)	17
Health costs from air pollution (Estes, 1996)	228
Air pollution, excluding damage to health (Cobb et al., 1995)	35
Health care and lost productivity costs as a result of air pollution (Cannon, 1985)	40
Health problems related to air pollution (Hawken, 1997)	100
Air pollution from motor vehicles (Chafee, 997)	30-200
Loss of old-growth forests (Cobb et al., 1995)	61
Economic losses caused by alien plant and animal species (Nature Conservancy, 1997)	97
Loss of farmland (Cobb et al., 1995)	81
Loss of wetlands (Cobb et al., 1995)	155
Depletion of stratospheric ozone layer (Cobb et al., 1995)	214
Other long-term environmental damage from non-renewable energy use (Cobb et al., 1995)	645

In this chapter we have discussed the concept that values, perceptions, and ethics permeate and directly impact regulatory risk decisions. For us to ignore their impact would be folly. What we need is recognition of the role they play, a different view of the current reality this view provides, and a movement towards understanding the "big picture." All aspects of regulatory decision making involve values and value judgements. These may appear as assumptions, default positions, and policies.

Are regulatory decisions too complex? Are there too many details, pressures, and components for any one person, or group of people, to understand, weigh, and analyze? Is it possible for even a single issue to be understood enough to be explained to those involved, the public and other interested parties? All of this demands that a simplifying model be developed that will facilitate understanding and move closer to producing better regulations and a safer and more healthy environment. We have attempted to do that here.

Ethical issues pervade regulatory decision making. This new area of practical ethics (Shirk, 1988) is an important component of decision making that helps guide us towards better decisions in protecting equity and value.

Is it time to develop an environmental ethic? Or is the situation so complex that a set of ethics needs to be developed in several areas? The answer to these questions could go a long way towards simplifying our understanding of environmental problems.

It seems essential that we work towards grasping the big picture in a more complete and better way. The use of modeling could help.

If we are going to deal with reality, we must be more honest. The scientific value of truth should be central to regulatory decision making. These values should be central to the development of a new paradigm in the field of regulatory decision making.

References

Brunk, C. G., L. Haworth and B. Lee. 1991. *Value Assumptions in Risk Assessment: A Case Study of the Alachlor Controversy*. Waterloo, Ontario: Wilfrid Laurier University Press.

Byrd, D. M., and L. B. Lave. 1988. Significant Risk is not the Antonym of De Minimus Risk. C. Whipple Ed., *De Minimus Risk*. New York: Plenum Press, p. 41–60.

Calabrese, E. J., Ed. 1992. *Biological Effects of Low Level Exposures:Chemicals and Radiation*. Boca Radon, FL: Lewis Publishers.

Calabrese, E. J., Ed. 1994. *Biological Effects of Low Level Exposures: Dose-Response Relationships*. Boca Radon, FL: Lewis Publishers.

Cannon, J. S. 1985. *The Health Costs of Air Pollution*. New York: American Lung Association.

Chaffee, J. 1997. Statement. *The Washington Post*. p. A4.

Cobb, C., T. Halsted, and J. Rowe. 1995. *The Genuine Progress Indicator: Summary of Data And Methodology*. San Francisco: Redefining Progress.

Coleman, M. E., and H. M. Marks. 1999. Quantitative and quantitative risk assessment. *Food Control* 10:289–297.

Corson, W. 1998. *Recognizing Hidden Environments; and Social Costs and Reducing Eologicol and Society Damage Through Tax, Price, and Subsidy Reform.* Unpublished Manuscript. Washington, D.C.: the Graduate School of Political Management, George Washington University.

Cothern, C. R. Ed. 1996. *Handbook For Environmental Risk Decision Making: Values, Perceptions and Ethics.* Boca Radon, FL: Lewis Publishers/CRC Press.

Cothern, C. R. and D. W. Schnare. 1984. The Limitations of Summary Risk Management Data. *Drug Metabolism Reviews* 17:145–159.

Cross, F. B., D. M. Byrd, and L. B. Lave. 1991. Discernable Risk—A Proposed Standard for Significant Risk in Carcinogen Regulation. *Administrative Law Review* 43: 61–88.

Davies, J. C., S. Gusman, and F. Irwin. 1979. *Determining Unreasonable Risk under the Toxic Substances Control Act.* Washington, D.C.: Conservation Foundation.

DeLong, J. V. 1997. *Property Matters: How Property Rights Are Under Assault—and Why You Should Care.* New York, NY: The Free Press

Estes, R. 1996. *The Tyranny of the Bottom Line: Why Corporations Make Good People Do Bad Things.* San Francisco: Berrett-Koehler.

Fiorino, D. J. 1989. Technical and democratic values in risk assessment. *Risk Analysis* 9 3: 293-299.

Gough, M. 1981. Unreasonable risk. *Assessment of Technologies for Determining Cancer Risks from the Environment,* Appendix B, Washington, D,C,: Office of Technology Assessment.

Graham, J. D., and J. B. Wiener. 1995. *Risk Vs. Risk: Tradeoffs in Protecting Health and the Environment.* Cambridge, Mass: Harvard University Press.

Hand, L. 1947. *U.S. v. Carroll Towing Co.* 158 F. 2nd 169 (2nd Cir.).

Hawkin, P. 1997. Natural Capitalism. *Mother Jones.* March April: 40–45

Jasanoff, S. 1993. Bridging the two cultures of risk analysis. *Risk Analysis* 13: 123–129.

Kopp, R. J., A. J. Krupnick, and M. Toman. 1997. *Cost-Benefit Analysis and Regulatory Reform: An Assessment of the Science and Art.* Discussion Paper 97-19, Washington, DC: Resources for the Future [Downloadable from www.rff.org].

Latai, D. 1980. *A Risk Comparison Methodology For the Assessment of Acceptable Risk.* Disertation. Cambridge, MA: Massachusetts Institute of Technology.

Lave, L. B. 1981. *The Strategy of Social Regulation: Decision Frameworks for Policy.* Washington, DC: Brookings Institution.

Lowrance, W. W. 1976. *Of Acceptable Risk: Science and the Determination of Safety.* Los Altos, CA: William Kaufmann.

Marsick, D. J., and D. M. Byrd. 1990. A Review of Information Sources for Right-to-know Purposes. *Fund. Appl. Toxicol.* 15: 1–5

National Research Council. 1975. *Decision Making for Regulating Chemicals in the Environment*. Washington, D.C.: National Academy Press.

National Research Council. 1996. *Understanding Risk: Informing Decisions in a Democratic Society*. Washington, D.C.: National Academy Press.

National Research Council. 1982. *Risk and Decision Making: Perspectives and Research*. Washington, D.C.: National Academy Press.

National Research Council. 1983. *Risk Assessment in the Federal Government: Managing the Process*. Washington, D.C.: National Academy Press.

National Research Council. 1997. *Preparing for the 21St Century: The Environmental and Human Face*. Washington, D.C.: National Academy Press.

Nature Conservancy. 1997. *Americas Least Wanted: Alien Species Invasions of U.S. Ecosystems*. Arlington, VA: Nature Conservancy.

Presidential Commission on Risk Assessment and Risk Management. 1997. *Risk Assessment and Risk Management in Regulatory Decision-Making*. Vol. 2, Washington, DC: Office of the President.

Raiffa, H. 1970. *Decision Analysis: Introductory Lectures on Choices under Uncertainty*. Reading, MA: Addison-Wesley Publishing Company.

Rosenbaum, W.A. 1991. *Environmental Politics and Policy*. Washington, DC.: CQ Press., A Division of Congressional Quarterly, Inc.

Russell, M., W. Colglazier, and B. E. Tonn. 1992. The U.S. hazardous waste legacy. *Enviroment* 34:12-5, 34-9.

Shrink, E. 1988. New dimensions in ethics. *J. Value Inquiry* 22:77-85.

Tsengs, T. O., M. E. Adams, J. S. Pliskin, D. G. Safran, J. E. Siegal, M. C. Weinstein, and J. D. Graham. 1995. Five-Hundred Life-Saving Interventions and Their Cost-Effectiveness. *Risk Analysis* 15: 369–90.

Weinstein, M. C. 1979. Decison making for toxic substances control. *Public Policy* 27(3): 333–383.

Westra, L. 1994. *An Environmental Proposal for Ethics: The Principle of Integrity*. Lanham, MD: Rowman and Littlefield Publishers, Inc.

Wilson, J. D., and A. Holbrow. 1997. *Default and inference options: Use in recurrent and general risk decisions*. Discussion Paper 97-46, Washington, D.C.: Resources for the Future.

Chapter Twelve

Risk Communication

Introduction to Risk Communication

If all else fails, a spectacular failure often insures lasting fame.
—John Maynard Keynes

Having assessed a risk and decided what you want to do to manage it, you will often find your efforts incomplete. You have made yourself an expert in these other areas, but now you have to communicate what you have done and why you have done it. You will have to interact with various groups that lack your in-depth knowledge about the risk.

If you don't explain yourself, the public will ignore your solution. In some situations, the public is not involved or a lack of public involvement is unimportant. For example, you may analyze a risk in a chemical process at a plant and find that the workers have a good technical grasp of what you have done. For the most part, however, public involvement is crucial.

Our society mostly depends on voluntary compliance. When a stakeholder group involved in a risk scenario disagrees with your risk management plan, your ability to communicate to all of the stakeholders and the public in order to obtain compliance becomes crucial. If only few understand, only a few will care, and little will change. Most importantly, some involved groups, perhaps even the general public, may know something you do not. You will have to understand their perspective, and perhaps even revise your assessment and management plan accordingly.

How do you explain a regulatory risk assessment? Our perspective on risk communication is fairly easy to understand. First, many people perceive risks differently than risk assessors do. The factors involved in risk perception are open to scientific study through psychometric research, and a substantial body of information is now available to the regulatory analyst to understand how and why experts often will view risks differently from the public. Second, risk communication is a skill that you can learn, just like playing tennis. Sufficient practical guidance is available to enable you to get started by doing some reading. Third, good risk communication is difficult, and your attitude toward it will be crucial to your success. If you take it seriously, anticipating hard work ahead, you probably will succeed. If you think that risk communication is nonsense and not worth any effort, you probably will fail, perhaps dramatically.

Try to keep four factors in mind when attempting to communicate risk: trust, benefit, control, and fairness. You cannot succeed with any group until you establish some level of trust. It may take time and thoughtful planning, but obtaining some trust will be well worth the effort. It may even save you time in the long run, because you will not have to keep explaining yourself to hostile audiences.

Some of the prerequisites to trust are honesty, consistency, and a willingness to change your mind as you learn. The easiest approach is to tell the truth, as you understand it, patiently and consistently, even when a little spin or bending of the truth would make your efforts much easier. In the regulatory arena, risk analyses have long histories, and an earlier falsehood, exposed later, will poison risk communications efforts.

In contrast, consistency and flexibility are more difficult to achieve in a regulatory process, and they even can conflict with each other. Having learned new information, if you make a major change in the risk assessment or risk management plan, opponents of the new decision will trumpet your apparent "inconsistency." Clinging to a poor risk management plan to maintain consistency in the face of new, contradictory information will generate even more problems. Preparing to explain changes by keeping a careful, open, and thorough history will help some, but the tension between consistency and flexibility will tend to undermine trust.

Try to understand what a group has to gain or lose in the relevant risk scenario. What benefits them? What are their concerns? Because our society values individual rights, it places control in the hands of individuals. Reasonably, control is important to them. They cannot depend on "big brother" to solve their problems through central planning. If the assumption of risk is involuntary, important stakeholders lose control, and you will have a considerable problem to overcome.

Our society also strongly values fairness. If a group does not view the distribution of risk as equitable, you will face another difficult communications problem. For example, if a plant generates products of great benefit to society, but the risk of operating the plant falls on a few workers, you can anticipate their skepticism and their unwillingness to go along with the conclusions from your analysis.

Different definitions and different points of view plague the field of risk communication. The general problems of ambiguous terminology and "fuzzy" meanings that we outlined earlier become even more acute in this area. From our point of view, the regulatory analyst usually wants to interact with the general public but lacks appropriate skills and experience. Often communications will take place indirectly—for example, through the media or elected representatives.

Adding even more complexity to the process, bureaucracies often are involved. Governmental communications channels often involve specialists, creating secondary problems of accurate information transfer within an organization. If different people carry out the functions of risk assessment, risk management, and risk communication, miscommunication between these people means that the public cannot possibly understand what is going on.

Our working definition is that regulatory risk communication involves a two-way exchange of information that meets the interests and needs of both senders and receivers. If the process is to work well, both senders and receivers must participate actively.

Before the 1950s, risk communication amounted to informing audiences; it was a one-way process largely practiced by public relations specialists. The difficulty that the professionals in this group have in the health and safety area is that they think they can solve any problem by "spinning" the message. Thus, they thought that the details of risk assessment and managerial decision making were mostly irrelevant to risk communications. This approach often failed spectacularly. We have a diametrically opposite view: before you can communicate risks, you need to understand them; before you can communicate regulatory decisions, you need to understand how these decisions were made.

By the 1960s, the risk analysis community began to learn about the importance of allowing the audience to be heard, although we really did not listen well. By the 1980s we generally realized the importance of consensus building, a two-way process in which the involved groups listen to the other. The listening process in risk communication is not just passive. It involves working hard to hear what others are trying to communicate. By the 1990s, risk communications skills had generally gotten much better. Some firms now even specialize in this area. Unfortunately, analysts and stakeholder groups are burning out from too much information and overly persistent involvement with some problems, such as dioxins.

The field of risk communication can be organized in different ways. One way is to divide it topically into areas like finance, environment, safety, health, or transportation. Other ways to organize the topics are by care, crisis and consensus communication, perception and decision making, and the audience (media, public, scientists, government regulators, industrialists, and so on).

Knowing the purpose of your communications effort is important. In planning a risk communication effort, you need to consider your resources, including time, space, funds, personnel, data, and the quality of the preceding steps. In many cases, regulatory and institutional policies will limit your efforts. Finally, you need some awareness of your audience, such as their background knowledge, preferred means of communication, and cultural values.

Try to avoid making promises that you may not be able to keep. In government agencies, it is an all too common practice for a manager to send a representative out to meet with a stakeholder group and then veto the representative's commitments. Do not agree to provide information, meet again, or broker changes, unless you command these internal processes. Your audience will not find your deference to some higher level in the organization very convincing. Instead, such statements indirectly communicate that you represent an organization that is not really involved. This cost is much less, however, than the long range consequence of not meeting commitments. Betraying promises will break any trust you have established and can create overt hostility in your audience.

Risk Perception

The way you present a risk estimate partially communicates values. If you state that the risk of cancer, given some management option, will be 10^{-6}, you communicate something different from saying that the average decrease in life expectancy will be six seconds per person, or saying that 260 people will die of cancer every year. Yet, we learned earlier in this book that these three expressions are all risk equivalent. They represent exactly the same risk.

Even when you express estimates of different risks in the same way for purposes of comparison, the public often perceives them differently than experts. Beginning in the 1970s, Slovic, Fishoff, and their colleagues undertook experimental studies to understand the basis of these perceptual differences. They now have provided the risk analysis community with an extensive body of descriptive information.

Governmental organizations like EPA are constantly torn between the rankings of risk from technical experts and the perceptions of the public (see Chapter 9). Congress is elected to represent the public and does so in the laws they pass. However, in developing regulations mandated by those laws, regulatory agencies endeavor to find a balance between technical estimates of risk and those perceived by the public. Knowledge of both is important in developing regulations. An example of this problem is the Superfund program, which is perceived by the public as addressing one of the most serious environmental problems. Most risk analyzers place it much lower on the scale of importance based on risk estimates.

If you are going to engage in risk communication, an important principal to remember is that the audience may understand the risks differently than you do. We do not mean that your perception, based on a technically correct risk assessment, is somehow better or more important than the public's perception. That would imply a value judgement on our part about whether your understanding is intrinsically better. Psychometric studies show that the public often brings additional information into play, not less. You do need to understand that a difference in perception is highly likely. If you fail to acknowledge this difference, your communications efforts probably will fail.

Often, simple approaches to risk perception illuminate the public-expert difference. For example, Littai and Rasmunsen (1980) categorized the risks of various technologies in terms of a group of dichotomous factors (see Table 12.1).

Table 12.1 Risks of various technologies in terms of a group of dichotomous factors (from Littai and Rasmunsen, 1980)

FACTOR	VALUES	
*Origin	Natural	Man-made
*Volition	Voluntary	Involuntary
*Manifestation	Immediate	Delayed
*Severity	Ordinary	Catastrophic
Controllability	Controllable	Uncontrollable
Mitigation	Practical	Impractical
Reversibility	Reversible	Irreversible
Manageability	Manageable	Unmanageable
Benefit	Clear	Elusive
Benefit	Enrich	Impoverish
Exposure	Continuous	Occasional
Exposed group	Large	Small
Necessity	Necessary	Unnecessary
Alternatives	Available	None
History	Old	New
Familiarity	Common	Unusual/Novel
Susceptibility	Average person	Sensitive subpopulation
Involvement	Occupational	Environmental
Triggering	Natural	Man-made
Identification	Statistical	Specific
Society	Developed country	Third world
Economic setting	Wealthy country	Poor country
Political setting	Democratic	Totalitarian
Social group	Civilian	Military
Circumstance	Normal	Emergency
Concern	Physical	Societal

They found that they could classify many technological risks into similarly perceived groups with just four factors: origin, volition, manifestation, and severity. Because each factor was binary, they created 16 groups in theory, but six contained no technology. Lay groups tended to regard the technologies in each group similarly. For example, consider the following causes of death:

(1) immediate, voluntary, man-made, catastrophic risk: airplane crashes;

(2) immediate, involuntary, man-made, catastrophic risks: dam failures, sabotage, nuclear energy;

(3) immediate, voluntary, man-made, ordinary risks: sporting accidents, car crashes, surgery;

(4) delayed, voluntary, man-made, ordinary risks: smoking, saccharin;

(5) immediate, involuntary, man-made, ordinary risks: homicides, crime;

(6) delayed, involuntary, man-made, ordinary risks: food additives, pesticides, coal energy, industrial pollution;

(7) immediate, involuntary, natural, catastrophic risks: earthquakes, hurricanes, epidemics;

(8) immediate, involuntary, natural, ordinary risks: lightening, animal bites, and

(9) delayed, involuntary, natural, ordinary risks: chronic diseases.

An analyst trained in risk assessment might well calculate the lifetime risk of death from each of these sources and rank them according to magnitude. Doing so will miss social attitudes toward them, and correspondingly, cause the analyst to have a misunderstanding of the audience when going into a session that includes the public.

Most importantly, because the risk perception studies have an empirical, scientific basis, they played a key role in persuading the overall risk analysis community that risk communication is an important component of the overall process.

Fulton's List

Fulton developed a set of 'musts' and 'shoulds' that provide a good idea of the requirements of risk communication:

- A communication message must be concise. For example, the conclusion should be stated in 12 to 15 words, and not contain shoptalk or jargon.
- Your message must be positive. Negative connotations divert attention and discourage action.
- Your message must address underlying concerns, or the public will think that you are not listening.
- A successful message also must be repeated. A well-known prescription is to tell the audience what you are going to tell them, tell them, and then tell them what you told them.

- Your message also must provide a source for more information. Some individuals will want to know more in order to follow up on suggestions.
- Your message must be in plain English. Most newspapers write at a sixth grade level; this is a good level for you to ensure that your message will transmit.
- The risk communication message should be memorable. If your audience has to write it down, you probably did not give them information that they will remember.
- Your message should include analogies. Anything that can relate the message to something the group understands will help them. Similarly, a successful message will involve third-party references. This broadens the scope, gives a better perspective, and starts to build trust. Thus the message should be personalized. If the listener does not realize that this message involves them personally, it will not impress them.
- Your message should be qualitative and not numerical. Millions, billions, orders of magnitude, and Greek prefixes seem designed to confuse the general public. Also, try to avoid vague terms like big, small, inconsequential, minimal, etc. Different groups may interpret them differently. Speak about the substance in terms of your concern for the health and welfare of the listener and do not make them feel ignorant with numbers or jargon.
- Your message should include stories and anecdotes. A personal element often will be the most remembered part of your message. (An effective sermon usually has a similar personal touch.)
- Your message should acknowledge major uncertainties. All risk estimates involve uncertainty. The more you acknowledge them honestly, the more the audience will trust you.

The Seven Cardinal Rules of Risk Communication

Vincent Covello and Derry Allen developed seven rules that also help to explain the range and quality of risk communication (1988):

(1) Accept and involve the public as a legitimate partner; they have a right and a responsibility to participate. Both the risk communicator and the public have rights and responsibilities. We are responsible for communicating all that we know, for building trust and providing opportunities for public input to the decision-making process. It is most important for us to listen carefully and actively to the public. They will only continue to be involved in the process in a positive way as long as they know that they are heard. The public has the right and responsibility to be an active part of the process and they have a right to be heard. They have a responsibility to listen and learn enough of the technical details to be conversant and to make a valid decision.

(2) Plan carefully and evaluate your efforts. The more carefully you plan, the better your communication efforts will be. Later in this chapter we will discuss some of the ingredients in the plan. They should include gathering information about the audience, deciding on the types of communications (written, oral, or visual), anticipating problems, and evaluating the process. Some wide ranges of 'publics' exist, each with different interests, needs, concerns, priorities, and preferences. Also, they are part of a wide variety of different organizations. All of these must be addressed for the most effective communication, even though they require different goals and strategies.

(3) Listen to the public's specific concerns. Active listening involves focusing on what the public is saying, hearing their concerns (especially their emotional concerns and perceptions), and reflecting to be sure that we understand what they are saying. Trust, credibility, competence, control, voluntariness, fairness, caring, and compassion are more important than the details of quantitative risk assessment.

(4) Be honest, frank, and open. This is the best and only way to build trust and credibility. Make sure that there are multiple ways for the public to give and get information. Provide open forums, leaflets, pamphlets, press releases, as well as advertisements on TV and radio. It is important to provide information about where more information can be obtained. Hotlines are useful in being perceived as open.

(5) Coordinate and collaborate with other credible sources. Third party references are essential for trust building. When the public accepts and trusts someone who agrees and supports your facts, information, and technical details, the public will begin to trust you by association. Build bridges, consult with others, and issue communications jointly. Possible local sources include clergy, physicians, university scientists, and local officials.

(6) Meet the needs of the media. Be aware of reporters' deadlines and prepare advance background materials for them in press kits. It is important to follow up on stories that praise or criticize.

(7) Speak clearly and with compassion. Do not try to say too much. Limit your points to those that are truly important. You will not satisfy everyone; however, when people are sufficiently motivated, they are quite capable of understanding complex risk information.

Other Approaches to Risk Communication

Lundgren suggested a different way to approach risk communication. She uses three levels, relating to care, crisis, and consensus (1994).

Care communication involves situations in which research has already determined risk and a way to manage it. The audience has accepted the assessment and plan. In the health care area, smoking and AIDS are examples. The main goal of care communication then becomes to inform and advise the audience. For example, risk communication provides information to choose among a variety of preventive and mitigative measures.

Crisis communication applies in situations involving extreme or sudden dangers, like emergencies and accidents. These situations include tornadoes, hurricanes, pending dam collapses, and influenza outbreaks.

Consensus communication involves informing and encouraging groups to work together to reach a decision. This involves audience interaction where the goal is to conform and not necessarily to persuade.

Lundgren's approach is useful because it focuses the communication process on the consensus part where most of the controversy, conflict, and dissension is.

Public outrage is another component of risk that you need to recognize in your communications efforts. Sandman (1987) has popularized his observation that risk = hazard + outrage. Outrage includes such factors as catastrophic, dreaded, fatal, memorable, invisible, involuntary, uncertain, uncontrollable, undetectable, unethical, unfair, unfamiliar, and unnatural risks. It also involves untrustworthy sources of information.

Thus emotional, perceptional, and gut level themes lead to outrage. When it comes into play, the audience may direct frustration and anger at you in direct and personal ways. These situations can be very unpleasant. The potential for outrage is an important reason why some risk analysts try to avoid risk communication.

The simplicity and elegance of Sandman's observation are some prime examples of the kind of message that Fulton advocates. It has the same characteristics as above. We are dealing with feelings and emotions.

Richard Peters tested various determinants of credibility in environmental risk communication in empirical studies that established preconceptions that the public has about organizations and their representatives. The results were expressed as stereotypes. Thus, the public typically believes that industry has expertise but lacks concern and care. The public also believes that government has concern but lacks commitment and involvement. The public usually believes that advocacy groups lack knowledge and expertise. Peters suggests that a good risk communication program will counter the public's stereotypes. Thus, a government representative should try to demonstrate commitment and involvement to improve credibility. Reinforcing a stereotype often is the worst thing to do. An industry representative who initially exhibits a lack of concern should expect to have severe problems gaining public trust.

Public Interactions

For members of the public, perception is the same as reality. In communicating with the public, we thus need to tune into these perceptions, acknowledge them, and include them in any communicating we do. Some important perceptual components of risk are natural versus man-made, voluntary versus involuntary, ordinary versus catastrophic, immediate versus delayed, continuous versus one time, controllable versus uncontrollable, old versus new, clear versus unclear, and necessary versus luxury.

For example, if a risk is natural, such as from indoor levels of radon gas, it is considered to be less risky than if it involves radioactivity from a man-made source, such as a nuclear power reactor. Generally speaking, the public will perceive a naturally occurring risk as one to two orders of magnitude less risky than an equally risky man-made risk. Similarly, the public will accept a voluntary risk such as smoking cigarettes, hang-gliding, or bungee jumping more readily than they will accept an involuntary risk such as passive smoking, being exposed to toxic fumes from an industrial site, or toxic contamination in the drinking water. Control is an important dimension here. In our modern age, many feel less in control of their lives and any sense of uncontrollability is a serious threat.

Morals and values are important to the public. Most consider pollution to be morally wrong, no matter what the cost. However, many have a limit to what they will individually pay. Values of fairness, equity, honesty, and safety are important to many. In risk communication situations, we need to listen to these ideas and address them.

The public finds dealing with uncertainty a major obstacle. To them it is either risky or safe. Risk communicators have an important role that includes helping the public understand that a risk can be overestimated to err on the safe side when the uncertainty is large.

Two of the most important issues to be addressed with the public are trust and control. The first objective of a risk communicator is to establish trust. Without this trust, any message will likely be ignored. Our society values control, and its loss is a grave threat. These issues must be addressed and met to start any communication process.

Risk decisions are more likely to succeed when the public shares the power and the responsibility. It helps to involve them in the decision-making process. Additionally, there have to be feedback routes from the public; they have to have both actual openness and a perception of it.

The key to explaining risk information is motivation on the public's part. When motivated, they will participate, learn, and enter the decision-making process in an active, supportive, and positive way. Without this motivation, the process will likely fail.

Risk communication is easier when the emotions of the public are seen as legitimate. This can be accomplished by acknowledging them and accepting them as real, important, and involved in the process.

There are particular words that should be avoided, and there are others that are helpful in communicating with the public. Table 12.2 contains a partial suggested list of both.

There are many ways to communicate with the public, including direct communication, indirect communication, and information gathering techniques.

Table 12.2 Lists of negative and positive words in communicating with the public

Bad Connotations, Bad Sounding or Negative Words		
absolutely	estimation (under- or over-)	poisonous
acute	evaluation	precarious
always	extrapolate	probability
anxiety	fear	provocation
average	guesstimate	radiation
below		
regulatory		
concern	hazard(ous)	regulations
carcinogen	insecurely	risky
causality	interpretation	scientific
controversy	judgement	statistical
chancy	law suit	statistics
cost	lifetime exposure	technical
danger	maximum-tolerated dose	toxic
decision	minimal risk	uncertainty
de minimus	never	unsafe
dispute	nuclear perilous	virtually safe dose

Good Connotations, Good Sounding or Positive Words		
acceptable	health	strength
admissible	intact	threshold
assurance	no effect	tolerable
benefit	predictable	truth
choice	prevention	trust
common sense	reliable	unforced
dependable	responsible	voluntary
faith	safety	willing
free	security	zero risk
freedom	stable	

(1) Direct Communication Techniques:

briefings	guest speaking	open houses
brochures	handbills	personalized letters
direct mailings	information fairs	purchased advertising
door-to-door visits	information hot line	slide shows
drop-in center	mobile office	telephone
fact sheets	newsletters	videos
flyers	newspaper inserts	volunteers

(2) Indirect Communication Techniques:

feature stories	press conferences and interviews
guest editorials	press kits
news releases	public service announcements

(3) Information Gathering Techniques:

advisory group	
brainstorming	focus groups
information contact person	
interview community leaders	
and key individual	door-to-door surveys
mailed surveys or	
questionnaires	open forums
	telephone surveys

Following are some of the common myths of risk communication that we hear all too often from communications naysayers:

(1) We don't have enough time and resources to do risk communication.

(2) Communicating with the public about risk is more likely to alarm people than keeping quiet.

(3) If we could only explain risks clearly enough, people would accept the information.

(4) We shouldn't go to the public until we have complete solutions to problems.

(5) These issues are too tough for the public to understand.

(6) Technical decisions should be left in the hands of the technical people.

(7) Risk communication is not my job.

(8) If we give this group an inch, they'll take a mile.

(9) If we listen to the public, we will devote scarce resources to issues that are not a great threat to public health.

(10) Activist groups stir up unwarranted concerns.

(11) If we respond to this statement, we will give it credibility. If we just ignore it, the issue will disappear.

Risk Communication Plans

Many risk analysts are surprised when they first learn that expert risk communicators have structured plans. Many conceive of the communications process as offhand and spontaneous. Thus, the idea of a systematic process that justifies commitments of time, money, and personnel to specific objectives can come as a revelation. Some also think that the communications process is too indefinite and murky to attempt any planning. Our view is diametrically opposite. Planning is most needed when the environment is most uncertain.

The main components of a risk communication plan are the objectives, the type of audience, the proper media channels, the issues, the available resources, and a timeline.

Many factors contribute to the objectives of a communications plan. Regulations or laws may compel the communications, and their requirements may shape the plan. Your organization will have policies, public relations procedures, and sensitivities to certain stakeholder groups. The objectives must include these considerations. In addition, the nature of the risk may shape the communication objectives. Is this a crisis, care, or consensus situation? Does the situation require time to establish trust and provide information for the public, or is the risk well known? The newness and visibility of the risk may become factors in the plan.

Knowing your audience is essential. Ideally you may have time to analyze the linguistic, cultural, educational, social, and psychological character of your audience. Factors that are important include age, gender, experience with the risk, trust level, jobs or occupations, and how the risk specifically affects them. There are several ways to gather this information. The best way is through face-to-face conversations or interviews and surveys. Government and local organizations such as the scientific or medical media can supply this kind of information.

Reading the local newspaper's op-ed columns as well as listening to local talk shows on radio and TV may be useful. Other sources include staff sociologists in large organizations. Sometimes market analysis information is available from polls taken by organizations such as opinion research organizations. Other sources include government dockets and environmental impact statements.

Different situations require different media for providing information. The main media to select from include written, oral, and visual.

Written messages can include a lot of data and information and are relatively inexpensive. However, they can be difficult to comprehend, contain jargon, or be overly technical. Examples of written messages include fact sheets, newspapers, newsletters, and pamphlets.

Oral messages can be used if the presenter is trusted, knowledgeable, and credible to the audience. However, these messages can be easily misunderstood. This medium is more flexible and can fit into nearly any schedule. For long term efforts, you will need a series of presentations.

Visual messages include graphical presentation, displays, direct advertising, tours, demonstrations, videotapes, and TV.

An emerging media is the Internet. This is new and some find it intimidating. It also only reaches a select audience—mainly young males.

It is essential to identify the controversial aspects of any risk communication before starting. Problem areas should be carefully researched and a strategy for dealing with controversial issues should be developed before any communication is attempted, especially for large public meetings.

Other important components of a risk communication plan are the resources needed (personnel and cost) and the timeline. These factors often limit the amount of communication that can be done.

Risk communication is not negotiation, arbitration, or mediation. Risk communication is listening to the other side, letting them know you are listening, and acting on that knowledge to work towards understanding, consensus, and finally action. When the risk communicator misunderstands the transmission and receiving role, and instead tries to become independent of the process, to shape or control it, many things will go wrong.

Some General Tips

Do not underestimate the shock that comes when you first get thrust into a public arena. If you acquire risk communications responsibilities, find out early in the process what kind of media training your organization will provide. People who think of themselves as good speakers often misperceive how they will sound or appear on camera. Many organizations afford opportunities to build public speaking skills by sponsoring organizations like Toastmasters.

While practice in private is a good idea, it cannot substitute for exposure to live audiences. Once you gain the attention of others, you will become self-conscious, and paradoxically, the very distraction of your self-awareness can befuddle any efforts to respond to your audience. As an example, many speakers can achieve a smooth delivery in private, but when others are listening, they sprinkle their speech with frequent phrases like "and-uh" that interrupt the listeners' chain of thought. Afterward, the listeners may attribute qualities like lack of knowledge and inexperience to the speaker.

In situations where there is a low level of trust and a high level of concern, body language can be as much as 75% of the message delivered. In presenting, you will do well to stand erect with your hands at your side. Try to gesture naturally. Avoid swaying and signs of nervousness such as foot tapping, clenched fists, or touching your face. Do not nod your head, especially when listening, as it may send the wrong signal. Practicing your presentation in front of a mirror or in front of critical colleagues will help.

In situations of low trust and high concern, humor is inappropriate; empathy and caring are important. A good management of time would use a ratio of between one to one and three to one of presentation time to open discussion. In this case, a third party endorsement is essential. Try to involve some local, well-known person who is trusted.

Constraints on Risk Communication

Many constraints can interfere with risk communication. These include both message (time and place) and source (credibility) problems.

The most common constraints are the organizational ones. In most organizations, there are scarce or inadequate resources such as funding and staff. All too often risk communication gets a low priority on the list of needs. Review and approval procedures delay and sometimes prevent successful risk communication. These may be overcome in a crisis situation where it can be shown that there is community support. Sometimes the structure of the organization can be a constraint—for example, when technical information must go through the public affairs section or the legal group, which thinks only positive and vague information should go out and therefore important details are omitted. Insufficient information is a perennial problem. If information is given out and later shown to be wrong, it can destroy trust. However, if you wait too long before releasing information, the public will be suspicious that you are hiding something.

Deficiencies in scientific understanding, data, models, and methods resulting in large uncertainties in risk estimates are a particularly difficult constraint. Uncertainty involves a large variety of problems that contribute to complexity (see Chapter 2). To the general public, uncertainty means we do not know. They are suspicious of ranges of numbers.

Disagreements between scientists can constrain delivery of the correct picture. Often managers ask for "one-handed scientists." The problem is to give the full picture with the information needed without overloading the audience or confusing them.

If an audience is angry or feels guilty, this and related emotions must be acknowledged and dealt with before any scientific and technical information can be received or any decision making can be entered into.

Regularly we read in newspapers about the technical or scientific ignorance of the public. This can lead to the erroneous conclusion that they cannot understand the science. When motivated, most audiences are capable of understanding enough of the scientific and technical detail to make a responsible decision.

Hostile audiences present a particularly important and common constraint. Components of the approach to such groups are important. First you establish some ground rules, such as setting a very short amount of time for speaking and then taking questions and comments. Ask the audience to speak into the microphones on the floor so everyone can hear. If you do not have a microphone, it is hard to argue with the one who does.

Often it helps to require the audience to submit written questions. This can temper the anger and also provides a copy of the questions for further study and use. Another ground rule that is helpful is the use of a moderator, especially of a person who is trusted, is a good communicator, and is knowledgeable about the topic.

Another suggestion in dealing with a hostile audience is to stay calm, even if they don't. Most important are these three things; listen, listen, and listen. It is important not to interrupt any speaker. Be caring and sensitive, and let the audience know that you are listening to them. Active listening is an important skill for dealing with hostile audiences. The idea is to relate personally, build trust, and be open. Attack the problem, never the person.

Communications Media

Several aspects must be kept in mind when attempting to provide risk communication information for the media. Remember that a risk is seldom a big story. The media is looking for news, not education. They are trained to ask where, what, when, why, who, and how. Unless the story is news and exciting, it is likely not to make it past their editors. In general, politics is more newsworthy than science.

Reporters cover viewpoints, and are not seeking 'truths.' They will try to get at least two opposing viewpoints, even if one represents a slim minority. Reporters, in response to public demand, will simplify the story by asking whether or not the source of risk is considered hazardous or safe. Further, they will personalize the risk. For example, would you let your family live there? This approach is just the opposite of a scientist's training, which is to keep oneself out of one's research and speak in the third person. Reporters often do their jobs with limited time and expertise (often environmental reporting is added to the reporter's main job).

Most individual risk stories are about blame, outrage, fear, anger, and other nontechnical issues, and little space is devoted to the risk. The technical information included usually has little, if any, impact on the audience. Reporters mainly use official sources, adding industry or activist experts when needed.

The motivation of reporters is to get their stories past their editors, who are mostly concerned with what will sell newspapers. The reporters themselves are trained to use mostly six types of questions (who, what, when, where, how, why).

Some Common Traps

In serious situations of risk communication, humor should, in general, be avoided. It is tempting to lighten the situation, but this usually only makes things worse. Avoid irony, sarcasm, cartoons, and even laughing at a preposterous question. Take it all seriously.

Repetition of a negative, even when refuted, will usually result in more support and reinforcement of the negative idea. A negative idea is three times more powerful than a positive one, and will often harm your position rather than supporting it. As pointed out earlier, it is important to avoid negative words and phrases like "no," "not," "never," or "can't."

Also try to avoid hedging words like "maybe," "perhaps," "I think," or even long pauses. These kinds of vacillating phrases make it appear that you lack knowledge, although this may not be the case. Try to present a positive, upbeat mood.

Jargon is a common trap among government employees. Acronyms, such as "DOD," "OPPTS," "RCRA," and words and phrases such as "remediation," or "ten to the minus sixth" put the audience on the defensive. Inadvertently, the speaker is communicating that membership in a special, expert group is required to understand the subject and the audience is ignorant. Try to avoid making a statement like "leaching is migrating off the site into the groundwater and is safely dispersed via a plume distribution."

Comparing risks seldom provides useful communications information. To say that the risk that concerns your audience has approximately the same probability as being hit by lightning, getting stung by a bee, or being hit by a meteor has little meaning and suggests that you think their concerns are trivial. Recall that your audience already has a stereotype of you—perhaps that you lack involvement. If you feel you must compare risks, try to compare similar risks. For example, compare the risk of lung cancer from smoking cigarettes to the risk of lung cancer from radon.

Summary of Risk Communications

Risk communication is one of the three major elements of risk analysis. After you assess a risk and decide how to manage it, you need to communicate this information with other groups. The important aspect is two way communication, which requires both sides to listen to each other. Only by working together can good communication be achieved and the goals of all parties involved be met. In general this is a continuous process and involves patience, endurance, and hard work.

References

Cohn, Victor. 1990. *Reporting on Risk: Getting It Right in an Age of Risk.* The Media Institute, Washington, D.C.

Cothern, C. R. 1990. Widespread Apathy and the Public's Reaction to Information Concerning the Health Effects of Indoor Air Radon Concentrations. *Cell Biology and Toxicology* 6:315–322.

Cothern, C. R., Ed.1996. *Values, Perceptions and Ethics in Environmental Risk Decision Making*. Boca Raton, FL: Lewis Publishers.

Covello, V. T. 1992. Trust and Credibility in Risk Communication, *Health and Environment Digest* 6:1–3.

Covello, V. T., and F. Allen. 1988. *Seven Cardinal Rules of Risk Communication*. Washington, DC: USEPA, Office of Policy Analysis.

Fulton, Keith. Fulton Communications, Houston, Private Communication.

Harrison, E. Bruce. 1992. *Environmental Communication and Public Relations Handbook*, (2nd). Rockville, MD: Government Institutes.

Lundgren, R. 1994. *Risk Communication: A Handbook for Communicating Environmental, Safety, and Health Risks*, Columbus. OH: Battelle P.

National Research Council. 1989. *Improving Risk Communication*. Washington, DC: National Academy Press.

Sandman, P. M.. 1987. Risk Communication: Facing Public Outrage, *EPA Journal*, 13:21–22.

Chapter Thirteen ....

Case Studies

Introduction

We have chosen several case studies to present real-world examples to illustrate the problems encountered in applying the principles of this book. We discuss here some of the features of each case study to show our rationale for selecting these particular examples.

Ecological Risk Assessment—Acidic Deposition

This is an example of a risk that mainly affects the ecological world and has only limited effects on human health and welfare. This example considers regions showing effects primarily in the northeast U.S. and a few other areas. We show that acid deposition has the character of a broad, general, and qualitative risk analysis. There is some quantitative information, but as is characteristic of ecological risk, we do not have an exposure-response curve.

Arsenic and Cancer

Our arsenic example illustrates a nonlinearity in the exposure-response curve. This nonlinearity has important consequences for the regulation of drinking water and other sources. The occurrence in the U.S. is limited, but there are serious health problems due to significant exposure for the drinking water supplies in many other parts of the world. We show that this is an example of a single contaminant that has numerous health endpoints. There are both cancer and non-cancer health endpoints.

We find this to be an interesting example because there are a lot of human epidemiology data, but animals seem to be much less sensitive. This has important implications for the default assumption for other contaminants where it is assumed that if animals are not affected, then humans must be similarly unaffected. The mode of action in humans is unclear for arsenic exposure. In addition, arsenic may be an essential element. If that is so, there can be a risk of having too much arsenic or having too little.

Electromagnetic Fields

We include this case study because it provides an example of a contaminant that we are all exposed to but where the health risks, if they exist, are very small. This is also an example of a controversial communication problem in that the main potential evidence is for childhood leukemia. The emotional dimension of it affecting mainly children clouds the risk assessment and creates risk communication problems. It is an example of the difficulty of proving that something does not exist. This case shows the statistical approach of pooling the existing epidemiology studies using a technique called meta-analysis.

Environmental Tobacco Smoke

Here we have another example of a contaminant to which we are all exposed. It has the complexity of being a mixture of many individual contaminants. We have direct human data in 30 epidemiology studies. There are both cancer and non-cancer health endpoints. Finally, we show that the estimated health effects for lung cancer can yield risks in the thousands for a lifetime. Besides the epidemiology studies, there is the surrogate of active smoking, where the data is even more extensive.

Indoor Air

We include this example because its complexity makes most of the normal approaches to risk assessment difficult. Because of the mixtures of toxic fumes and biological contaminants, the large range of sensitivities in humans, and the wide range of health complaints in sick buildings, the normal approach of monitoring for contaminants does not work. The best approach is to take detailed information from those suffering from a range of ailments. Thus we find that the quantitative approach does not work well here. To make matters worse, there are few cures. The best approach is to determine what the cause is and eliminate it. Over half of the cases reported so far involve inadequate ventilation—an easy matter to address.

Radon

Extensive information exists about the extent to which lung cancer results from exposure to radon. There are over 17 studies of hard rock miners and epidemiology studies for lung cancer resulting from exposure to radon. Many of these have produced exposure-response curves. However, there are none for exposure levels commonly found in homes. Thus there are lots of data at higher levels (only an order of magnitude or so above home levels), but none where the levels are those found in homes. This is an example of the difficulty of developing suitable risk estimates to support the regulation of ambient exposures. We examine here the policy assumption of a linear, non-threshold exposure-response curve. The estimated risks are quite high (tens of thousands per lifetime), but the public reaction is widespread apathy.

Case Study: Ecological Risk Assessment—Acidic Deposition

You will find several case studies of ecological risk assessment in the following volumes: USEPA (1993 and 1994), Cairns (1992), and Suter (1993). We will examine one of these here that involves acid deposition. This example is based on a discussion in USEPA (1993), and the references are listed there if you wish to continue in more detail.

Introduction

Acid deposition is a phenomenon that directly affects living things other than humans. Although we are affected, the main effect is on the ecology. We have the results of many research papers, but the primary source of information is a ten-year research program called the National Acid Precipitation Assessment Program (NAPAP). The goal of this extensive study was to assess the effects of acid deposition on the environment. The National Surface Water Survey (NSWS) determined the percentage and extent of lakes and streams that became acidic or were potentially susceptible to this acidic deposition. We will find this a useful example because there is a wealth of information. Although there are extensive data, the character of the risk assessment is different from those you might find in human health risk assessments. The main difference we will encounter is that there are no individual or population quantitative risk estimates.

In our language, the stressor is acidic deposition. This deposition takes the form of both wet and dry deposition. Although the acid content of lakes and streams may come from other sources, we will focus on the deposition from industrial sources, primarily from fossil fuel combustion.

The NAPAP research program assessed adverse effects on aquatic systems, forests, agricultural crops, construction materials, cultural resources, atmospheric visibility, and human health. We will focus here on the effects on lakes and streams.

The rationale for the NAPAP research program was to provide Congress, the President, and federal and state policy officials with relevant information to use in formulating policy, legislation, and regulations. EPA had the main responsibility for the assessment of effects on lakes and streams. In this effort, they were guided by four questions: (1) how extensive is the damage to aquatic resources due to current and historical levels of acidic deposition; (2) what is the anticipated future damage to these resources; (3) what levels of damage to sensitive surface waters are associated with various rates of acidic deposition, and (4) what is the rate of change or recovery of affected systems if acidic deposition rates decrease?

We will use here the ecological risk assessment guidelines discussed in Chapter 10. An overview of this process is shown schematically in Figure 13.1.

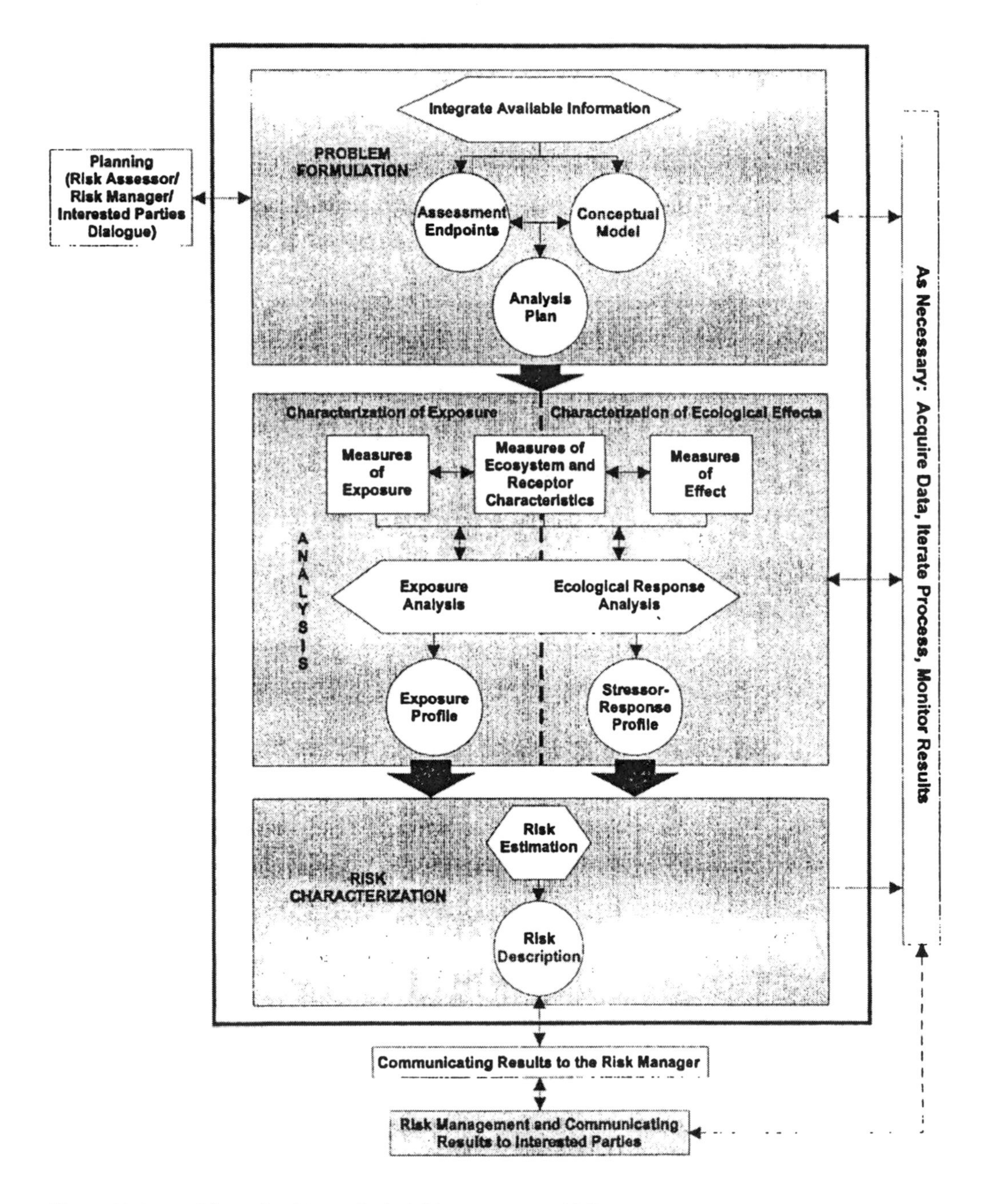

Figure 13.1 Schematic of the ecological risk assessment guidelines

Problem Formulation

As you are aware, the primary components of acidic deposition that are found in lakes and streams are sulfuric and nitric acids. Also we know that the largest sources of acid are sulfur compounds linked to fossil fuel combustion in electric power-generating and industrial facilities. These compounds become airborne and move with the generally eastward flow of air. Thus we will find the eastern lakes and streams are the most affected. The higher hydrogen ion content (lower pH) can be directly toxic to aquatic species. Low pH can also mobilize metals, such as aluminum, that are toxic to aquatic species, especially fish.

The potential of lakes and streams to buffer against deleterious changes is called the acid-neutralizing capacity (ANC), which is used as a measure to characterize the systems. This buffering capacity varies widely among aquatic systems, so we will have to consider both the amount of acidic deposition and the buffering capacity of the lakes and streams. The effects we will consider occur over decades rather than years and over large geographic areas such as New England. Thus we will consider regional effects rather than those in any specific lake or stream.

To limit the study to a practical level, only larger aquatic systems were considered in the NSWS survey. The target population was composed of lakes with surface areas >4 hectares (ha) in the East, >1 ha in the West. Lakes were stratified by alkalinity or ANC class (e.g., <100, 100–200, >200 microequivalents/liter). The NSWS target steam population contained stream reach segments with drainage areas less than 150 km^2. The target populations for evaluation were lakes in the Northeast, Upper Midwest, Southeast, Florida, and Mountainous West, and streams in the Mid-Atlantic Highlands and Coastal Plains, Southeastern Highlands, and Florida regions.

We can recognize the enormity of this study and appreciate that it had to be limited to the vulnerability of fish species to limit the cost. The potential effects were based on laboratory mortality data or field data of fish presence or absence as a function of water chemistry. The primary data included ANC, pH, sulfate, dissolved organic carbon (DOC), calcium, and aluminum.

This research study established criteria for screening lakes and streams to identify those with low resistance to changes in their acidic status. The study also provides us with an extensive background or baseline for evaluating future changes. Since our assessment is limited to fish and other components such as zooplankton are not included, the picture we get will not be complete.

Analysis

We can determine the relationship between surface-water acidity and acute effects by combining the data and information from both laboratory and field studies. The effects are seen in Figure 13.2 in relation to the pH levels shown at the top. As we can see, the small mouth bass start to die below 5.5 and are gone by 5.0. Similarly, clams and snails are gone for pHs below 5.0.

We will find useful an index called the acidic stress index (ASI). This index is directly related to the mortality of fish on the scale of 0 to 100. ASI>10 would describe sensitive species, while ASI>30 would be representative of more tolerant species.

We next turn our attention to the characterization of the exposure. As we can see in Table 13.1, 4% of an NSWS target population of about 28,000 lakes were acidic, and 8% of 60,000 streams. Almost no lakes were acidic in the West; virtually all acidic lakes and streams were in the eastern U.S. We can see that about half of these lakes and streams had ANC < 200 meq/L, indicating that they were potentially susceptible to acidic deposition, and 20% of the aquatic resources were considered sensitive to acidic deposition (ANC < 50 meq/L). Note that in footnote a of Table 13.1, there is not uniform agreement among scientists about the level of ANC and pH where stress enters. After eliminating systems that were acidic because of nonatmospheric sources, it was found that about 75% of the acidic lakes and about 50% of the acidic streams were that way primarily due to acidic deposition.

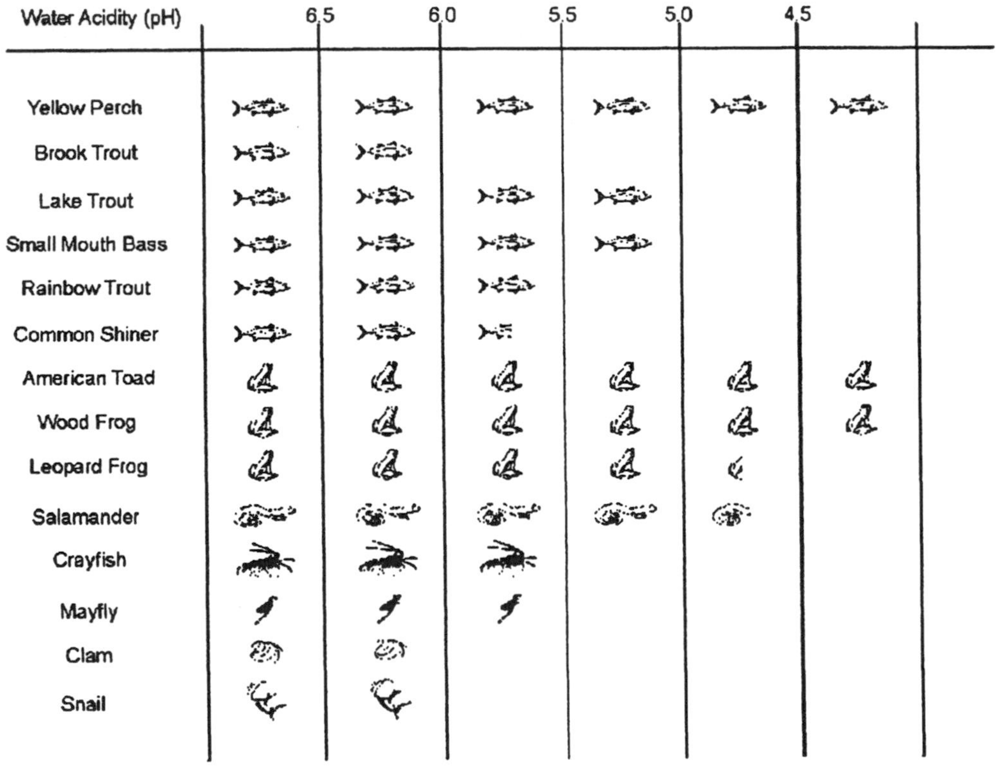

Water Acidity (pH) 6.5 6.0 5.5 5.0 4.5

Yellow Perch
Brook Trout
Lake Trout
Small Mouth Bass
Rainbow Trout
Common Shiner
American Toad
Wood Frog
Leopard Frog
Salamander
Crayfish
Mayfly
Clam
Snail

Figure 13.2 Effect of pH on some Aquatic Species

Table 13.1 Percentage estimates of number of lakes and streams with ANC and pH below three arbitrary reference values[a] (from USEPA, 1993)

	Percentage of Lakes and Streams							
	Lake or	Total		ANC			pH	
Region	Stream[b]	Number	≤0	≤50	≤200	≤5	≤5.5	≤6
New England	L	4330	4	40	64	2	6	11
Adirondacks	L	1290	14	38	73	10	20	27
Mid-Atlantic	L	1480	6	14	41	1	6	8
Highlands	S	27,700	8	22	48	7	11	17
Mid-Atlantic Coastal Plain	S	11,300	12	30	56	12	24	40
Southeastern	L	258	<1	1	34	<1	<1	<1
Highlands	S	18,900	1	8	52	1	2	9
Florida	L	2100	23	40	50	12	21	33
	S	1730	39	70	78	31	50	72
Upper Midwest	L	8500	3	16	41	2	4	10
West	L	10,400	<1	16	66	<1	<1	1
All NSWS	L	28,300	4	19	56	2	5	9
	S	59,600	8	20	51	7	12	22

[a] The chemical definition of an acidic system is ANC<0. Most scientists agree that acid-sensitive fish are stressed at pH < 5. However, some scientists believe that acidic episodes can occur if ANC < 200 and that fish can be stressed if pH < 6. Therefore, three reference values are given for ANC and pH.

[b] The stream estimates in this table are based on the chemistry measured at the upstream end of each surveyed stream reach.

Risk Characterization

In the final step of ecological risk assessment we integrate information on exposure, exposure-effects relationships, and the refined target population of lakes and streams to characterize the risk from acidic deposition to that population. As new technology replaces old, the emissions of sulfur and nitrogen would be expected to decrease. The regions and aquatic systems projected to show the greatest increase in unsuitable fish habitat are the Adirondack lakes and Mid-Atlantic Highland streams. The Adirondack region has a large proportion of the lakes with low ANC. Without emission controls, this estimate was projected to increase to 22% and then decrease to 8%.

This ecological risk assessment shows how models can be combined and provides a quantitative estimate of the consequences. However, it is regional in character and only gives a broad picture. This is the nature of most ecological risk assessments. Since the assessment only involves fish, it may underestimate the extent of the problem. Also, with only a few categories of ANC and pH, the information contains a bias.

References

Cairns, J. C., Jr., B. R. Niederlehner and D. R. Orvos, Eds. 1992. *Predicting Ecosystem Risk.* Princeton, NJ: Princeton Scientific Publishing.

Suter, Glenn W. II. 1993. *Ecological Risk Assessment.* Boca Raton, FL: Lewis Publishers.

USEPA. 1993. A Review of Ecological Assessment Case Studies from a Risk Assessment Perspective, *Risk Assessment Forum,* Volume I, EPA/630/R-92/005. Washington, D.C.

USEPA. 1994. A Review of Ecological Assessment Case Studies from a Risk Assessment Perspective, *Risk Assessment Forum,* Volume II, EPA/630/R-94/003. Washington, D.C.

Case Study: Arsenic

Introduction

A series of international conferences provides us with much of the scientific details of the health effects and exposure to arsenic (Chappell et al., 1994; Abernathy et al., 1997).

Several aspects of the exposure and health effects of arsenic make it an important and interesting case study. There is evidence that arsenic may be an example of a contaminant that exhibits a nonlinear exposure-response curve for cancer. This may be in the form of a sublinear or hockey stick shape, or a kind of threshold. This possibility has important regulatory implications. Currently the drinking water standard (Maximum Contaminant Level or MCL) for arsenic is 50 μg/L and the associated cancer rate is of the order of one in a thousand for this concentration based on linear fits to epidemiological data. EPA is considering lowering the standard to a concentration in the range of 2 to 5 μg/L. The WHO provisional guideline is 10 μg/L. EPA's estimate of the annual cost of compliance for drinking water treatment is $24 million for the current MCL of 50 μg/L, but would rise considerably to $2.1 billion for an MCL of 2 μg/L. However, if as we suspect there is a significant nonlinearity at this exposure, the risk may not be as high as previously thought, reducing the pressure to lower the standard. Some modes of carcinogenic action support linear exposure-response curves. However, arsenic's mode of action is not known. Thus the shape of the exposure-response curve in the range from 2 to 50 µg/L has large implications for our risk analysis.

There have been suggestions by some scientists that arsenic is an essential element for humans. The information relating to this possibility is limited, but the implications would be large.

There are some important questions raised by the health effects data for arsenic exposure. Animals appear to be much less sensitive. Only recently have tumors been seen in animal studies. The reason for this is poorly understood.

It has been reported that arsenic produces several cancers (bladder, kidney, liver, lung, and skin) as well as a long list of non-cancer health effects. For us it is an example of an environmental contaminant that can lead to many serious health endpoints.

Occurrence and Exposure

Arsenic (As) is a naturally occurring element. It exists as a metalloid in the -3, +3, and +5 valence states. It occurs in sulfitic ores and is released during smelting of ores as a trioxide. Other releases include burning of fossil fuels, manufacturing, and weathering of rocks. Recently it has found use in gallium-arsenide (GA-As) semiconductors. It moves into environmental compartments by oxidation and reduction, volatilization, methylation, and absorption.

Arsenic has been known to mankind for about 4000 years and was originally combined with copper to make bronze. We all know of its notoriety as a poison. It has been used extensively in medicine, agriculture, and industry. Fowler's solution (1% potassium arsenite) is used to treat dermatological problems. Lead and copper arsenates have been used as insecticides, and methylated arsenicals as herbicides. These uses have been largely discontinued.

We are exposed to arsenic in food, water, air, soil, and occupational settings. The average person in the U.S. is exposed to a range of 1–50 µg/day from food. Drinking water concentrations are below 10 µg/L for 98% of the U.S. population, while the remaining 2% may be exposed to higher concentrations.

Besides the exposures to us in the U.S., arsenic exposures have been reported in the drinking waters of Argentina, Bangladesh, Chile, Finland, Hungary, Mexico, inner Mongolia, Thailand, Taiwan, and West Bengal, India. Major exposures due to coal burning have been reported in China, India, and Slovakia while mining and smelting exposures have occurred in Ghana, Montana, Portugal, Sweden, and Thailand.

In order for us to achieve a holistic and integrated exposure scenario, several characteristics are important, including chemical form, soil chemistry, bioavailability, source (air, water, soil, food, dust, medicines), route of exposure (oral, respiratory, dermal), and distribution in the body via physiologically based pharmacokinetics.

Arsenite (As^{+3}) is known to be more toxic than arsenate (As^{+5}). Arsenate reduction to arsenite requires glutathione (GSH). Other forms of arsenic whose toxicity is lesser known are methylarsonic acid (MMA(V)), monomethylarsonous acid (MMA(III)), dimethylarsinic acid (DMA(III)), trimethyl arsine oxide, tetramethyl arsonium cation, arsenocholine, arsenic lipids, and arsenosugars.

Non-Cancer Health Effects

Acute toxic exposures to arsenic in humans cause one or more of the following:

- burning and dryness of the oral and nasal cavities;
- gastrointestinal disturbances;
- muscle spasms;
- vertigo, delirium, and coma; and
- edema about face and eyelids.

Chronic, toxic exposures to arsenic cause one or more of the following non-cancer effects in humans:

- malaise and fatigue;
- gastrointestinal disturbances;
- hyperpigmentation or hypopigmentation of skin;
- pale bands in the nails of fingers and toes;
- slightly hypochromic anemia with basophilic stipling; and
- red cell disruption, decreased red cell production, and leukopenia.

A partial list of human diseases reported to be associated with arsenic exposure include:

- blackfoot disease;
- conjunctivitis;
- diabetes mellitus;
- EEG abnormality;
- hyperkeratosis;
- hypertension;
- ischemic heart disease;
- lens opacity;
- melanosis;
- mental retardation;
- peripheral neuropathy; and
- Raynold's syndrome-hyperkeratosis.

Also, weak evidence has been reported for teratogenic, reproductive, and developmental effects.

Cancer

Cancers reported from arsenic exposures include bladder, kidney, liver, lung, and skin.

Modes of Action (Mechanisms)

It is important for us to realize that the methylation of arsenite is a detoxification mechanism for acute effects. This detoxification occurs in two stages. Arsenite is methylated by enzymatic transfer of the methyl group from S-adenosyl methionine (SAM) to monomethylarsonic acid (MMA) and further methylated to dimethylarsinic acid (DMA). There is a wide variation in the rate at which different animals methylate arsenic. At low exposures of arsenic, almost all is methylated to DMA. At higher exposures the DMA levels fall to about 60% as a limiting value, and inorganic and MMA levels rise to about 40%.

The possible modes of action resulting from exposure to arsenic are not fully known. We have no evidence that arsenic or its metabolites form DNA adducts, which are used as a marker of genetic toxicology. Also they do not induce point mutations. However, such exposure does induce abnormalities such as sister chromatid exchange, and change in chromosome structure and numbers. We suggest that although there is not direct DNA damage, exposure to arsenic likely leads to a decrease in the fidelity of DNA replication or an inhibition of DNA repair or perhaps a decrease in the efficiency of the cell division process that can eventually result in chromosomal abnormalities. Arsenite has been shown to enhance the mutagenicity of ultraviolet light, indicating that it may be a co-carcinogen. Thus arsenite is not strictly a non-genotoxic carcinogen. Limited information also supports other mechanisms, including oxidative stress and cell proliferation.

Nutrition

Nutritional essentiality of arsenic for humans has been suggested, but there is no definitive information. There is substantial evidence in animals that arsenic is an essential trace element. Thus on the curve in Figure 13.3, it is possible that deficiency of arsenic could be deleterious to our health.

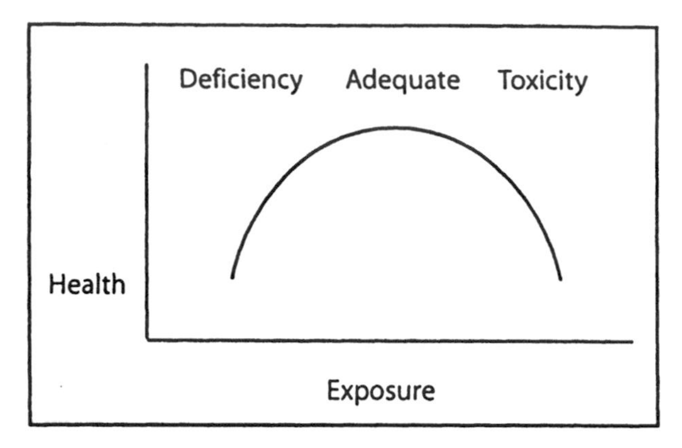

Figure 13.3 Exposure-Response Curve

Exposure-Response Curve

When we fit the multistage model to epidemiology data we get a risk estimate of 5 x 10^{-5} lifetime risk of cancer per microgram per liter of arsenic in drinking water. Thus at a concentration of 50 μg/L, the lifetime risk is 2.5 per thousand. Other drinking water standards of EPA correspond to risk in the range of one in ten thousand (10^{-4}) to one in a million (10^{-6}). Most standards imply lower risks, only those for radionuclides are as high as one in ten thousand. Thus EPA might lower the standard by an order of magnitude to provide consistency. However, analysis of the nonlinearity does not support increased skin cancer below 120 μg/L (Byrd et al, 1996). Adequate data are not available for other cancer sites, such as bladder. Thus there is an intense interest in the shape of the exposure-response relationship at these lower levels.

Figure 13.4 shows the range of possible shapes in the 1 to 100 μg/L range. The metabolism and possible modes of action of arsenic in producing a cancer suggest a nonlinear exposure-response curve. Consider methylation; we start at low doses where the arsenite is detoxified by the methylation process. However, as we increase the concentration, the methylation capacity is reached and at higher exposures more is not detoxified. This would lead to a bend or hockey stick curve. We could describe this as a practical threshold.

Conclusions

Why are humans so sensitive? Why is there no evidence for tumors in animals?

Some argue that we need to first answer these questions and determine the mode of action before we can address the shape of the exposure-response curve. Instead, we suggest that the available information poses a problem in risk management: considering the best public policy, given either interpretation of the exposure-response relationship.

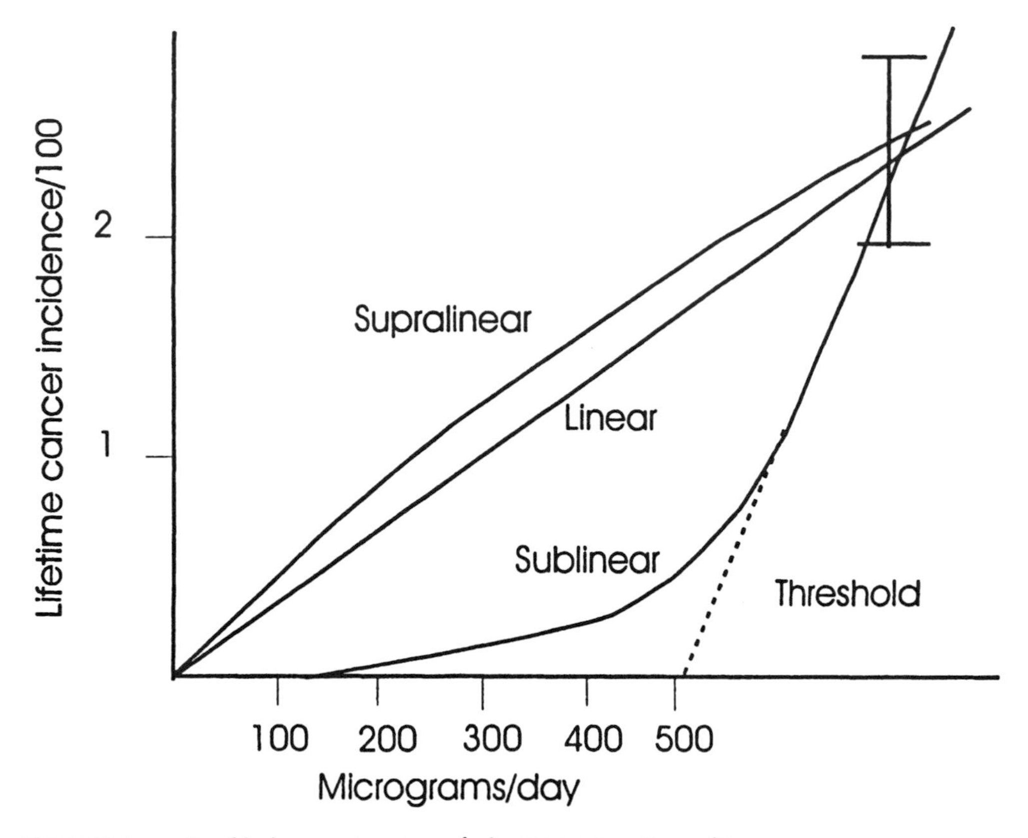

Figure 13.4 Possible dose-response curves for human exposure to arsenic

Drinking Water Standard for Arsenic

We have two likely possibilities for the arsenic drinking water standard. We could leave it at the current value of 50 μg/L or lower it.

Our arguments for leaving the standard where it is are:

(1) we do not have enough information to change it and there is too much uncertainty;
(2) the suggestion of a nonlinearity in the exposure-response curve indicates that the risk could be less than previously believed;
(3) we need much more data and could propose a research plan; and
(4) we do not know the causative mechanism.

Our arguments for lowering the standard an order of magnitude are:

(1) the risk is perceived as unacceptably high, as seen in data and information from other countries;
(2) the possible nonlinearities in the exposure-response curve may not relate to all sites, only skin cancers. Even a lowering of the standard might leave a total lifetime risk on the order of 10^{-4}, and
(3) a lower standard may protect against the wide range of non-cancer health effects.

Conclusion

As you can see, the possibility of a nonlinearity in an exposure-response curve can have important implications for regulatory decision making.

References

Abernathy, C. O., R. L. Calderon, and W. R. Chappell. 1997. *Arsenic: Exposure and Health Effects,* Second International Conference on Arsenic Exposure and Health Effects. New York: Chapman and Hall.

Byrd, D. M., M. L. Roegner, J. C. Griffiths, S. H. Lamm, K. S. Grumski, R. H. Wislon, and S. Lai. 1996. Carcinogenic risks of arsenic in perspective. *International Archives of Occupational and Environmental Health.* 68: 484–494.

Chappell, W. R., B. D. Beck, K. G. Brown, R. Chaney, C. R. Cothern, K. J. Irgolic, D. W. North, I. Thornton, and T. A. Tsongas. 1997. Inorganic Arsenic: A Need and an Opportunity to Improve Risk Assessment, *Environmental Health Perspectives.* 105:1060–1067.

Chappell, W. R., C. O. Abernathy, and C. R. Cothern. 1994. *Arsenic: Exposure and Health; Science and Technology Letters,* First International Conference on Arsenic Exposure and Health Effects, Northwood, UK.

Rossman, Toby G. 1998. Molecular and Genetic Toxicology of Arsenic, Environmental Toxicology: Current Developments, Ed. J. Rose. Amsterdam: Gordon and Breach Science Publishers, 171–187.

USEPA. 1998. *Special Report on Ingested Inorganic Arsenic: Skin Cancer, Nutritional Essentiality, Risk Assessment Forum.* EPA/625/3-87/013, Washington, D.C.

Case Study: Electromagnetic Fields

The possibility that electromagnetic fields (emfs) cause childhood leukemia or any other deleterious effect presents us with a case where almost everyone is exposed, but the risk, if it exists at all, is quite small.

The problem is proving conclusively that something does not exist. From the evidence, can we say that there is zero risk? Can we ever say there is zero risk with authority, conviction, and complete confidence? Likely not.

General

First let us consider the characteristics of electric and magnetic fields. Electric fields are measured in volts/m. They can be easily shielded by conducting objects like trees and buildings and are reduced in strength by distance (they diminish inversely as the square of the distance from the source). Near the earth the strength is about 200 volts/m and under electrical storms can be as high at 50,000 volts/m. Directly under high voltage transmission lines the strength can be in the range of 1 to 10 volts/m and 30 m away can be in the range of 0.1 to 1 volts/m.

Magnetic fields are measured in gauss (G) or tesla (T), where one milligauss = 0.1 microtesla. Magnetic fields are not easily shielded and are reduced in strength by the inverse square law with increasing strength from the source. The magnetic field of the earth is about 500 mG. In normal offices we are exposed to magnetic fields in the range of a few mG. Typical appliances such as ranges, ovens, toasters, and irons produce magnetic fields in the range of a few to a few hundred mG, depending on how far away we are. We are exposed to fields in the range of a few mG from power lines like those in our homes and offices. Electric blankets produce magnetic fields that can be as high as 20 to 40 mG. Directly under high voltage transmission lines the magnetic fields can be in the range of 30 to 100 mG, diminishing to 2 to 10 mG at a distance of 30 m.

Risk Analysis

"There is no known mechanism by which magnetic fields of the type generated by high voltage power lines can play a role in cancer development. Nevertheless, epidemiologic research has rather consistently found associations between residential magnetic field exposure and cancer. This is most evident for leukemia in children" (Feychting and Ahlbom, 1995) (see Chapter 6, Epidemiology).

In this case study, we need to focus on the data displayed in Figure 13.5 (from Kheifets et al., 1995). In this figure we see that if the risk estimate is greater than one, there is an effect; if it is less than one, there is no effect. As you can see, 28 measurements were published over a 20-year period. Of these measurements, only five have error ranges that clearly put them above one. The data have been pooled using meta-analysis in the data point to the right called "pooled." With what confidence can you say that there is or is not an effect? If there is a risk, what is the quantitative level of that effect?

We see clearly in Figure 13.5 that if there is an effect, it is very small. However we are unable to completely reject the possibility that there is an effect.

Conclusion

We are all taught that the dodo bird is extinct; can you prove conclusively that there are no dodo birds? The scientific challenge of showing that something does not exist is huge. To many, including the authors, a risk of zero has no meaning. If you accept that premise,

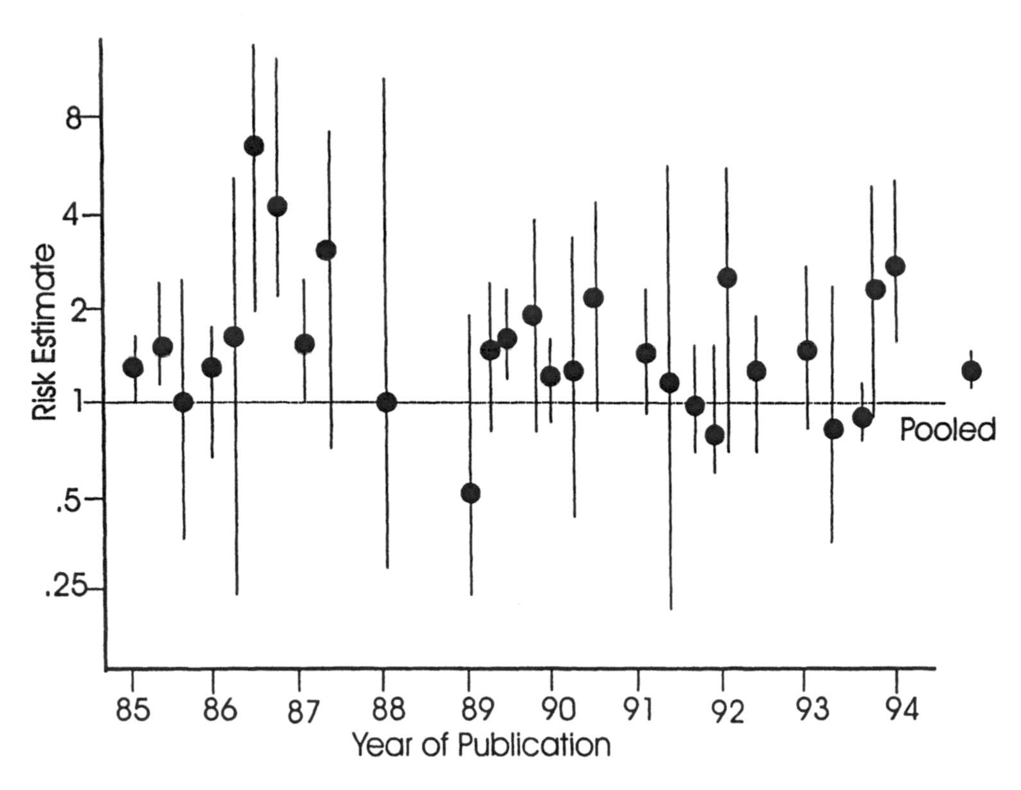

Figure 13.5 Results of epidemiology studies of the effect of electromagnetic fields on humans and a point to the right representing the sum of all the others

you cannot completely and fully reject that there is any deleterious effect from electromagnetic fields. This important point is, for us, the dilemma in assessing the risk of electromagnetic fields. We do know that there is not enough energy in the electric and magnetic fields to which we are exposed to break the bonds in cells. However, we cannot say for sure that there is not some mechanism of which we are unaware that can break bonds in DNA and cause cancer and that does not involve energy.

This is clearly a case where we need to make a policy decision. It would seem prudent to avoid high exposures to electric and magnetic fields and to continue to conduct some research into this problem since we are all exposed.

References

Feychting, M. and A. Ahlbom. 1995. Childhood Leukemia and Residential Exposure to Weak Extremely Low Frequency Magnetic Fields, *Environmental Health Perspectives* 103:59–62.

Goodman, E. M., B. Greenebaum, and M. T. Marron. 1995. Effects of Electromagnetic Fields on Molecules and Cells, *International Review of Cytology* 158:279–338.

Kheifets, L. I., A. A. Afifi, P. A. Buffler, and Z. W. Zhang. 1995. Occupational Electric and Magnetic Field Exposure and Brain Cancer: A Meta-Analysis, *Journal of Occupational and Environmental Medicine* 37:1327–1341.

National Institute of Environmental Health Sciences and U.S. Department of Energy. 1995. *Questions about EMF: Electric and Magnetic Fields Associated with the Use of Electric Power.* Washington, DC: U.S. Government Printing Office.

Washburn, E. P., M. J. Orza, J. A. Berlin, W. J. Nicholson, A. C. Todd, H. Frumkin, and T. C. Chalmers. 1994. Residential Proximity to Electricity Transmission and Risk of Childhood Leukemia, Childhood Lymphoma, and Childhood Nervous System Tumors: Systematic Review, Evaluation and Meta-Analysis, *Cancer Causes and Control* 5:299–309.

Case Study: Environmental Tobacco Smoke

Introduction

First of all, some nomenclature. We are exposed to tobacco smoke in an active way (those who smoke tobacco products) and a passive way (those who are exposed to the tobacco smoke of others). The smoke from a cigarette often is referred to as main stream (MS) smoke, which passes through the cigarette and into the lung and is exhaled, and side stream (SS) smoke, which leaves the end of the cigarette without being inhaled. Environmental tobacco smoke (ETS) refers to tobacco smoke of others that we are exposed to. The ETS label is ambiguous, as it sometimes refers to SS smoke exposure and sometimes refers to the cooled, recirculated mixture of MS and SS smoke.

This case study has some unusual or unique qualities. The analysis of the health effects of environmental tobacco smoke (ETS) uses the EPA's concept of weight-of-evidence (USEPA, 1986). This concept involves weighing all the available evidence to reach an overall, systematic analytic conclusion (see Chapter 7). ETS is also a mixture and we do not have a good way to estimate risks for mixtures that includes synergisms and antagonisms of all the individual components. However, for ETS there are epidemiological studies (that is, data for human beings) for this mixture in exposure studies of the general public. Thus, for ETS we have some of the best kind of data (epidemiology data) and a range of health endpoints that are accepted by the public as being adverse (including lung cancer, respiratory illnesses and irritation, and aggravation of childhood asthma) for a situation where there is exposure to a mixture of constituents, many of which are known carcinogens by themselves. Thus we have three ways to estimate carcinogenic potency: (1) from epidemiology data about ETS exposure, (2) from the analogy to effects seen among cigarette smokers, and (3) by attempting to estimate the effects of the constituents. Another important issue is the statistical meanings of one and two tailed distributions.

The health impacts of ETS might include 3,000 lung cancer deaths annually for non-smoking adults (for the likely range of this estimate, see discussion of uncertainty range later in this chapter). For children, there is a range of deleterious health outcomes from exposure to ETS. Between 150,000 and 300,000 children could have respiratory tract infections such as bronchitis and pneumonia due to exposure to ETS. ETS is associated with increased prevalence of fluid in the middle ear of children and between 200,000 and 1,000,000 cases of increased severity of symptoms in children of asthma. Using the weight of the available scientific evidence, you might conclude that the widespread exposure to environmental tobacco smoke in the U.S. presents a serious and substantial public health impact (USHHS, 1993).

Background

Tobacco smoking has long been recognized as a major cause of mortality and morbidity and is responsible for over 400,000 premature deaths per year in the U.S. (USHHS, 1993), or more than one in every six deaths. Health effects include lung cancer, respiratory diseases, and heart disease. It has long been recognized that exposure via passive smoking involves approximately the same chemical contaminants as found in exposure for active smokers.

As we will see in Table 13.2, most of the constituents of main stream smoke are found in side stream smoke. Thus, health effects that occur from main stream smoke are likely to occur as a result of exposure to side stream smoke or environmental tobacco smoke.

The weight-of-evidence approach for evaluating the likelihood of lung cancer from ETS involves epidemiology studies for both active and passive smoking, animal studies, genotoxic studies, as well as the biological measurements of human uptake of tobacco smoke. The evidence was sufficient for EPA to attempt to classify ETS as a class A carci..ogen under the EPA guidelines; thus it is a known human carcinogen. EPA's weight-of-evidence for non-cancer respiratory effects is primarily based on epidemiology studies.

Exposure

You are most likely aware that almost everyone has been exposed to tobacco smoke, even though many sit in the non-smoking section and do their best to avoid exposure. The same contaminants found in main stream tobacco smoke are, not surprisingly, found in side stream smoke or in environmental tobacco smoke. We find that in many cases the amounts in side stream smoke are higher than those found in main stream smoke (see

Table 13.2). Of the constituents listed in Table 13.2, five are known human carcinogens, nine are probable human carcinogens, and three are animal carcinogens.

Table 13.2 Distribution of constituents in fresh, undiluted mainstream smoke and diluted sidestream smoke from non-filter cigarettes (from USHHS, 1993)

Constituent	Amount in MS	Range in SS/MS
Vapor phase		
Carbon monoxide	10-23 mg	2.5-4.7
Carbon dioxide	20-40 mg	8-11
Carbonyl sulfide	12-42 mg	0.03-0.13
Benzene[1]	12-48 mg	5-10
Toluene	100-200 mg	5.6-8.3
Formaldehyde[1,2]	70-100 mg	0.1-~50
Acrolein	60-100 mg	8-15
Acetone	100-250 mg	2-5
Pyridine	16-40 mg	6.5-20
3-methylpyridine	12-36 mg	3-13
3-vinylpyridine	11-30 mg	20-40
Hydrogen cyanide	400-500 mg	0.1-0.25
Hydrazine[2]	32 ng	3
Ammonia	50-130 mg	3.7-5.1
Methylamine	11.5-28.7 mg	4.2-6.4
Dimethylamine	7.8-10 mg	3.7-5.1
Nitrogen oxides	100-600 mg	4-10
N-nitrosodimethylamine[2]	10-40 ng	20-100
N-nitrosodiethylamine[2]	ND-25 ng[4]	<40
N-nitrosopyrrolidine[2]	6-30 ng	6-30
Formic acid	210-490 mg	1.4-1.6
Acetic acid	330-810 mg	1.9-3.6
Methyl chloride	150-600 mg	1.7-3.3
1,3-buradiene[2]	69.2 mg	3-6

[1]Known human carcinogen
[2]Probable human carcinogen
[3]Animal carcinogen
[4]ND means not detected

Table 13.2 (continued) Distribution of constituents in fresh, undiluted mainstream smoke and diluted sidestream smoke from non-filter cigarettes (from USHHS, 1993)

Constituent	Amount in MS	Range in SS/MS
Particulate Phase		
Particulate matter	15-40 mg	1.3-1.9
Nicotine	1-2.5 mg	2.6-3.3
Anatabine	2-20 mg	<0.1-0.5
Phenol	60-140 mg	1.6-3.0
Catechol	100-360 mg	0.6-0.9
Hydroquinone	110-300 mg	0.7-0.9
Aniline[2]	360 ng	30
2-toluidine	160 ng	19
2-naphthylame[1]	1.7 ng	30
4-aminobiphenyl[1]	4.6 ng	31
Benz[a]anthracene[3]	20-70 ng	2-4
Benzo[a]pyrene[2]	20-40 ng	2.5-3.5
Cholesterol	22 mg	0.9
g-butyrolactone[3]	10-22 mg	3.6-5.0
Quinoline	0.5-2 mg	3-11
Harman	1.7-3.1 mg	0.7-1.7
N-nitrosonornicotine[3]	200-3,000 ng	0.5-3
NNK	100-1,000 ng	1-4
N-nitrosodiethanolamine[2]	20-70 ng	1.2
Cadmium[2]	110 ng	7.2
Nickel[1]	20-80 ng	13-30
Zinc	60 ng	6.7
Polonium-210[1]	0.04-0.1 pCi	1.0-4.0
Benzoic acid	14-28 mg	0.67-0.95
Lactic acid	63-174 mg	0.5-0.7
Glyolic acid	37-126 mg	0.6-0.95
Succinic acid	1 pg	2
PCDD's and PCDF's	1 pg	2

[1]Known human carcinogen
[2]Probable human carcinogen
[3]Animal carcinogen

Hazard Assessment—Lung Cancer

Active smoking has been causally associated with lung cancer. You need to understand the importance of this statement. It does not mean that active smoking causes lung cancer. The mechanism of lung cancer is not known. Until the exact mechanism is known we cannot say that active smoking causes lung cancer. However, this statement is stronger than saying that there is an association or correlation. The evidence moved a step closer to causation in the finding that benzo[a]pyrene or structurally related compounds are involved in the transformation of human lung tissue (Dinissenko et al., 1996).

In order to test the strength of the causal association of ETS and lung cancer, you would need to review the 30 epidemiology studies that are the basis of the USHHS report (1993) concluding that there is a "causal association." There are some problems in this kind of analysis. The epidemiologist calls these confounding factors. In this case you will realize that it is very difficult to find a group of humans that have not been exposed to some tobacco smoke. Thus, a true and independent control group does not likely exist. It turns out that the best comparison is between females who have never smoked (female never-smokers) and are married to smokers, and female never-smokers who are married to nonsmokers.

Another analytical problem that you will meet in any epidemiology study is that of statistical power. That is, are there enough people involved in the study to make a statistically meaningful statement? In the case of female never-smokers, there is barely enough statistical power. Thus, the confidence level is low.

Consider the epidemiologist's statements in the USHHS report. First, 9 of the 30 studies show a causal association between exposure and lung cancer that is unlikely to occur by chance with a probability of less than 10^{-4}. You should appreciate the strength of this statement. The likelihood that this association exists only by chance is one in 10,000 (or it is 99.99% sure). Next, it appears that only 17 of the studies have exposure level data. However, every one of these 17 shows an increased risk for the highest exposure group. You should realize that is a very strong statement. That every one of 17 studies has the same result is truly remarkable. Only nine of these are significant at the $p<0.05$ level despite having low power, and this result is highly unlikely to occur by chance at the $p<10^{-7}$ level. The more common epidemiology study results support a conclusion of causal association and others, although they do not suggest this is an incorrect conclusion, do not support it. However, that 10 of the 14 showed a statistically significant exposure-response trend is a powerful concluding statement. The agreement among the studies is remarkable for most epidemiology studies of environmental contaminants at environmental exposure levels.

By convention, statistical significance usually means that there is less than a 5% probability of the results occurring by chance. Normally, this is applied at both ends of the distribution. However, when there is independent evidence, as there is for ETS, that the substance is harmful, a one tailed approach is taken, which allows a 10% threshold. This is a standard statistical technique found in all introductory statistics courses.

We next turn to the weight-of-evidence analysis in order to determine what we can learn from this approach. There are nine parts to this analysis.

(1) There is biological plausibility. We can clearly understand that ETS is taken up in the lungs in a way similar to that for MS tobacco smoke. The presence of the same carcinogens is a valuable piece of evidence in the analysis. It establishes the plausibility that ETS is a lung carcinogen.

(2) The carcinogenicity of tobacco smoke has been demonstrated in inhalation studies in the hamster, intrapulmonary implantation in the rat, and skin painting in the mouse. Although these demonstrations do not reproduce lung cancer directly, if one type of cancer results from animal studies, perhaps another can result after human exposure. Positive results of genotoxicity testing for both MS and ETS provide corroborative evidence for their carcinogenic potential.

(3) There is a consistency or response in the epidemiology studies. All four of the cohort studies and 20 of the 26 case-control studies showed a higher risk of lung cancer among the exposed never-smokers.

(4) The epidemiology studies are broadly based. The 30 studies are from eight different countries, employing a wide variety of study designs and protocols, and were conducted by many different research teams. There is no alternative explanation or variable given for the observed associations.

(5) There are exposure-related statistical associations between passive smoking and lung cancer. Of the 14 studies that provided sufficient data for a trend related to exposure, ten were statistically significant despite most having low statistical power.

(6) The studies were performed at environmental exposure levels and not the very high levels often found in animal studies.

(7) The association remained even after adjustment for possible misclassification.

(8) Pooling the highest exposure groups yielded a relative risk in the range of 1.04 to 1.35 ($p<0.05$) after adjusting for possible smoker misclassification.

(9) Confounding factors such as history of lung disease, home heat sources, diet, and occupation did not explain the association.

The combination of factors provides a strong argument based on this weight-of-evidence principle that a causal association exists between ETS and lung cancer.

Population Risk Estimates—Lung Cancer

Since so many people are exposed to ETS, the risk does not have to be very large to translate into a significant health hazard for the U.S. We need to adjust the relative risk for misclassification and the fact that we are all exposed to some ETS. The suggested relative risk resulting from this analysis for ETS is 1.59. However, 1.59 does not allow for the difference, if any, between SS smoke and ETS. This translates into approximately 1,500 annual lung cancer deaths for female never-smokers, 500 male never-smokers, and 1,000 former smokers of both sexes for a total estimate of 3,000 lung cancer deaths annually. These components have varying degrees of confidence. The estimate of 1,500 for female never smokers has the highest confidence, 500 for male never-smokers is less certain, and 1,000 for former smokers has the lowest confidence. An estimate by another pair of scientists is 5000 ± 2400 (Repace and Lowrey, 1990). An alternate analysis (USHHS, 1993) gives an uncertainty range of 2,700 to 3,600. The estimate of uncertainty used in the report of the USHHS, based on sensitivity analysis, yielded a range of 400 to 7,000 annually; however, they used very conservative assumptions.

Conclusions

Other reports from the National Research Council (1986) and the Public Health Service (DHHS, 1986) independently assessed the health effects of exposure to ETS. Both reports concluded that ETS can cause lung cancer in adult nonsmokers and that children of parents who smoke have increased frequency of respiratory symptoms and acute lower respiratory tract infections, as well as evidence of reduced lung function.

We find here an example of an environmental contaminant, ETS, that might cause deleterious effects based on weight-of-evidence. We also find that the estimated risks reveal a potentially serious public health problem. This is an example not only of the power of the weight-of-evidence approach, but the importance of epidemiology data.

References

Dinissenko, M. F., A. Pao, M. Tang, and G. Pfeifer. 1996. Preferential Formation of Benzo[a]pyrene Adducts at Lung Cancer Mutational Hotspots in P53, *Science* 274:430–432.

National Research Council. 1986. *Environmental Tobacco Smoke: Measuring Exposures and Assessing Health Effects.* Washington, DC: National Academy Press.

Repace, J. L. and A. H. Lowrey. 1990. Risk Assessment Methodologies for Passive Smoking-Induced Lung Cancer, *Risk Analysis* 10:27–37.

USEPA. 1986. Guidelines for Carcinogen Risk Assessment, *Federal Register* 51:33992–34003.

USHHS. 1986. *The Health Consequences of Involuntary Smoking: A Report of the Surgeon General*, DHHS Publication No. (PHS) 87-8398, Public Health Service, Office of the Assistant Secretary for Health, Office of Smoking and Health.

USHHS and USEPA. 1993. *Respiratory Health Effects of Passive Smoking: Lung Cancer and Other Disorders*, NIH Publication No. 93-3605, Monograph 4.

Case Study: Indoor Air

Introduction

Most of us are city dwellers. We spend about 90% of our time in indoor atmospheres. In these situations we are exposed to a mixture of toxic fumes from a variety of sources. All too often we are in indoor air spaces that have inadequate ventilation or too little fresh air. Historically we have closed in our homes and offices to conserve energy. Few of our buildings have windows that open. We generally turn off the ventilation overnight, and this allows toxic fumes to build up. In addition we put new sources of toxic fumes into our offices, including particle board furniture, carpets, drapes, photocopying machines, fax machines, and computers. It should not surprise us that we are getting sick in these spaces.

Into our homes, offices, and factories we have brought a wide variety of chemicals. We now use about 80,000 man-made chemicals. In addition, a variety of biological materials circulate in indoor air, such as dust mites and molds. Our wide range of sensitivities as humans mask health problems because some of us seem immune to the effects of toxic fumes, while others are very sensitive.

The resulting health effects are given a wide variety of names. The two most common are sick building syndrome (SBS) and building related illness (BRI). As a general rule, a building is considered "sick" if more that 20% of the occupants are complaining of health problems that interfere with their work. Generally speaking we find that over half of the health complaints are related to inadequate ventilation. Approximately 800,000 to 1.2 million commercial buildings in the U.S. have produced building related illness, which translates to 30–70 million people exposed to potential building related health problems.

U.S. medical care costs due to indoor air quality problems have been estimated at over one billion dollars annually.

To most scientists and risk assessors, the immediate reaction to the above problem is to measure the levels of the toxic fumes. However, because many of us have developed sensitivities to very low levels, this approach generally is not useful. The most productive and successful approach to this kind of a problem is for us to start with a visual inspection of the facility and then do a thorough questioning of those complaining. This will be covered in more detail later.

One label for building related illnesses that has brought considerable controversy is multiple chemical sensitivity. This disease was first recognized in the 1950s. The controversy involves the relationship between mind and body. Some think that problems like anxiety and depression are a consequence of physical ailments, while others think these physical ailments are the result of mental problems. We advise using this label carefully, if at all.

Common Symptoms and Responses

Besides sick building syndrome and building related illness we find several other names given to the illnesses found in indoor air, among which are those listed in Table 13.3.

Table 13.3 Some of the more common names given to illnesses found in indoor air circumstances
-BUILDING RELATED ILLNESS
-CEREBRAL ALLERGY
-CHEMICAL AIDS
-CHEMICAL HYPERSENSITIVITY SYNDROME
-CHEMICAL PNEUMONITIS
-CHEMICAL SENSITIVITY
-CHEMICALLY-INDUCED IMMUNITIES REGULATION
-ECOLOGIC ILLNESS
-ENVIRONMENTAL ILLNESS (IL)
-ENVIRONMENTAL MALADAPTION SYNDROME
-HUMIDIFIER FEVER
-HYPERSENSITIVITY PNEUMONITIS
-MULTIPLE CHEMICAL SENSITIVITY (MCS)
-SICK BUILDING SYNDROME
-TOTAL ALLERGY SYNDROME
-TOXIC HEADACHE
-TWENTIETH CENTURY DISEASE
-UNIVERSAL ALLERGY

The common symptoms we experience in indoor air environments fall in the general categories of nervous system, respiratory system, psychological, and other. Some of the many ailments reported include those listed in Table 13.4.

Table 13.4 indicates to us the complexity of trying to find the cause of a building being sick. This complexity is the reason that it is more efficient to interview those suffering rather than try to find a cause or several causes by monitoring.

In general we respond in several ways to ailments produced in indoor air. Our first response is one well known to psychologists. We usually first deny that there is a problem. Males especially try to "tough it" by thinking they can show their strength in the face of some adversity. In many of the cases of indoor air diseases, denial is the worst thing we can do. In some cases our bodies become sensitized permanently. An emotional tie that is important in our homes is our belief that our home is our castle. Thus we believe that there is nothing in our home that can make us sick. This is another fantasy that we need to overcome. To many of us our job is our identity and we will suffer a lot before we give that up. Too often we do not realize the price that we are paying in terms of our health. Finally, we trust that those responsible for watching over our health and safety are doing their job. In many cases, building managers are ignorant of the consequences of inadequate ventilation, water collecting in low areas, and toxic fumes from furniture, clean-

Table 13.4 Some of the more common ailments reported as a result of exposure to indoor air circumstances

-HEAD AND CENTRAL NERVOUS SYSTEM RELATED
-LETHARGY
-FATIGUE
-HEADACHES
-MEMORY DIFFICULTIES AND LOSS
-DIZZINESS, FAINTING
-DEPRESSION
-DIFFICULTY CONCENTRATING
-GROGGY
-TENSE, IRRITABLE
-DIFFICULTY MAKING DECISIONS
-SLEEP DISTURBANCE
-UPPER RESPIRATORY SYSTEM
-BRONCHITIS
-ASTHMA
-SHORT OF BREATH
-RHINITIS
-SINUS CONGESTION
-ALLERGIES
-WHEEZING
-IMMUNE SYSTEM
-SUSCEPTIBILITY TO INFECTION
-ALLERGY PROBLEMS
-COLDS
-OTHER
-EYE IRRITATION
-UPSET STOMACH, NAUSEA
-PROBLEMS DIGESTING FOOD
-JOINT PAIN
-CONSTIPATION
-DIARRHEA
-CRAMPS
-GAS
-BLOATING
-PSYCHOLOGICAL
-DEPRESSION
-ANXIETY
-SUDDEN ANGER
-IRRITABILITY
-CONFUSION
-LETHARGY
-SUICIDAL

ing supplies, and other sources. On the other side of this picture, those of us who are lazy can use sick building syndrome as an excuse for not working. The problem is to sort out legitimate cases from the illegitimate ones.

Possible Sources of Agents that May Cause Disease

One major problem we face in analyzing the risk due to indoor air is the myriad of sources. These fall in several categories and include a wide range of individual sources. In Table 13.5 we list some of the many possible sources of indoor air health risks.

There are other conditions found in indoor air that may not be as serious as the above, but we must consider them in any indoor air risk analysis. These are environmental conditions such as thermal comfort, humidity, lighting, and noise.

As you can see from Table 13.5, there are so many potential sources that it is easier to start the risk assessment with a simple walk through, looking for those materials listed above.

Table 13.5 Some of the potential sources of health problems in indoor air

•	asbestos
•	carbonless carbon paper
•	carpeting
•	cleaning agents
•	computers
•	drapes and window coverings
•	dust
•	dust mites
•	fax machines
•	glues
•	man-made mineral fibers
•	microbial contaminants: viruses, fungi, bacteria
•	mold
•	paints
•	particle board furniture (formaldehyde and vocs)
•	pesticides
•	photocopying machines
•	printers (laser, ink jet)
•	solvents
•	tobacco smoke
•	video display terminals

Diagnosing Problem Buildings

Historically, the industrial hygienist investigated indoor air quality complaints. We find this approach relatively unsuccessful due to the approach of measuring contaminant levels to determine compliance with federally permissible exposure limits (PELs), or Threshold Limit Values (TLVs) of the American Conference of Government Industrial Hygienists. If the levels measured are below these limits or values, it is assumed that there is no problem despite the presence of health problems. We find the industrial hygiene approach inappropriate because sources of indoor air contaminants are numerous and diffuse, levels of potential contaminants are several orders of magnitude below PELs or TLVs, and contaminant control is usually achieved by dilution, not local exhaust ventilation

The features of successful assessments of risks in indoor air quality investigations include: walk through; detailed interviews of the occupants; assessments of the heating, ventilation, and air conditioning systems (HVAC), and some limited measurements. The most successful risk assessments involve stages of investigation to narrow down the possible problem areas and a team of two or three specially trained investigators.

We find that the best first step is to do a walk through looking for sources of toxic fumes, dust, and mold (see list above), and a preliminary inspection of the HVAC system. The next step is a thorough interview with each person who is complaining. You need a detailed questionnaire to ensure that you systematically ask the same questions and obtain the same information so you can consistently compare notes. Next, make a more rigorous inspection of the HVAC system. You start with measurements of air flows into and out of rooms and notations whether the fans are running constantly when the building is occupied. By measuring the carbon dioxide level you can determine if adequate outdoor air is being brought into the system. Other measurements you may find useful are carbon monoxide, respirable particulates, and total volatile organic compounds.

Some potential problems in the HVAC system are: insufficient outdoor air for the effective control of bioeffluent and other contaminations generated in the building; migration of contaminants from one building or zone to another; reentry of building exhausts; entrainment into intake air of contaminants generated outdoors; generation of man-made mineral fibers from disintegration of sound liners in air handling units; microorganisms and organic dust contamination in condensate drip pans, humidifiers, filters, and porous thermal/acoustical insulation; inadequate dust control; inadequate control of temperature, relative humidity, and air velocity; and inadequate air flows into building spaces due to system imbalances

If there is an inadequate outdoor air supply, you can look to the following areas: lack of design to include it; design inadequate for building; increase in building occupancy beyond design; malfunction of inlet dampers; operating changes to conserve energy; and reduced flows due to poor maintenance of filters, fans, and other HVAC system components.

Sometimes we find that the HVAC system may itself be a source of indoor air contaminants due to microorganisms proliferating in poorly maintained condensate drip pans, contaminated air filters, porous acoustical ducts, and air-handler insulation.

For most indoor air quality problems, the above protocol will locate the problem and also indicate the solution.

Why Are Indoor Air Quality Problems Not Better Understood and Generally Not Addressed?

Most of us are ignorant of the problems brought about by inadequate ventilation, sensitivity to toxic fumes, and the importance to our health of living in clean, fresh air. Why is this?

One reason is that medical doctors are reluctant to deal with the problem area. This reluctance is due to there being no clinical definition and no direct tests for many of the ailments. As we pointed out above, there are no known causes for most of the health problems. We see that many of the symptoms are subjective, such as headaches and difficulty concentrating. Since most of the symptoms are self-reported by the individuals, the information is anecdotal, and few systematic studies exist. Finally, physicians are trained to look at specific symptoms and often do not take the holistic approach needed to deal with these complex problems.

We find a reluctance to deal with these problems because there is a paucity of cures or treatments. For those who get sick in buildings, the alternatives are: spend time in a new environment and change the diet to flush the body of the toxics; use medications that can help in some cases, such as for headaches; try alternative medicine such as biofeedback training; or avoid the agent that seems to be causing the symptoms.

What Can We Do?

Why should we be concerned about indoor air quality? Our concern should result from our needs to preserve our health, sustain the health of others, and avoid liability. A few suggestions for dealing with indoor air quality problems are listed in Table 13.6.

Table 13.6 Some suggestions for dealing with indoor air quality problems

- Lower exposure levels; most diseases are dose related.
- Restrict the use of pesticides, especially petrochemically based ones.
- Eliminate perfumes, colognes, and after shave lotions, or only use natural ones.
- Provide adequate ventilation, such as the minimum required by the American Society of Heating, Refrigeration and Air-Conditioning Engineers-ASHRAE which is 20 cubic feet per minute per person and introduction of 20% outside air.
- Do not use air fresheners, deodorizers, and disinfectants.
- Avoid exposure to photocopying machines, laser printers, and fax machines.
- Open the windows.
- Avoid cleaners that are petrochemically based.
- Minimize exposure to exhaust from heating systems and appliances that use natural gas, oil, or wood.
- Do not smoke.
- Steam clean instead of dry clean.
- Avoid use of petrochemicals for maintenance, repair, construction, or remodeling.
- Keep areas well vacuumed and free of dust (including your automobile).
- Avoid urea formaldehyde carpeting.
- Avoid synthetic carpeting.
- Use only products that come from plants and try to avoid those that sound like they come from a chemistry lab.

Conclusion

Our risk assessment of indoor air quality has a very different character than other health risk assessments. This risk involves mixtures, a range of human sensitivities, and a wide variety of symptoms and diseases. Thus it requires a different approach:

- let your nose be your guide about what chemicals to avoid or eliminate;
- pay attention to clues from your body—aches, pains, dizziness, etc.;
- let the building air-out after painting at 30-35°C for at least 48 hours; do the same for newly-installed particle board furniture; and
- look for problems like entrainment, reentry, and cross-contamination.

References

Ashford, N. and C. Miller. 1991. *Chemical Exposures: Low Levels and High Stakes.* New York: Van Nostrand Reinhold.

Bower, J. 1993. *Healthy House Building: A Design and Construction Guide.* Bloomington, IN: Healthy House Institute.

Bower, J. 1995. *Understanding Ventilation.*Bloomington, IN: Healthy House Institute.

Department of Labor, Occupational Safety and Health Administration, Proposed Rule for Indoor Air Quality, *Federal Register*, 15968–16039, Tuesday, April 5, 1994. (This is a good general reference for indoor air problems and information; the other references are included for further reading.)

Godish, T. 1995. *Sick Buildings: Definition, Diagnosis and Mitigation.* Boca Raton, FL: Lewis/CRC Press.

Good, Clint. 1989. *Healthful Houses: How to Design and Build Your Own.* Bethesda, MD: Guaranty Press.

National Technical Information Service. 1995. *Sick Building Syndrome: Annotated Bibliography.* Springfield, VA, 703-487-4650.

Rea, W. 992, 1994, and 1995. *Chemical Sensitivity*, Volumes 1-4. Boca Raton, FL: Lewis Publishers.

Sparks, P. J., W. Daniell, D. W. Black, H. M. Kippen, L. C. Altman, G. E. Simon, and A. I. Terr. 1994. Multiple Chemical Sensitivity Syndrome: A Clinical Perspective, I. Case Definition, Theories of Pathogenesis and Research Needs, *J. Occupational and Environmental Medicine* 36:718–730.

Sparks, P. J., W. Daniell, D. W. Black, H. M. Kippen, L. C. Altman, G. E. Simon, and A. I. Terr. 1994. Multiple Chemical Sensitivity Syndrome: A Clinical Perspective, II. Evaluation, Diagnostic Testing, Treatment and Social Considerations, *J. Occupational and Environmental Medicine* 36:9–17.

Wallace, Lance. 1987. The Total Exposure Assessment Methodology (TEAM) Study: Summary and Analysis: Volume I, Office of Research and Development, US Environmental Protection Agency, Washington, D.C., EPA/600/6-87/—2a.

Case Study: Radon

Introduction

It is important to know that radon is a colorless, odorless, naturally occurring, and inert radioactive gas (in the series from the periodic table of helium, neon, argon, krypton, xenon, and radon). There are 27 known isotopes of radon, of which three occur naturally, ^{219}Rn, ^{220}Rn, and ^{222}Rn. The latter has the longest life (half-life of 3.8 days) and is commonly called radon (Cothern and Smith, 1987).

It is also important to know that radon occurs naturally and contributes about 1/2 of the background radiation to which we are all exposed (Cothern and Lappenbusch, 1986).

The units describing exposure to radon are new to most (see appendix to this chapter), but it is not necessary for you to understand all their relationships to appreciate the importance and effect of environmental radon. What is important is that above an exposure level of approximately[1] 100 WLM (see Figure 13.6), there are statistically valid epidemiological data indicating that radon is a health hazard. Below that level there are measurements in mines and homes, but almost all fail to show any statistically valid effect.

The exposure-response curves for radon have been derived from epidemiological studies (primarily for hard rock miners) that span almost three orders of magnitude of dose. These curves cover exposures above about 1/3 Jhm^{-3} (about 100 WLM) and provide the best known curves for any environmental contaminant. However, no one knows the shape of the curve below about 1/3 Jhm^{-3}, which includes the exposure levels commonly found in homes. In order to protect public health and safety, we assume and state as policy that the shape of the exposure-response curve is proportional[2] (that is if the exposure increases by two, the response increases by two) and that there is not a threshold in the region relating to homes (from a few WLM to around 100 WLM[3]). In addition, the linear (proportional) curve is the simplest to explain and easiest to understand.

Why Is Radon Important?

We consider exposure to radon a major public health problem involving the possible opportunity to prevent thousands, even tens of thousands, of lung cancer deaths each

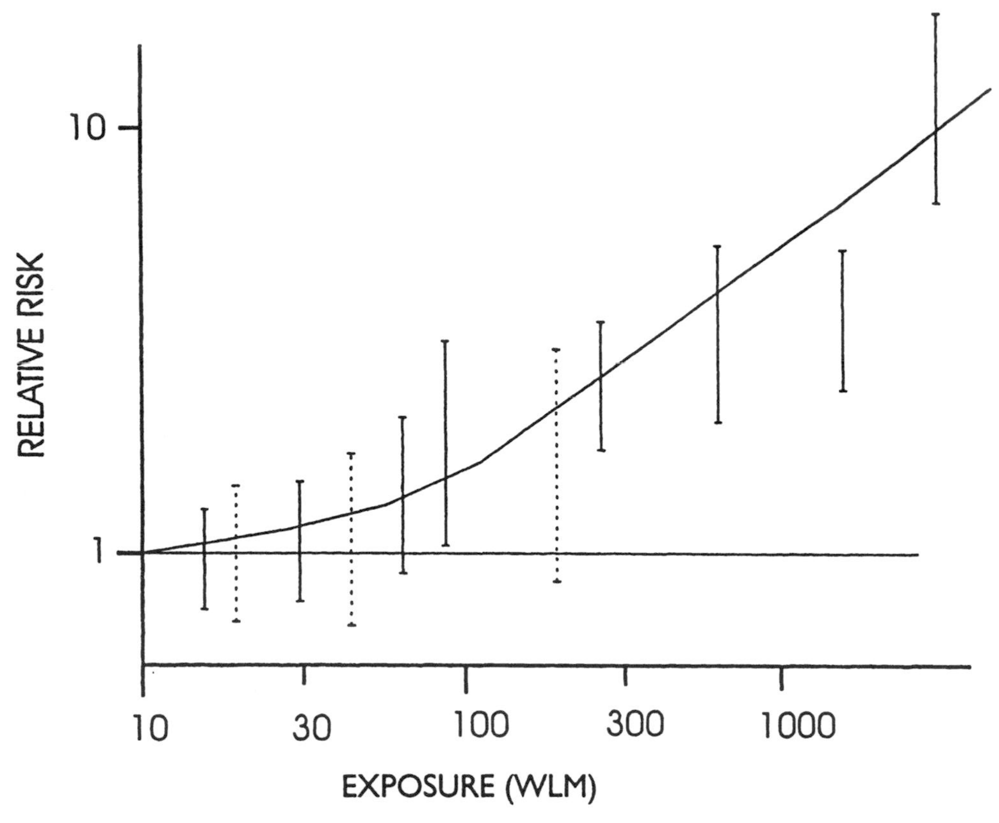

Figure 13.6 A schematic representation (the exposure is in WLM for conceptual simplicity) that compares the epidemiology of hard rock miners and of residential homes with respect to relative risk (RR). The proportional curve (Risk Ratio = 1 + 0.0049 WLM) represents the analysis of 11 hard rock miner epidemiology studies (Lubin et al, 1994a). The solid error bars are data from these studies. The miner data is reported for exposures of WLM, while the residential studies use the radon activity of pCi/L. These two units are only approximately comparable. The dashed error bars are representative of results of studies of homes from Lubin et al. 1994b. All of the error bars below 100 WLM include both the relative risk of one and the proportionally extrapolated curve from the miner studies. Thus, the results of the studies in homes cannot distinguish between the extrapolated miner curve and a no-effect level.

year. These risk estimates are larger than estimates for other environmental contaminants. For example, the estimates for the risk of cancer from radon in drinking water are larger than those for all other drinking water contaminants combined.

As shown in Figure 13.7, EPA sets environmental standards for a lifetime risk (70 years of exposure) in the range of one in ten thousand to one in a million. The risk due to radon exposure at the EPA suggested guidance level (4 pCi/L or about 150 Bq/m^3) corresponds to a lifetime risk level of about two in a hundred (2%). Thus the risk (assuming a proportional exposure-response curve) for radon far exceeds that for other environmental contaminants, while all other environmental risks represent only a few percent of the total number of cancer fatalities.

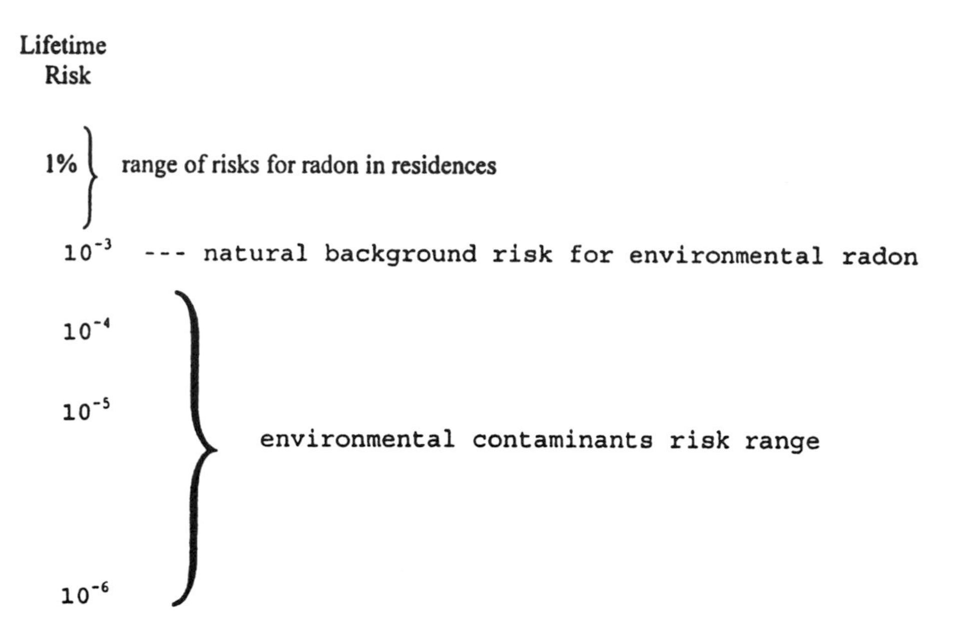

Figure 13.7 Comparison of lifetime risks from exposure to radon showing the risk due to most environmental contaminants in the range of one in ten thousand to one in a million while that for radon in residences is in the range of a few percent

[1] The units of exposure for Figure 13.7 are expressed as WLM for conceptual simplicity.

[2] The exposure-response relationship in Figure 13.6 is linear, but the appearance may cause some confusion in a semilog plot such as Figure 13.6 where the "linear" curve appears to have curvature.

[3] Based on current scientific judgement, this appears to be a conservative position; however, there is evidence that very low levels of radioactivity are essential for life (Polycove, 1994).

A problem in our analysis is that radon and tobacco smoke interact synergistically, exacerbating the health risks or exposure to both. The risk level for a smoker exposed to radon is in the range of two to four times that of the same individual who is not smoking.

Exposure: Mainly from Soils and Drinking Water

Soil is the main source of indoor radon. The radon gas enters homes through cracks in the basement, areas around loose fitting pipes, and holes in the foundation. The source of this radon is the naturally occurring parent, ^{238}U, which is found in minerals, including granite. Radon also enters homes through the drinking water. The radon gas then is released when the water is exposed to air through showers, baths, washing clothes and dishes, and flushing toilets.

When we inhale radon with indoor air, the radon remains in the lungs for only a short time, and then most is exhaled. When radon decays in air it loses a positively charged alpha particle from its nucleus and produces another element, ^{218}Po, which has an electrostatic charge for a very short period of time (milliseconds). This highly radioactive ^{218}Po ion is electrostatically attracted to tiny airborne particulates. The most important of these particulates are those with dimensions less than 0.7 micrometers, which may be small enough to move past the ciliary defenses of the body into the airways and be deposited in the lungs, usually at the bifurcations where passages split. The progeny of radon (the major ones being, ^{214}Pb, ^{214}Bi, and ^{214}Po) then decay in our lungs sequentially, releasing damaging nuclear (alpha, beta, and gamma) radiation. Thus it is ^{218}Po and its progeny that actually cause the damage to the brachial epithelium and induce lung cancer. Only they will remain in the lungs for an adequate time to result in significant decay.

Epidemiological Studies

You will begin to realize that there are more epidemiological studies of the health effects of exposure to radon than any other single environmental contaminant. The studies that produce an exposure-response relationship are for hard rock miners. Studies of homes, in general, do not statistically show any excess lung cancer. However, the concentrations in homes are so low that there are not enough people exposed to find an effect. At environmental levels, the estimated relative risk for indoor air radon may be approximately 1. Some journals do not accept epidemiological studies with relative risk under two (Taubes, 1995), and for radon, if the proportional extrapolation is valid, this would preclude studies for exposures less than about 400 WLM.

There are 17 studies of the health effects of exposure to radon in a variety of hard rock mines. Eleven of these yield statistically valid exposure-response curves and have been combined to produce the curve shown in Figure 13.6 (Lubin et al., 1994a). The curves from the different mining conditions generally agree and show a similar proportional (linear) shape. Two are from the U.S., four from Canada, and one each from China, West Bohemia, France, Australia, and Sweden. These studies provide reproducible points on the exposure-response curve from below 1/3 J h m^{-3} to above 20 J h m^{-3} (or about 100 WLM to above 5000 WLM). Exposure conditions for these studies involve a range of conditions, ventilation rates, smoking histories, and exposure to other contaminants.

As we have learned, it is usually the case with epidemiology that there are confounding factors, uncertainties, and problem areas. The most universal problem with the existing studies is the problem of exposure. Many estimates were made retroactively, long after the exposure occurred, and many exposures were estimated from measurements at nearby mines. Some of the values are guesstimates. Other problems include exposure to tobacco smoke, previous unknown exposures, losing track of older miners, only males being exposed, exposure to other contaminants (such as asbestos, arsenic, chromium, nickel, or diesel fumes), cohorts of mainly older men, exposure rate effects (thought to be less than a factor of two), some histopathological differences between radon-induced lung cancers and cigarette-induced lung cancers, health endpoints other than lung cancer, and thresholds (Cothern, 1989). There are also problems in interpretation due to differences between household dust and hard rock mine dust, age and sex of those exposed, and time of exposure.

In spite of all these confounding factors, the epidemiological studies agree and show a statistically valid exposure-response relationship in the range of 1/3 Jhm^{-3} to 20 Jhm^{-3}.

Animal and Cellular Studies

Another important question concerns the shape of the exposure-response curve and the possible existence of a threshold. There are two main areas of investigation that relate to a threshold for radon. First, there are multiple steps involved in the sequence of transformation from a normal cell to a cancer cell (Cohen and Ellwein, 1991). Large exposures can induce changes in all steps, while low radiation exposures, such as those from residential levels of radon, may only change one step. Second, there are possible repair mechanisms as determined by studies of chromosomal damage (see, for example, Conforth and Goodwin, 1991).

Simulations based on cellular theories provide insight into possible exposure rate effects at low exposures (Crawford-Brown and Hofmann, 1993). Also there may be a promotional effect on irradiated cells where cancer initiated cells are killed (Crawford-Brown and Hofmann, 1990). It has been demonstrated that for some living creatures, ionizing radiation at low levels is beneficial (Polycove, 1994).

Quantitative Risk Assessment

No one knows what the shape of the exposure-response curve is in the exposure range found in homes (below about 1/3 J h m^{-3} or about 100 WLM) (*Health Physics*, 1995). Regulatory agencies have developed a policy of erring on the safe side and assuming a linear non-threshold curve. This is based on the assumption that such a curve yields the highest estimated risks at low levels. The bottom line is that there is no scientific, statistically valid proof that a threshold does (or does not) exist or that the shape of the curve is (or is not) linear.

The present information is shown in Figure 13.9. In the region below about 1/3 J h m^{-3} (about 100 WLM), the data from the miner studies (Lubin et al., 1994a) and the indoor (home) studies (Lubin et al., 1994b) cannot distinguish between no effect (a relative risk of one) and the extrapolated curve from the miner studies.

We estimate that the risk of developing lung cancer from exposure to radon in indoor air is nearly equivalent to smoking a pack of cigarettes a day. About 130,000 people in the U.S. die of lung cancer due to smoking cigarettes each year; about 400,000 die from all diseases caused by smoking cigarettes. In addition, approximately 3000 lung cancer fatalities occur each year as a result of the effects of passive smoking (USEPA, 1992).

Estimates of the annual U.S. lung cancer deaths from exposure to radon include: 4,000 to 30,000 (Cothern, 1989), 6,000 to 36,000 (Lubin et al., 1994a), and 3,000 to 32,000 (NRC, 1998). This latter estimate also includes the estimate that perhaps 1/3 of these fatalities can be avoided by reducing radon levels in homes below the four pCi/L action guidance level of EPA.

Public Reaction Is Widespread Apathy

The public reaction to the information about exposure to radon in indoor air has been widespread apathy (Cothern, 1990). There are a number of reasons for this response. Radon is not detectable by the human senses and thus many doubt its seriousness, while others react with a fear of the unknown. In general, environmental contaminants are considered bad if they are man-made and not so bad if they are natural, like radon. For those who look to the courts to solve environmental problems, natural contaminants like radon present a unique problem in finding a defendant. To understand the importance of radon requires mastering some technically complex facts and a completely new set of units. This is beyond the endurance of many. With so many environmental problems being described in the media, perhaps we all have reached a kind of environmental-problems burnout. For many, money is the driving factor. They do not want to spend their limited resources for radon mitigation; they are looking for reassurance that it is not a serious problem.

Comparison between Risk Estimates for Exposure to Radon and Other Environmental Contaminants

Radon has several unique aspects when compared to those for other environmental contaminants. Perhaps the most important characteristic for us is that the supporting data arises from human epidemiology, and not from experiments on rodents. This eliminates the bothersome assumption that what happens in a mouse will happen in a human.

You will note that the epidemiological data span almost three orders of magnitude in exposure, which gives us much more confidence in the shape of the dose-response curve. Remember that for most animal studies there are only two or three exposures that are very close together, making interpolation tenuous. You may have observed that the uncertainty range for the combined epidemiological studies is about one order of magnitude, whereas the range for other contaminants shown in Chapter 2 is often six orders of magnitude or even more.

We have already noted that radon is a naturally occurring contaminant, whereas many environmental contaminants are man-made. This unique aspect leads many to downplay the importance of radon.

The risk from radon exposure is much higher than the risks for many other environmental contaminants. Individual risk from radon is in the range of one in a hundred to one in a thousand, while the maximum risk at the highest allowed exposure under most environmental standards is in the range of one in a hundred thousand to one in a million.

Finally, we note that the cost of remediation is in the range of $5 million per case avoided, while the cost per case avoided for most environmental contaminants is in the range of $200 million to $1 billion. In general we think that it is more cost effective to remediate radon than most other environmental contaminants.

References

Cohen, S. M. and L. B. Ellwein. 1991. Genetic Errors, Cell Proliferation and Carcinogenesis, *Cancer Res.* 51:6493–6505.

Conforth, M. N. and E. H. Goodwin. 1991. The Dose-Dependent Fragmentation of Chromatin in Human Gibroblasts by 3.5 MeV Alpha Particles from ^{238}Pu: Experimental and Theoretical Considerations Pertaining to Single Track Effects, *Radiation Res.*, 127:64–74.

Cothern, C. R. and W. L. Lappenbusch. 1986. Drinking-Water Contribution to Natural Background Radiation, *Health Physics* 50:33–47.

Cothern, C. R. and J. E. Smith, Jr., Eds. 1987. *Environmental Radon.* New York: Plenum Press.

Cothern, C. R. and P. A. Rebers, Eds. 1990. *Radon, Radium and Uranium in Drinking Water.* Chelsea, MI: Lewis Publishers.

Cothern, C. R. 1989. Health Effects of Inhaled Radon Progeny, *Envir. Carcino. Revs., J. Envir. Sci. Hlth.* C7(1):75–108.

Cothern, C. R. 1990. Widespread Apathy and the Public's Reaction to Information Concerning the Health Effects of Indoor Radon Concentrations, *Cell Biology and Toxicology* 6:315–322.

Crawford-Brown, D. J. and W. Hofmann. 1990. A Generalized State-Vector Model for Radiation-Induced Cellular Transformation, *Int. J. Radiat. Biol.* 57:407–423.

Crawford-Brown, D. J. and W. Hofmann. 1993. Extension of a Generalized State-Vector Model of Radiation Carcinogenesis to Consideration of Dose Rate, *Math. Biosci.* 115:123–144.

Health Physics Society. 1995. Radiation Dose-Response Model, *Health Physics Society Newsletter*, Special Issue, 23:3–17.

Lubin, J. H., J. D. Boice, Jr., C. Edling, R. W. Hornung, G. R. Howe, E. Kunz, R. A. Kusiak, H. I. Morrison, E. P. Radford, J. M. Samet, M. Tirmarche, A. Woodward, S. X. Yau and D. A. Pierce. 1994a. Radon and Lung Cancer Risk: A Joint Analysis of 11 Underground Miners Studies, NIH Publication No. 94-3644, U.S. Department of Health and Human Services, Public Health Service, National Institutes of Health.

Lubin, J. H. 1994b. Invited Commentary: Lung Cancer and Exposure to Residential Radon, *J. of Epidemiology*, 140:323–332.

Lubin, J. H., J. D. Boice, Jr., C. Edling, R. W. Hornung, G. R. Howe, E. Kunz, R. A. Kusiak, H. I. Morrison, E. P. Radford, J. M. Samet, M. Tirmarche, A. Woodward, S. X. Yau, and D. A. Pierce. 1995. Lung Cancer in Radon-Exposed Miners and Estimation of Risk From Indoor Exposure, *JNCI*, 87:817–827.

National Research Council. 1990. *Health Effects of Exposure to Low Levels of Ionizing Radiation; BEIR V.* Washington, DC: National Academy Press.

National Research Council. 1998. *Health Effects of Exposure to Radon; BEIR VI*. Washington, DC: National Academy Press.

Pollycove, M. 1994. Positive Health Effects of Low Level Radiations in Human Populations IN *Biological Effects of Low Level Exposures: Dose-Response Relationships.* Ed. E. J. Calabrese. Boca Raton, FL: Lewis Publishers.

Smith-Sonneborn, J. 1994. Stress Proteins and Radiation IN *Biological Effects of Low Level Exposures: Dose-Response Relationships.* Ed. E. J. Calabrese. Boca Raton, FL: Lewis Publishers.

Taubes, G. 1995. Epidemiology Faces Its Limits, *Science* 269:164–169.

USEPA. 1992. *Respiratory Health Effects of Passive Smoking: Lung Cancers and Other Disorders*, EPA/600//-90/006F, Washington, DC.

Appendix

Units

For describing the concentration of toxic and hazardous pollutants, units such as mg/liter, microgram/liter, ppm, and similar units are generally used. However, certain unique properties of radioactive substances require different units to directly compare the health effects of different radionuclides.

Two important units in the analysis of radon exposure are the activity expressed in becquerels per liter of air (Bq/L), historically as picocuries per liter of air (pCi/L), and the energy of alpha particle energy emitted expressed as $J\ h\ m^{-3}$ (historically as working levels, WL). Units for activity are named after the winners of the 1903 Nobel Prize in physics for the discovery of radioactivity, Henri Becquerel, Marie Curie, and Pierre Curie. One becquerel (1 Bq), one disintegration per second, is the unit of activity in the international SI system; one curie (1 Ci) is the historical unit being the activity of one gram of radium. For conversion purposes, 1 Bq = 27 pCi.

Historically the measure of the energy emitted by the collection of isotopes resulting from the decay of radon (the progeny) was the working level (WL). At the time of the invention of the unit, one working level (1 WL) was thought to be a safe exposure level for miners. It represents the sum of the total energy released by all the progeny of radon (the major ones being, ^{214}Pb, ^{214}Bi, and ^{214}Po). At equilibrium, one working level of exposure is numerically equal to a radon activity of 100 pCi/L. Due to differences in equilibrium between the progeny, this could be as high as 200 pCi/L. A unit of total exposure is the product of the energy and time with a common unit being working level months (WLM). Modern measurements are in the units J h m^{-3}, where 1 WLM = 0.0035 J h m^{-3}.

Index

Page numbers followed by "t" indicate tables.

L

M

U

V

CPSIA information can be obtained at www.ICGtesting.com
Printed in the USA
BVOW052354251112

306182BV00005B/39/P

9 780865 876965